Alchemy in Byzantium

Medical Traditions

Edited by
Alain Touwaide

Volume 12

Alchemy in Byzantium

Edited by
Gerasimos Merianos

DE GRUYTER

ISSN 2567-6938
ISBN 978-3-11-157254-3
ISBN 978-3-11-158961-9 (PDF)
ISBN 978-3-11-159017-2 (E-PUB)
DOI https://doi.org/10.1515/9783111589619

Library of Congress Control Number: 2026942307

Bibliographic information published by the Deutsche Nationalbibliothek
The Deutsche Nationalbibliothek lists this publication in the Deutsche Nationalbibliografie; detailed bibliographic data are available on the Internet at https://dnb.dnb.de.

De Gruyter and Walter de Gruyter GmbH are part of De Gruyter Brill.
www.degruyterbrill.com

Questions about General Product Safety Regulation: productsafety@degruyterbrill.com

Cover illustration: Collage of illustrations in medical treatises from the 1st to the 16th century CE, from Greece and Rome to the Arabic World and China (1. peaches from a Japanese edition from the Li Shi Zhen, Bencao Gangmu, A.D. 1714; 2. plant from the manuscript of Padova, Biblioteca Seminario, 194, Constantinople, ca. A.D. 1430; 3. scene of uroscopy from the manuscript of Paris, Bibliotheque nationale de France, grec, 2294, 15th century;
4. a man climbing on a tree from a manuscript of Dioscorides, Arabic, Bologna, Biblioteca universitaria, 2954, A.D. 1244).
Printing: CPI books GmbH, Leck

MIX
Papier | Fördert gute Waldnutzung
FSC® C083411
FSC
www.fsc.org

Contents

Acknowledgments

This volume is, to a great extent, the offspring of the international workshop "Alchemy in Byzantium: A Research Field on the Rise", which was held virtually on 9–10 November 2020. The event was organized by the internal project *Alchemy in Byzantium* of the Institute of Historical Research of the National Hellenic Research Foundation (Athens), which I coordinate. It was funded by the project *ANAVATHMIS. Historical Research and Digital Applications* (MIS 5002357).

I wish to express my gratitude to all the speakers for their valuable contributions and stimulating discussions, as well as to the participants from around the world for their insightful questions. In the bleak atmosphere of the COVID-19 pandemic, it was invigorating to engage with subjects of shared interest, even if only through virtual means. Not all of the workshop speakers are represented in this volume, while some contributors were subsequently invited to participate.

I owe special thanks to all authors for our impeccable cooperation, especially to those who first presented their contributions and patiently waited for their publication, demonstrating remarkable goodwill and collegiality. I would also like to thank my friends and colleagues Marina Koumanoudi and Nikolaos Livanos, with whom I shared the initial idea of organizing such an event and who were always beside me, even in the smallest details. Last but not least, my thanks are due to Alain Touwaide, who, as Series Editor, kindly invited me to publish this volume within his series and with whom I had close and fruitful cooperation throughout the preparation of this collection of essays for publication.

Athens, October 2025 — Gerasimos Merianos

https://doi.org/10.1515/9783111589619-203

List of Contributors

Bojidar Dimitrov completed an MA in Byzantine and Oriental Studies at the University of Münster (2004–2012) and a PhD in Semitics at the University of Marburg, with a research fellowship at the Bavarian Academy of Sciences and Humanities in Munich within the *Ptolemaeus Arabus et Latinus* project (2013–2017). In the framework of the AlchemEast project (University of Bologna) he has worked on a critical edition, translation, and commentary of Ǧābir ibn Ḥayyān's *Rectifications to Plato*.

Vangelis Koutalis is a postdoctoral researcher at the Department of Pedagogy and Primary Education of the National and Kapodistrian University of Athens. He also teaches physics in secondary education. His research focuses mainly on Byzantine alchemy and the emergence of chemistry in Greek-speaking communities in the Ottoman Empire.

Matteo Martelli is Professor of the History of Science at the University of Bologna. His research focuses on Graeco-Roman and Byzantine science – particularly alchemy and medicine – and on its reception in the Syro-Arabic tradition. He served as Principal Investigator of the ERC project AlchemEast and is currently Co-Principal Investigator of the ERC Synergy project MOSAIC, in which his research focuses on Byzantine and Syriac occult traditions.

Maria Mavroudi is Professor of Byzantine History at the University of California, Berkeley. Her research pioneered the study of the Byzantine translations from Arabic into Greek by placing them centrally within Byzantine literary culture and its reception in "East" and "West" during the medieval and early modern period. This highlights the intellectual exchanges between Byzantium and its "Eastern" and "Western" neighbors as a dialogue of contemporary civilizations rather than a transfer of fossilized ancient Greek traditions from the Byzantine to the Western Christian and the Islamic worlds. It also begs reconsidering the position of the ancient Greek classics within the Byzantine, Arabic, and Latin intellectual traditions, as well as the supposed marginality of Byzantium within a broader medieval intellectual universe.

Gerasimos Merianos is Research Director in Byzantine History and the History of Byzantine Science at the Institute of Historical Research (IHR) of the National Hellenic Research Foundation, Athens. He coordinates the research project *Alchemy in Byzantium* at the IHR and has published extensively on Byzantine alchemy. In addition to the present volume, he is currently preparing, together with Athanasios Rinotas, an edited volume on the connections between Byzantine and medieval Latin alchemy (Routledge), as well as a monograph in Greek on the history of the "sacred art" in Byzantium (Kaminos).

Michèle Mertens is Associate Professor and Curator at the University of Liège. Her research focuses on Greco-Egyptian alchemical texts, with particular emphasis on Zosimus of Panopolis, whose *Authentic Memoirs* she has edited in the *Collection des Universités de France* series on Greek alchemists (Belles Lettres), with extensive commentary. She has published several studies on Zosimus' writings and their textual transmission and is currently preparing an edition of Zosimus' other works.

Maria K. Papathanassiou is Professor Emerita in the Department of Mathematics at the National and Kapodistrian University of Athens. Her research and publications focus on the history of exact and occult sciences in antiquity and the Byzantine period. She is currently working on the edition of Stephanus' astrological treatise on the horoscope of Islam and participates in the iMOUSEION project of the Center for Hellenic Studies at Harvard University.

https://doi.org/10.1515/9783111589619-204

Günter Prinzing is Professor Emeritus of Byzantine Studies at the Johannes Gutenberg University Mainz, where he taught from 1986 to 2009. His research focuses on the history of the Middle and Late Byzantine periods, with a special interest in the relations between Byzantium and its Slavic neighbors, but also in Byzantine vernacular literature. He is co-editor (with Andrea Schmidt) of *Das Lemberger Evangeliar* (1997) and editor of the *Ponemata diaphora* of Demetrius Chomatenus (2002).

Athanasios Rinotas is a Guest Researcher at IMT Lucca and an Affiliated Researcher at the De Wulf-Mansion Centre, KU Leuven. He received his PhD in Philosophy from KU Leuven in 2023. His research focuses on medieval philosophy and science, particularly the interplay between magic, alchemy, and natural philosophy in the work of Albertus Magnus. He has published widely on these topics in journals such as *The British Journal for the History of Science, Early Science and Medicine*, and *Ambix*. Together with Gerasimos Merianos, he is currently editing a collective volume on the connections between Byzantine and medieval Latin alchemy (Routledge).

Alexandre M. Roberts is Associate Professor of Classics and History at the University of Southern California, in Los Angeles, California. His publications include *Reason and Revelation in Byzantine Antioch: The Christian Translation Program of Abdallah ibn al-Fadl* (University of California Press, 2020) and a series of articles on Byzantine and Arabic alchemy, most recently "Disentangling Alchemy". He is currently writing a book on chemistry, especially chemical theory, in Byzantium and the Islamic world.

Gerasimos Merianos

Introduction

Is there Byzantine alchemy? The question is not as trivial as it might first appear. One might be tempted to reply that “Byzantine alchemy is simply the alchemy studied and practiced in Byzantium”, yet such a definition merely states rather than demonstrates its existence. There is no doubt that manuscripts and distinct works from the period have survived, and that alchemy has left traces in non-alchemical Byzantine literature. In fact, apart from the two chemical papyri of Leiden and Stockholm, our knowledge of Greek alchemy derives, for the most part, from Byzantine manuscripts. This implies that we have inherited a distinctly Byzantine perspective on the Greek alchemical tradition – one significantly shaped by Byzantine choices about what to preserve and transmit, as Michèle Mertens eloquently demonstrates in her case study on the transmission of Zosimus of Panopolis’ writings within the context of Byzantine florilegic culture in this volume. However, the historiographical challenge lies in determining whether the alchemy practiced in Byzantium should be seen merely as a perpetuation of the late antique tradition.

1 Byzantine Alchemy in Recent Historiography

Let me, then, take the question a step further: is there really a need to consider Byzantine alchemy a field of study in its own right? Could it not be treated simply as a continuation of late antique alchemy, its principal achievement being the preservation of and commentary on earlier texts? These rhetorical questions reflect the conventional, linear accounts of the history of alchemy that, until very recently, have dominated modern historiography. In these narratives, Byzantine alchemy is viewed as little more than an echo of its Graeco-Egyptian antecedent, an intermediary rather than a creative agent.

Allow me to employ a theatrical metaphor. If “Alchemy” were a play, then behind the Graeco-Egyptian protagonist there would stand a secondary character, barely visible to the audience. This character also speaks Greek, though with a different pronunciation. She perfectly understands the Graeco-Egyptian protagonist but cannot enter into dialogue, because her role is incomplete: the playwright lets her mumble clichés and nothing more. Then, once the Islamicate character enters the stage, she vanishes without notice. In most versions of the scenario, we do not even hear her name; in some, we learn that she is called “Byzantine”, a peculiar scholarly name, neither fully ancient nor properly medieval, neither Eastern nor Western.

This metaphor captures how Byzantine alchemy has often been treated in modern historiography. When mentioned at all, it appears only briefly and superficially in

https://doi.org/10.1515/9783111589619-001

histories of alchemy, usually with discussions ending around the seventh century CE.[1] This is not because the authors of otherwise excellent surveys deliberately excluded Byzantium, but because, until recently, there had been no sufficient body of research on Byzantine alchemy to build upon. This situation created a vicious circle: since Byzantine alchemy had not attracted a critical mass of scholarship, it remained largely absent from broader histories of alchemy, and its absence, in turn, reinforced the perception that there was little to study.

The issues discussed above are associated, by and large, with a persistent historiographical view cultivated within the political context of European colonialism in the late nineteenth and early twentieth centuries, in which Byzantium, as a whole, is often viewed as the "subaltern".[2] Consistent with this perspective, it is portrayed as playing only a minor role in the development of Western thought and science. Its main contributions are perceived as limited to two functions: (1) preserving the scientific knowledge of Greek antiquity, and (2) disseminating Greek manuscripts brought by Byzantine immigrants to the West, before or especially after the fall of Constantinople in 1453. Through these manuscripts, Europe gained access to ancient Greek heritage without relying on Arabic translations. Consequently, Byzantium is deemed to have played a significant role only as a preserver and transmitter of ancient inherited knowledge – first to the Islamic East and later to the West – gradually contributing to the Renaissance.[3] According to this historiographical narrative, Byzantium's independent contribution to scientific thought is negligible; Byzantine science appears as an adapted or, at most, an updated version of its Greek and Roman predecessors. The place ascribed to alchemy within this framework is in great part a direct reflection of that same historiographical bias.

As a Byzantinist, I must acknowledge that the disregard of Byzantine alchemy has also stemmed, at least in part, from Byzantine Studies itself. For a long time, Byzantinists hesitated to engage with the subject, often conflating it with magic and other "occult sciences",[4] and dismissing it as marginal or non-serious.[5] As a result, alchemy remained an essentially nonexistent field within Byzantine Studies until quite recently.

1 See, e.g., Principe 2013: 24–26.

2 Cameron 2014: 9, 115; Stouraitis 2022. See now Anderson and Ivanova 2023.

3 The topic is treated in greater depth by Maria Mavroudi in her contribution to this volume; see also Mavroudi 2013; 2015; 2023.

4 This stereotype – by no means limited to Byzantine Studies – has its roots in nineteenth-century historiography; see Principe and Newman 2001. For a discussion of the complex and contested term "occult", see, e.g., Hanegraaff 2012: 177–191; Monod 2013: 1–19.

5 There are, however, notable exceptions, such as Schreiner and Oltrogge 2011, which examines eighty Byzantine recipes for inks, pigments, and related substances as part of the Byzantines' engagement with practical chemistry (p. 13). The authors acknowledge that some of the recipes appear in alchemical manuscripts and note the influence of alchemy in certain cases. Of course, this once again brings to the fore the issue of the fluid boundaries between alchemy and a range of crafts.

It is therefore not surprising that most work on what is conventionally referred to as the Greek Alchemical Corpus has focused on texts up to the seventh century – a chronological horizon that is often considered to mark the end of Late Antiquity. It should be noted that there is no unanimous answer as to when the Roman Empire in the East – or Late Antiquity – ended and when what we now call "Byzantium" (a historiographical term introduced in the sixteenth century by Hieronymus Wolf)[6] took shape. Averil Cameron has argued that the scholarly expansion of Late Antiquity has created a crisis of periodization that threatens to blur or even erase Byzantium's identity. She shows how the growing focus on material culture, archaeology, and the eastern Mediterranean – often ending the narrative around the rise of Islam – has led many scholars to place the beginning of Byzantium only in the seventh century. She also observes that the traditional (usually, though not solely, from an Anglo-Saxon perspective, I would say) hinge of the sixth century (the age of Justinian) has been neglected or absorbed into Late Antiquity. Yet, such shifting of period boundaries reflects historiographical choices: Byzantium did not suddenly appear after 600 CE, nor did it simply mark the end of antiquity. Instead, Cameron calls for recognizing the continuity between Late Antiquity and Byzantium.[7] This uncertainty about when Byzantium emerges mirrors the uncertainty about when Byzantine alchemy does so.

I would add that the issues of periodization become even more complex in this context, since what is otherwise labeled "proto-" or "early Byzantine"[8] often tends to be studied more readily under the rubric of antiquity. With regard to Greek alchemy, it can be said that interest in it persists only so long as the boundaries of antiquity can be stretched forward. This problem of chronological boundaries is further complicated by the fact that most of the alchemical authors known to us between Olympiodorus the alchemist (sixth century) and the time of Michael Psellus (eleventh century) are notoriously difficult to date, their chronology often oscillating between centuries. A characteristic example is provided by the four iambic poems on chrysopoeia, whose dating is discussed in Maria Papathanassiou's contribution as well as in my own in this volume. The aim is not to downplay the Graeco-Egyptian origins of Byzantine alchemy, or to limit the scope of scholars of Late Antiquity, or to draw rigid boundaries between the two branches of Greek alchemy – *far from it*, since we are constantly reminded by the texts themselves that late antique and Byzantine alchemy form a continuum. Rather, it is to underscore the persistence of historiographical stereotypes that diminish Byzantium's role and significance in the history of science, and of alchemy in particular.

6 Kazhdan 1991: 344.

7 Cameron 2016.

8 Depending on disciplinary or national historiographical traditions, such as the Greek one, to which I belong, which traditionally places the beginning of Byzantium in the fourth century CE – a choice not without its own drawbacks, as Stouraitis 2022: 26 demonstrates.

The study of alchemy in Byzantium is a multidisciplinary enterprise that depends on its Graeco-Egyptian foundations but must also account for its Byzantine identity. Nevertheless, modern scholarship has often approached the Greek Alchemical Corpus primarily with a "salvage" attitude – an effort to recover or edit late antique texts from Byzantine manuscripts. This is sometimes an inevitable undertaking. Yet, when this salvaging tendency is detached from interest in the Byzantine context or content of the manuscripts, it unintentionally reinforces the stereotype of Byzantium as merely a grand depository of knowledge. In this framework, the reasons for the transmission, reception, and cultural function of these texts in Byzantium have remained largely unexplored, and works that cannot be tied to antiquity are often regarded as marginal or of lesser significance.

On the other hand, when Byzantine texts are studied for their own sake, the approach has sometimes been overly literary, often at the expense of historical and contextual interpretation. Ideas and practices specific to Byzantium have frequently been detached from their settings, limiting the depth of analysis. There is, therefore, an evident need for contextualization when studying alchemy in Byzantium. Recognizing what we customarily call Byzantine "alchemy" as an area of inquiry within the broader Greek alchemical tradition requires methodological dialogue between history, philology, the history of science, and material culture, among other disciplines. The many practices and ideas that we subsume under the collective term "alchemy" were continually reinterpreted and rearticulated within Byzantine social, political, economic, and cultural settings, as the papers in this volume demonstrate. Certain Byzantine intellectuals regarded the "sacred art" as an integral part of their own culture, while still acknowledging its earlier origins; "alchemy", in turn, interacted dynamically with multiple aspects of Byzantine culture. To uncover what is "Byzantine" in the "sacred art", one must therefore have a firm grasp of the contexts in which it was cultivated and evolved. For instance, as Bojidar Dimitrov's focus on the "Proclean shift" in this volume implies, and as earlier scholarship has likewise acknowledged,[9] Byzantine alchemy did not exist in isolation but engaged actively with prevailing philosophical currents.

In addition to methodological considerations, the question of terminology must also be addressed – above all, the use of the word "alchemy". As is well known, "alchemy" derives from the Latin *alchimia*, itself transmitted through Arabic *al-kīmiyā'*, and ultimately, from Greek *chym(e)ia* or *chēm(e)ia*. Yet, the semantic range of these terms varies considerably across different linguistic and cultural contexts. The long-standing conceptual separation between alchemy and chemistry is itself the product of early modern polemics, as William Newman and Lawrence Principe have demonstrated in their influential 1998 article, which traced the historiographical error underlying the modern

9 See, e.g., Merianos 2022.

dichotomy between the two.[10] Recent scholarship, such as that of Alexandre Roberts (see below), has further argued that the category of "alchemy" as such is problematic, since it imposes a retrospective and homogenizing interpretive framework upon premodern traditions of material transformation. These critiques have sharpened our awareness of the conceptual and historiographical complexities involved. Nonetheless, in practice many scholars studying the Greek Alchemical Corpus continue to use the term "alchemy", albeit with increasing caution in recent scholarship and often alongside Greek designations where appropriate. In the absence of a widely accepted alternative terminology, "alchemy" persists – not least because of its cross-disciplinary intelligibility – provided we remain aware of its limitations and of the conceptual distortions it can introduce. Rather than denoting a unified discipline, the term should be understood as a provisional shorthand for the heterogeneous activities and conceptual frameworks that Byzantine authors themselves variously described under the names *hiera technē* ("sacred art"), *chē(y)meia*, *chymeusis*, and chrysopoeia, among others.[11]

Although interest in the "sacred and divine art" in Byzantium (especially from the Middle Byzantine period onward) was limited in the past, the situation has gradually improved over the last few decades. In my view, two broad phases can be discerned in this heightened attention: the period of the 1990s–2000s and that from the 2010s to the present. This is not to suggest, of course, that the study of alchemical topics in Byzantium began only after the 1990s. Earlier discussions,[12] some in specialist Byzantinist journals and reference works,[13] attest to a sporadic scholarly interest in Byzantine alchemy well before its recent consolidation as a field. A comprehensive study of the modern historiography of Greek alchemy remains a desideratum, but this lies beyond the scope of the present Introduction.

What follows is a necessarily selective and subjective sketch of the development from the 1990s to the present. Some scholars, publications, or aspects of other scholars' work may have been omitted; I can only emphasize that such omissions are unintentional and that my aim here is simply to indicate research that touches on Byzantine alchemy to varying degrees.

Five figures particularly shaped interest in various aspects of Byzantine alchemy in the 1990s and 2000s, and most of them continue to play leading roles. However, it should be stressed that not all of them have an exclusive focus on Byzantine alchemy. I refer to them in alphabetical order.

---

10 Newman and Principe 1998; also Vangelis Koutalis' contribution to this volume.

11 For a similar situation of multiplicity of terms in Islamicate alchemy, see Moureau 2020.

12 To give an early example, Byzantine authors are discussed in Marcellin Berthelot's *Les origines de l'alchimie* (1885); however, he does not typically view them within a specifically Byzantine framework.

13 See, for instance, Stephanides 1927; Rehm 1939; and the overview of Byzantine alchemical literature in Hunger 1978: 2.279–282.

Andrée Colinet has given us magisterial editions of the *Anonymous of Zuretti*, a work written in Greek in southern Italy around 1300[14] (see Athanasios Rinotas' contribution to this volume), as well as of the alchemical recipes preserved in the fifteenth-century manuscripts *Parisinus graecus* 2419 and *Holkhamicus graecus* 109, together with the text attributed to Cosmas the Hieromonk (fifteenth[?] century).[15] She also wrote an influential study, in which she proved that the so-called *Work of the Four Elements* (possibly twelfth century) is closely related to a text attributed to Jābir ibn Ḥayyān, providing yet another instance of the connections between Byzantine and Islamicate alchemy.[16]

Among his prolific work since the late 1970s, Robert Halleux – one of the pioneers in the study of alchemy across different traditions and an outstanding connoisseur of the technical texts of the Greek tradition – edited the first volume of the *Les Alchimistes grecs* series, devoted to the chemical papyri of Leiden and Stockholm.[17] Any brief reference of this kind, however, cannot do justice to the breadth and impact of his scholarship. I will therefore limit myself here to his last volume in this series to appear during his lifetime: the first part of the ninth volume (*Traités des arts et métiers*), which contains the re-edition, French translation, and commentary of technical texts and recipe collections dealing with the manipulation of metals, minerals, and pearls, and with the making of hollow bronze molds for counterfeiting coinage.[18] This volume not only highlights the close interconnections between alchemy and metallurgy and the interaction of alchemy with specific artisanal sectors in Byzantium, but also preserves, in several texts, traces of Byzantine knowledge of Islamicate alchemy, as has already been noted in the scholarship.[19]

Michèle Mertens, an authority on Zosimus of Panopolis (late third–fourth century) and editor of his *Authentic Memoirs*,[20] among her work on Graeco-Egyptian alchemy, provided us in 2006 with the most influential for many later scholars, including myself, account of the reception and evolution of Graeco-Egyptian alchemy in Byzantium.[21] For years, this paper was the most reliable starting point of entry into Byzantine alchemy, as discussed below.

Maria Papathanassiou has most extensively engaged with Byzantine alchemy and, before the emergence of the next generation of scholars, represented the most explicitly Byzantine-focused approach to the subject. An expert on Stephanus of Alexandria (seventh century) – regarded by many as the first veritable Byzantine alchemical

---

14 Colinet 2000a; see also Colinet 2005.

15 Colinet 2010.

16 Colinet 2000b. See also the pertinent description of Masayo Watanabe's dissertation below.

17 Halleux 1981.

18 Halleux 2021.

19 Mavroudi 2002: 401–403; more recently, Roberts 2022a.

20 Mertens 1995.

21 Mertens 2006.

author – she has provided a much-needed re-edition of his text.[22] She has examined, and continues to explore, various aspects of Stephanus' work in relation to philosophy, religion, medicine, and the *quadrivium*.[23] In addition, she has addressed, among other topics, the intersections of alchemy with crafts, particularly metallurgy.[24]

Cristina Viano, an expert on Olympiodorus the alchemist (sixth century),[25] has illuminated the close relationship between Greek philosophy and alchemy, especially its engagement with Platonic and Aristotelian thought.[26] Her edited volume on alchemy's philosophical roots in the Greek and Arabic traditions remains unsurpassed.[27] In her work, she interprets alchemical writings not as isolated technical manuals but as expressions of a broader philosophical reflection on nature, transformation, and matter.[28] Through her studies of Olympiodorus and other commentators, she highlights how alchemical theory developed, by balancing theoretical inquiry with practical application.[29]

From the 2010s to the present, most of the scholars mentioned above have continued to work on related topics, and younger researchers have produced important new contributions. This collective activity has helped to establish Byzantine alchemy as a field of inquiry in its own right, while also expanding its thematic and cross-cultural horizons. It is worth recalling that prior to this phase the scholarly resources available to those seeking even a preliminary orientation were extremely limited. The dedicated, though already outdated in the 2000s, entry on alchemy in the *Oxford Dictionary of Byzantium* was one of the very few points of introductory guidance.[30] The situation improved significantly with the publication of Michèle Mertens' 2006 paper (noted above), together with Maria Mavroudi's lucid overview of Byzantine engagement with Arabic alchemy.[31] For years, these two surveys served as the most reliable points of departure for anyone wishing to begin exploring Byzantine alchemy.[32] It is in this con-

22 Papathanassiou 2017.
23 Papathanassiou 1990; 2005; 2006; 2018.
24 Papathanassiou 2002.
25 See, e.g., Viano 1995; 2021.
26 Viano 1996; 2005b; 2015a.
27 Viano 2005a.
28 Viano 2015b.
29 Viano 2018b.
30 Pingree and Cutler 1991.
31 Mavroudi 2002: 400–403. It should be noted that Maria Mavroudi (2006) positioned Byzantine alchemy as an integral and interconnected field within the Byzantine intellectual world – an aspect not so evident at the time – which was embedded in the broader Mediterranean context of knowledge exchange, essential for understanding how Byzantines conceptualized science and the natural world.
32 To these may be added the succinct overviews of alchemy as part of Byzantine science in Tihon 2008 and 2013. Interestingly, Tihon 2008 places the one-paragraph subsection on alchemy under the section "Pseudo-Sciences", whereas Tihon 2013 dedicates a two-paragraph section to the topic under the title "Alchemy and Chemistry".

text that the subsequent appearance of several new overviews, which are more or less focused on the Byzantine period, may be regarded as one of the most visible features of the current phase, reflecting the consolidation of the field.[33]

At a deeper level, this consolidation has encouraged a new phase of research characterized by collaboration, systematic publications, the participation in or supervision of research projects, and the organization of scientific events. In this environment, several scholars, three of whom are listed below in alphabetical order, have played a central role in shaping the contours of the field today.

Matteo Martelli, the editor of Pseudo-Democritus' work (first century CE) and Synesius' commentary (first half of the fourth century),[34] is an expert on late antique alchemy who has also explored its reception in the Syro-Arabic tradition, as well as aspects of the "sacred art" in Byzantium.[35] For example, he has co-authored a preliminary study of an important Byzantine alchemical lexicon.[36] Furthermore, he identified that a version of the recipe collection *On the Most Praiseworthy and Illustrious Art of Goldsmithing* – long thought to survive only in *Parisinus graecus* 2327 – is also preserved in codex *Meteora, Mone Hagiou Stephanou* 97[37] (see Günter Prinzing's contribution to this volume). He has brought to light other previously ignored recipe collections as well (e.g., the fourteenth–fifteenth-century recipe book found in *Parisinus graecus* 2314),[38] while his work on the reconstruction of (al)chemical recipes transcends Greek alchemy, extending from its Graeco-Egyptian to its Byzantine manifestations.[39] His research on the relationship between alchemy and medicine likewise crosses chronological and cultural frontiers.[40] His studies have illuminated the continuities between theoretical reflection and experimentation within the Greek alchemical tradition.

My research has sought to consolidate the study of Byzantine alchemy as a defined field of research. This work has examined the social, economic, and intellectual dimensions of Byzantine alchemical discourse, exploring how chemical knowledge intersected with artisanal practices, metallurgy, minting, and counterfeiting.[41] My investigations

33 The surveys devoted exclusively to Byzantine alchemy are Merianos 2017 and Viano 2018a. Martelli 2016 offers an overview of Graeco-Egyptian and Byzantine alchemy up to the seventh century. Koutalis et al. 2018 surveys Greek alchemy from antiquity through the Byzantine period to early modernity. Papathanassiou 2020 examines alchemy within the broader context of the "occult sciences" in Byzantium. Finally, Martelli et al. 2022 provides an overview of several concepts in Byzantine, Islamicate, and medieval Latin alchemy.

34 Martelli 2013.

35 E.g., Martelli 2019a.

36 Martelli and Valente 2013.

37 Martelli 2018.

38 Martelli 2024.

39 E.g., Martelli 2022; Manco and Martelli 2024.

40 Martelli 2019b.

41 Merianos 2017; 2024c; Merianos and Sakorrafou 2013.

have also addressed the relationship between Byzantine alchemy and Christianity,[42] the ethical and mathematical foundations (the "mathematization") of alchemical reasoning,[43] and the heuristic examination of alchemy in relation to concepts that we would now call "innovation".[44] Further research has focused on alchemical authors, such as the philosopher Christianus (late sixth–seventh century),[45] and writers who addressed alchemy, like John Canabutzes (fifteenth century).[46] Finally, studies of substances such as milk, cheese,[47] oil, wine, and the byproducts of wine (lees),[48] have shown how everyday materials acquired both practical and symbolic significance within Byzantine alchemical contexts. Altogether, this work has contributed to defining Byzantine alchemy as an integral part of Byzantine science and culture.

Alexandre Roberts has likewise sought to refine our understanding of alchemy in Byzantium and to situate it within its broader cross-cultural environment, particularly in relation to the Islamicate tradition.[49] He has examined several aspects of the seminal alchemical manuscript *Marcianus graecus* 299 (coll. 584; hereafter: **M**), focusing especially on its opening pages as evidence for Byzantine conceptions of alchemy and its place within learned culture.[50] His study of the alchemical epigram in **M** both convincingly dates the manuscript to the second half of the tenth century and situates it within the milieu of the Middle Byzantine elite.[51] He has also studied evidence for Byzantine engagement with Islamicate alchemy.[52] Moreover, he has offered a historiographical critique of the term "alchemy", arguing that it obscures the diverse intellectual, textual, and technical practices it seeks to describe. His relevant work has underscored the need to view Byzantine "sacred art" not as an isolated phenomenon but as part of a larger, evolving discourse on the transformation of matter.[53]

A further important contribution in this phase is the work of Vincenzo Carlotta, whose research focuses on the relationship between Greek alchemy and natural philosophy from Late Antiquity through the Byzantine period, with particular attention to the works attributed to Cleopatra and to Stephanus of Alexandria. Of special note is his forthcoming critical edition, English translation, and commentary on the *Dialogue of*

42 Merianos 2024a.
43 Merianos 2022.
44 Merianos 2024b.
45 Merianos 2022.
46 Sakorrafou and Merianos 2014.
47 Leontsini and Merianos 2016.
48 Merianos 2014; 2025.
49 Roberts 2023.
50 Roberts 2019; 2022b; 2022c.
51 Roberts 2020.
52 Roberts 2022a; 2022c.
53 Roberts 2022b; 2023; see also his contribution to this volume.

the Philosophers and Cleopatra (ca. seventh century).[54] He has also explored the integration of Greek alchemy within a Christian intellectual context in Byzantium, particularly in the case of Stephanus of Alexandria.[55] In addition, he has investigated how alchemical symbols in key manuscripts such as **M** functioned as vehicles for conveying operative knowledge in practice.[56] In a co-authored paper with Matteo Martelli, he examines conceptions of metals as living substances, especially the relationship between mercury and gold, showing how alchemical ideas intersected with metallurgical techniques across Greek, Byzantine, and Syriac sources.[57]

However, as important as these contributions may be, the most significant breakthrough of the period after 2010 lies in the appearance of doctoral dissertations devoted to topics more or less directly related to Byzantine alchemy. This development marks an important step that further legitimizes, within the community of historians of alchemy, the visible presence of Byzantine alchemy as a field of study. Before mentioning some of these dissertations, it should be noted that this moderate bloom is partly linked to a parallel development, the existence of successful funded projects, which in certain cases facilitated the undertaking of doctoral research. Three projects, now completed, stand out, based respectively (in chronological order of inception) in Athens,[58] Mainz,[59] and Bologna,[60] whose scope included Byzantine alchemy. All three have produced significant research findings. Yet, in terms of doctoral dissertations, the Bologna-based ERC project AlchemEast (PI: Matteo Martelli) is noteworthy, as it also encouraged the pursuit of topics related to Byzantine alchemy. Of course, this does not mean that relevant topics were not pursued independently of funded projects at universities worldwide.

First of all, we now have dissertations that have re-edited the texts of authors first published in the late nineteenth or early twentieth century, provided commentaries on them, and offered translations into major European languages.[61] Marco Bellini

54 Carlotta (Forthcoming).
55 Carlotta 2022.
56 Carlotta 2023.
57 Carlotta and Martelli 2023.
58 Thales Project *Digital Archive concerning Alchemy in Byzantium and in Greek-Speaking Communities of the Ottoman Empire* (DACALBO), PI: Efthymios Nicolaidis, National Hellenic Research Foundation, Athens, 2012–2015 (https://dacalbo.hpdst.gr/).
59 Leibniz-WissenschaftsCampus Mainz Project *Der griechische Traktat "Über die hochgeschätzte und berühmte Goldschmiedekunst" – Edition und interdisziplinärer Kommentar*, PIs: Susanne Greiff, Antje Bosselmann-Ruickbie, and Stefan Albrecht, Leibniz-Zentrum für Archäologie (LEIZA), Mainz, 2013–2022 (https://www.leiza.de/forschung/projekt/der-griechische-traktat-ueber-die-hochgeschaetzte-und-beruehmte-goldschmiedekunst).
60 ERC Consolidator Grant *Alchemy in the Making: From Ancient Babylonia via Graeco-Roman Egypt into the Byzantine, Syriac and Arabic Traditions (1500 BCE–1000 AD)* (AlchemEast), PI: Matteo Martelli, Alma Mater Studiorum – Università di Bologna, 2017–2023 (https://alchemeast.eu/).
61 In what follows, I refer only to dissertations that have some degree of relevance to Byzantine alchemy and that have been completed. Any omissions are unintentional and due to lack of information.

has studied the work of the philosopher Christianus;[62] Flavio Bevacqua has examined the writings of the Anonymous (*Anepigraphos*) Philosopher;[63] and María Belén Boned Fernández has analyzed the four iambic poems on chrysopoeia attributed to Heliodorus, Theophrastus, Hierotheus, and Archelaus.[64] The fact that we are now in a position to have modern editions of these seminal authors, whose thought is fundamental to the evolution and systematization of Greek alchemy in Byzantium, is in itself a milestone. Even before the appearance of these new editions, scholars had begun to recognize internal links and parallels among these texts, as well as connections between some of them and the work of Stephanus of Alexandria. These dissertations, especially once published, will play a fundamental role in uncovering the course and depth not only of the Christianization but also of the Byzantinization of Greek alchemy.

In the field of manuscript studies, Giorgia Pausillo has produced, in her dissertation, a comprehensive new catalogue of the Greek alchemical manuscripts preserved in Italian libraries, in accordance with modern standards, thereby replacing Carlo Oreste Zuretti's 1927 catalogue,[65] since Zuretti's work, particularly in its codicological and palaeographical descriptions, is inadequate. Each manuscript has now been carefully re-examined, and the findings provide fresh insights on the circulation of Greek alchemical texts and their readership.[66]

In the meantime, eye-opening dissertations have appeared on previously uncharted or undervalued topics. Alexandre Roberts has examined how scholars in eleventh-century Constantinople and Antioch understood matter and material transformation. He argues that matter remained a dynamic and culturally significant concept, reshaped through Christian exegesis and philosophical reflection, and that Byzantine engagement with alchemical texts in this period was active and interpretive rather than merely preservational.[67] Shannon Steiner's study of the relationship between Byzantine alchemy and cloisonné enameling shows that the latter embodied a broader system of esoteric learning. Furthermore, the expert handling of metal, glass, minerals, and heat reflects Byzantine knowledge of nature.[68] Vangelis Koutalis, for his part, has explored the presence of alchemy in the works of three Greek-speaking scholars active in Italy between the fifteenth

62 Bellini 2023. For Christianus' description – among the earliest in the Mediterranean world – of lac-dye and shellac production, see Bellini 2025. Furthermore, Bellini 2020 explores a case of petrifaction described by Michael Psellus in his chrysopoetic treatise *On How to Make Gold*, situating it within its broader philosophical context. On Psellus and chrysopoeia, see also Katsiampoura 2008; Merianos 2017: 243–246, 250; 2024a: 24, 40–43, 47; Roberts 2023: 611–615.

63 Bevacqua 2025.

64 Boned Fernández 2025.

65 Zuretti et al. 1927.

66 Pausillo 2022; for an English, revised monograph version of her dissertation, see Pausillo (Forthcoming).

67 Roberts 2015.

68 Steiner 2020. Steiner 2022 offers an English translation of Nicephorus Blemmydes' text on chrysopoeia.

and seventeenth centuries – Cardinal Bessarion, Angelo Forte, and Leone Allacci – who played an essential role in preserving the written testimonies of Greek alchemy. Koutalis shows that alchemy in that period was recognized as a potentially productive discipline of natural transformation, attracting scholars of high philosophical training and intellectual engagement.[69] Masayo Watanabe has examined the notion of "nature" in the Jābirian alchemical corpus, especially the key concept of the "fifth nature" and its Greek philosophical background. Building on Colinet's relevant paper,[70] Watanabe's study also affirms that the Jābirian corpus exerted an influence on Late Byzantine alchemy, as exemplified by the *Work of the Four Elements* (possibly twelfth century), a Greek text that draws on the Jābirian *Book of Thirty Words*, probably through its Latin translation.[71]

The preceding discussion shows that in recent decades, scholarship on Byzantine alchemy has made remarkable progress. We are now in a considerably more advanced position than we were in the not-so-distant past. Nevertheless, much remains to be explored, both in relation to the texts themselves and with regard to broader methodological and historiographical questions. It is evident that a number of works require modern critical editions – one such case is the text on chrysopoeia attributed to Nicephorus Blemmydes (1197–ca. 1269) – as do certain technical writings, since we remain partly dependent on the late nineteenth-century edition of texts by Marcellin Berthelot and Charles-Émile Ruelle.[72]

At the same time, our understanding of the reception, diffusion, and social appeal of alchemy within Byzantine society across different periods still requires further refinement through closer contextualization of the relevant texts, theories, and practices (an appropriate example is Alexandre Roberts' contribution to this volume). It is equally important to situate alchemical discourse in dialogue with contemporary developments in philosophy and theology and to trace the ways in which such intellectual currents are reflected – directly or indirectly – in the alchemical works of various eras. Continued attention should also be given to the transmission and transformation of alchemical knowledge across cultural and linguistic boundaries, particularly in relation to Syriac and Arabic traditions (as demonstrated by Matteo Martelli and Bojidar Dimitrov in this volume), as well as to the points of contact and divergence between Byzantine and medieval Latin alchemy (as Athanasios Rinotas shows in this volume). Finally, closer engagement with scholars working on other alchemical traditions – such as those focusing on early modern Europe – would be especially fruitful. This kind of

69 Koutalis 2022. Besides his dissertation, and among his many publications – primarily on early modern alchemy – Koutalis (Forthcoming) lists and describes a substantial number of manuscripts currently held in Greek libraries that were not included in the fifth volume, second part, of the *Catalogue des manuscrits alchimiques grecs* (Zuretti and Severyns 1928). This undertaking began as part of the DACALBO Project; see note 58 above.

70 See note 16 above.

71 Watanabe 2023.

72 Berthelot and Ruelle 1887–1888.

dialogue would enable us to identify both continuities and discontinuities in the questions posed to alchemy from Late Antiquity onward (as Vangelis Koutalis' contribution demonstrates), but also to reflect heuristically on how different scholarly communities have approached enduring issues within their respective historical frameworks, such as the relationship between alchemy and religion.

2 Structure and Scope of the Volume

The present volume originates from the international workshop "Alchemy in Byzantium: A Research Field on the Rise", held virtually on 9–10 November 2020 and organized by the Institute of Historical Research of the National Hellenic Research Foundation (Athens). It was the first scholarly event devoted exclusively to Byzantine alchemy and aimed at a systematic exploration of its many dimensions. Not all the papers presented there are included here; however, the volume has been enriched with contributions from scholars who, though unable to participate in the original event, were subsequently invited to contribute. After several years of preparation – which, I am afraid, has become almost customary for undertakings of this kind – the final outcome is now placed in the hands of the reader, in the hope that it will further advance the study of alchemy in Byzantium.

It is now time to turn to the papers that comprise this volume, which is divided into two parts. The contributions in Part I address alchemy in Byzantine contexts. They are arranged chronologically and examine key authors and the evolution of the Greek alchemical tradition, the presence of alchemy in non-alchemical literature, and the compilation of recipe collections.

Michèle Mertens, in her contribution "'I Could not Insert the Complete Chapters': Zosimus of Panopolis' Writings and Byzantine Florilegic Culture", examines the transmission of Zosimus of Panopolis' writings within the framework of Byzantine florilegic culture. Drawing on recent scholarship on compilation practices, Mertens situates the alchemical texts preserved in the tenth-century **M** within broader trends of excerpting, abridging, and reorganizing that are characteristic of Byzantine literary production. Through a detailed analysis of the *Authentic Memoirs* and the *Chapters to Eusebeia*, the study highlights the extent to which Zosimus' works were reshaped by compilers – sometimes preserving integral pieces, sometimes reducing them to excerpts, and occasionally interpolating later material. A preliminary edition to and commentary on one chapter to Eusebeia, namely the Περὶ ἀφορμῶν συνθέσεως (Compilation "On Principles"), illustrates these processes. Through this approach, the paper underscores the value of compilation studies for understanding the Byzantine reception of alchemical literature and the ways in which Zosimus' voice was mediated by florilegic practices. In doing so, it effectively suggests that studies of compilation in Byzantium should also take into account the "chrysopoetic" genre.

Maria Papathanassiou's contribution, "The Poet-Alchemists and Their Relation to Stephanus of Alexandria", explores the connections between Stephanus of Alexandria's alchemical lectures and the four Byzantine iambic poems on gold-making attributed to Heliodorus, Theophrastus, Hierotheus, and Archelaus. According to Papathanassiou, each poem reworks themes drawn from Stephanus, ranging from the transformation of the "living stone" and the symbolism of the *ouroboros* dragon to the description of an astronomical phenomenon in 617 CE and imagery reminiscent of Holy Week liturgy. With regard to the latter, this section of the paper is well embedded within the same interpretive framework as an earlier study by Papathanassiou, which examines the imprint of the Greek patristic tradition on Stephanus' work,[73] and thus touches upon aspects of alchemy and religion in Byzantium. Papathanassiou also addresses the question of authorship of the poems, considering whether they may be read as the work of four different authors or of a single disciple of Stephanus.

My paper, "Alchemy in Byzantine Hagiography: Menia the Wise and Heliodorus the Sorcerer", analyzes how Byzantine hagiography incorporated ideas of alchemy and metallurgical transformation, focusing on two figures: Menia the Wise in the *Life of Saint Pancratius of Taormina* and Heliodorus the Sorcerer in the *Life of Saint Leo of Catania*. Menia is portrayed as divinely endowed with metallurgical knowledge, producing both precious and base metals in abundance to enrich her city – an image that resonates with alchemy as a divinely sanctioned and beneficial art. By contrast, Heliodorus' counterfeit transformations of base materials into precious ones are depicted as diabolical deceptions. The paper develops a parallel reading of these episodes alongside selected alchemical texts, focusing on terminology and recurring motifs. Through these contrasting portrayals of Menia and Heliodorus, hagiography reflects broader Byzantine attitudes toward the "sacred art", distinguishing between "true alchemy" – associated with wisdom, divine inspiration, and communal benefit – and fraudulent practices linked to deception, approximating in this sense sorcery. The paper highlights how alchemical notions permeated popular genres, shaping moral discourse and cultural imagination well beyond the circles of specialist practitioners.

Alexandre Roberts' contribution, titled "Alchemy, History, and Myth in the *Suda* Lexicon", examines the alchemy-related entries in the *Suda*, a cornerstone of Middle Byzantine scholarship, and explores how the lexicon framed *chēmeia* and its place in history, myth, and culture. Roberts shows that the *Suda* presents *chēmeia* primarily as the legitimate craft of producing gold and silver, condemned only by tyrannical figures such as Emperor Diocletian, who is portrayed as burning alchemical books alongside Christian scriptures. The *Suda* further embeds alchemy within myth, identifying the Golden Fleece as a parchment book of chrysopoetic recipes. It also refers to Zosimus of Panopolis in such a way as to imply that his work had a practical chemical aspect. Taken together, these renowned and important entries – which seem even to have brought

73 Papathanassiou 2018.

later humanists into contact with the term *chēmeia* itself[74] – reveal that Byzantine intellectuals could regard *chēmeia* as both a rational *technē* and a practice of deep antiquity, integral to their cultural and historical imagination.

Günter Prinzing's paper, "On the Relationship between the Two Versions of the Late Byzantine Alchemical Recipe Collection on the Art of Goldsmithing", investigates the transmission of the Byzantine recipe collection *On the Most Praiseworthy and Illustrious Art of Goldsmithing*, preserved in *Parisinus graecus* 2327 (Chandax, 1478) and in *Meteora, Mone Hagiou Stephanou* 97 (1503/1504).[75] Prinzing compares the two versions in meticulous detail, examining differences in content, headings, and linguistic features, while tracing both back to a common ancestor. His analysis clarifies the composition and number of recipes, the role of omissions and later additions, and the scribal interventions shaping each manuscript. The study highlights the significance of these collections for understanding the preservation and circulation of artisanal knowledge in Byzantium, as well as their place within the broader manuscript tradition of alchemy and goldsmithing.

Part II of the volume turns outward, tracing the contact zones in which Byzantine alchemy intersected with Syriac and Arabic traditions and, in later centuries, engaged medieval Latin concepts, as well as early modern debates about the nature and history of chemical knowledge.

This part opens with Maria Mavroudi's paper, which sets the stage for the comparative reading of Byzantine and Islamicate alchemy and organically links the preceding part with the present one. Her contribution, "Alchemy, Philosophy, and Related Disciplines between the Greek and the Islamicate World, Seventh–Ninth Centuries: Towards a New Narrative on the Medieval Transmission of Philosophy and Science", challenges the entrenched linear narrative of intellectual transmission from antiquity to modernity by re-examining Byzantium's role during what have inaccurately been called the "dark centuries". Mavroudi argues that alchemy and related fields of knowledge (Hermetism, the science of letters, theosophy, astrology, medicine, among others) provide crucial evidence that Byzantine intellectual life did not vanish after Late Antiquity but evolved in tandem, and in dialogue, with early Arabic learning. Her study highlights the Byzantine contribution to the intellectual milieu of the seventh–ninth centuries, showing that Byzantine learning was contemporary with, and relevant to, the formation of early Islamicate philosophy and science rather than a mere antiquarian residue. In doing so, Mavroudi proposes a new model of transmission: not a unidirectional sequence of handovers between civilizations, but a dynamic, multidirectional continuum shaped by overlapping intellectual, religious, and social contexts. The paper

---

74 Matton 1995: 311–315.

75 This recipe collection constitutes the focus of the Mainz-based interdisciplinary project *Der griechische Traktat "Über die hochgeschätzte und berühmte Goldschmiedekunst" – Edition und interdisziplinärer Kommentar* (see note 59 above).

thus repositions Byzantium as an active participant in the medieval history of science and philosophy, with alchemy serving as a key case study.

Matteo Martelli, in "Homeric Alchemy between Byzantium and Baghdad", explores the intriguing intersections between Homeric poetry and alchemical traditions across Byzantine, Syriac, and Arabic contexts. While Byzantine alchemical anthologies scarcely mention Homer and only in general allegorical terms, Syriac and Arabic sources elevate the poet as an authoritative alchemical figure. The paper situates these developments within a broader late antique and Byzantine tendency to allegorize Homeric myth and epic as repositories of natural philosophy, technical knowledge, and transformative arts. Martelli shows that in the Islamicate world these allegorical strategies were extended, leading to a direct portrayal of Homer as an alchemical authority, whereas Byzantine sources remained more restrained. By tracing these parallel yet divergent receptions, the study sheds light on the dynamics of cultural transmission between Byzantium and the Islamicate world.

If Martelli reconstructs the pathways by which Homer acquired alchemical authority beyond Byzantium, Bojidar Dimitrov follows that thread within the *Corpus Gabirianum*. Dimitrov's paper, "Pseudepigraphical References to Homer in Some Ǧābirian Texts", analyzes Homer's pseudepigraphical role within the corpus attributed to Jābir ibn Ḥayyān, where the poet becomes a sage, whose authority encompasses multiple fields of knowledge. Dimitrov highlights passages linking Homer with Pythagorean number symbolism, theories of artificial generation, and with figures such as Pythagoras, Plato, and Galen. Through these attributions, Homer emerges as a polymathic authority, endorsing both theoretical and practical aspects of alchemy. The paper situates this reception within the intellectual milieu of late antique Alexandria, where Neoplatonic philosophy, medicine, and natural science intersected, and interpretative strategies turned classical authors into guarantors of technical and esoteric knowledge. Though less directly related to Byzantine alchemy, the study underscores Alexandria's role as a melting pot of ideas and demonstrates how the Alexandrian environment of the last pre-Islamic – hence Byzantine – centuries fostered the themes later associated with Homer in the *Corpus Gabirianum*. In doing so, Dimitrov highlights the importance of pseudepigraphy and cross-cultural exchange in shaping Homer's image as an alchemical master in Islamic thought.

The traffic of ideas, however, did not run only eastward. The final two papers redirect the focus westward: the first shows how Late Byzantine alchemy engaged with and appropriated concepts from the medieval Latin alchemical tradition, illuminating a medieval moment of cross-cultural interaction; and the second follows a single manuscript – **M** – into its early modern reception, shaping arguments about chemistry's origins.

Athanasios Rinotas' contribution, "Western Medieval Alchemical Influences in the *Anonymous of Zuretti*: With a Special Focus on the Case of Albertus Magnus", investigates the impact of medieval Latin alchemy on Byzantine chrysopoetic discourse through the case of the *Anonymous of Zuretti*. Rinotas demonstrates how the text incorporates key doctrines associated with Albertus Magnus, including debates on art and nature, the theory of the three sulphureous humidities, and Aristotelian concepts such as *symbola*. By identifying these influences, the study reveals the intellectual permeability of Late Byzantine alchemy. The analysis highlights the *Anonymous of Zuretti* as a significant witness to the integration and transformation of Western philosophical and scientific ideas within a Graeco-Byzantine alchemical framework.

Vangelis Koutalis, in his "Χημεία, *ars sacra, alchymia, philosophia naturalis*: *Marcianus graecus* 299 (coll. 584) among the Early Modern *literati*", traces the reception and intellectual uses of **M**, the earliest surviving witness to the Greek Alchemical Corpus, from Bessarion's fifteenth-century library to its early modern readers. Koutalis highlights how the manuscript, initially preserved as part of Bessarion's humanist legacy, became entangled in debates over the origins of chemistry, the status of alchemy, and the relationship between Byzantine χημεία and natural philosophy. Early modern *literati* such as John Dee, Domenico Pizzimenti, Leone Allacci, and Juan Andrés variously reframed **M** as a source of Greek "chemistry", a philosophical *ars sacra*, or a precursor to experimental science, while later historians of chemistry marginalized its significance. By recovering these shifting perceptions, the study underscores both the manuscript's role in shaping early modern conceptions of natural knowledge and the historiographical "blind spots" that have long obscured the Byzantine contribution to the history of alchemy and chemistry.

Collectively, the ten papers assembled in this volume demonstrate the breadth and depth of Byzantine alchemical discourse, tracing its evolution from Early Byzantium to the early modern period and its interactions with surrounding intellectual traditions. They reveal that alchemy in Byzantium is not an isolated curiosity but a significant component of both the Byzantine and the broader history of science, technology, philosophy, and culture.

Bibliography

Anderson and Ivanova 2023: Benjamin Anderson and Mirela Ivanova (eds), *Is Byzantine Studies a Colonialist Discipline? Toward a Critical Historiography* (ICMA Books – Viewpoints). University Park, PA: Pennsylvania State University Press, 2023.

Bellini 2020: Marco Bellini, "Psellos' Petrified Root: Transmutations and Natural Wonders from Classical Antiquity to Byzantine Times", *Physis* 55.1–2 (2020), pp. 223–241.

Bellini 2023: Marco Bellini, *L'opera alchemica del filosofo Cristiano. Edizione, traduzione e note di commento*. PhD Dissertation. Alma Mater Studiorum – Università di Bologna, 2023.

Bellini 2025: Marco Bellini, "The Byzantine Alchemist Christianos and His Procedures on Indian Lac", *Ambix* 72.2 (2025), pp. 127–142.

Berthelot 1885: Marcellin Berthelot, *Les origines de l'alchimie*. Paris: G. Steinheil, 1885.

Berthelot and Ruelle 1887–1888: Marcellin Berthelot and Charles-Émile Ruelle (ed. and Fr. transl.), *Collection des anciens alchimistes grecs*, 3 vols. Paris: G. Steinheil, 1887–1888 (reprint: Osnabrück: O. Zeller, 1967).

Bevacqua 2025: Flavio Bevacqua, *Les œuvres du Philosophe Anépigraphe entre alchimie, néoplatonisme et philosophie naturelle. Édition critique, traduction et commentaire*. PhD Dissertation. Sorbonne Université; Università degli Studi di Padova, 2025.

Boned Fernández 2025: María Belén Boned Fernández, *Los poemas alquímicos atribuidos a Heliodoro, Teofrasto, Hieróteo y Arquelao: edición, traducción y comentario*. PhD Dissertation. Universidad Complutense de Madrid, 2025.

Cameron 2014: Averil Cameron, *Byzantine Matters*. Princeton, NJ: Princeton University Press, 2014.

Cameron 2016: Averil Cameron, "Late Antiquity and Byzantium: An Identity Problem", *Byzantine and Modern Greek Studies* 40.1 (2016), pp. 27–37.

Carlotta 2022: Vincenzo Carlotta, "Introducing Greek Alchemy to Christianity. Inclusion and Exclusion of Religious Elements in Stephanus's *Lessons*", *ARYS* 20 (2022), Special Issue: *Chemical Arts and Religion in Antiquity*, pp. 323–348.

Carlotta 2023: Vincenzo Carlotta, "Translating Alchemical Practice into Symbols: Two Cases from Codex *Marcianus graecus* 299", in Katja Krause, Maria Auxent, and Dror Weil (eds), *Premodern Experience of the Natural World in Translation*. New York: Routledge, 2023, pp. 217–229.

Carlotta (Forthcoming): Vincenzo Carlotta, *The Alchemical Teachings of Cleopatra* (Sources of Alchemy and Chemistry) (Forthcoming).

Carlotta and Martelli 2023: Vincenzo Carlotta and Matteo Martelli, "Metals as Living Bodies. Founts of Mercury, Amalgams, and Chrysocolla", *Ambix* 70.1 (2023), pp. 7–30.

Colinet 2000a: Andrée Colinet (ed. and Fr. transl.), *L'Anonyme de Zuretti ou l'Art sacré et divin de la chrysopée par un anonyme* (Les Alchimistes grecs 10). Paris: Belles Lettres, 2000.

Colinet 2000b: Andrée Colinet, "Le *Travail des quatre éléments* ou lorsqu'un alchimiste byzantin s'inspire de Jabir", in Isabelle Draelants, Anne Tihon, and Baudouin van den Abeele (eds), *Occident et Proche-Orient: contacts scientifiques au temps des Croisades* (Réminisciences 5). Turnhout: Brepols, 2000, pp. 165–190.

Colinet 2005: Andrée Colinet, "L'*Anonyme de Zuretti*, un traité alchimique italo-grec de 1300", in Viano 2005a, pp. 135–152.

Colinet 2010: Andrée Colinet (ed. and Fr. transl.), *Recettes alchimiques (Par. gr. 2419; Holkhamicus 109); Cosmas le Hiéromoine, Chrysopée* (Les Alchimistes grecs 11). Paris: Belles Lettres, 2010.

Halleux 1981: Robert Halleux (ed. and Fr. transl.), *Papyrus de Leyde. Papyrus de Stockholm. Fragments de recettes* (Les Alchimistes grecs 1). Paris: Belles Lettres, 1981.

Halleux 2021: Robert Halleux (ed. and Fr. transl.), *Traités des arts et métiers* (Les Alchimistes grecs 9.1). Paris: Belles Lettres, 2021.

Hanegraaff 2012: Wouter J. Hanegraaff, *Esotericism and the Academy: Rejected Knowledge in Western Culture*. Cambridge: Cambridge University Press, 2012.

Hunger 1978: Herbert Hunger, *Die hochsprachliche profane Literatur der Byzantiner*, 2 vols (Handbuch der Altertumswissenschaft, Abt. 12, Byzantinisches Handbuch 5). Munich: C. H. Beck, 1978.

Katsiampoura 2008: Gianna Katsiampoura, "Transmutation of Matter in Byzantium: The Case of Michael Psellos, the Alchemist", *Science and Education* 17 (2008), pp. 663–668.

Kazhdan 1991: Alexander Kazhdan, "Byzantium", in Alexander P. Kazhdan (ed.-in-chief), *The Oxford Dictionary of Byzantium*, 3 vols. New York and Oxford: Oxford University Press, 1991, vol. 1, pp. 344–345.

Koutalis 2022: Vangelis Koutalis, *Η αλχημεία στο έργο των ελληνόφωνων λογίων της Ιταλίας: 15ος–17ος αιώνας (Βησσαρίων, Angelo Forte, Leone Allacci)*. PhD Dissertation. University of Ioannina, 2022.

Koutalis (Forthcoming): Vangelis Koutalis, *Graeco-Egyptian Alchemy: A Catalogue of Manuscripts in Greek Libraries* (Medicine in the Medieval Mediterranean). Berlin and Boston, MA: De Gruyter (Forthcoming).

Koutalis et al. 2018: Vangelis Koutalis, Matteo Martelli, and Gerasimos Merianos, "Graeco-Egyptian, Byzantine and Post-Byzantine Alchemy: Introductory Remarks", in Nicolaidis 2018, pp. 11–43.

Leontsini and Merianos 2016: Maria Leontsini and Gerasimos Merianos, "From Culinary to Alchemical Recipes. Various Uses of Milk and Cheese in Byzantium", in Ilias Anagnostakis and Antonella Pellettieri (eds), *Latte e Latticini. Aspetti della produzione e del consumo nelle società mediterranee dell'Antichità e

del Medioevo (Collana MenSALe, Documenta et monumenta 6). Lagonegro: Grafica Zaccara, 2016, pp. 205–222.
Magdalino and Mavroudi 2006: Paul Magdalino and Maria Mavroudi (eds), *The Occult Sciences in Byzantium*. Geneva: La Pomme d'or, 2006.
Manco and Martelli 2024: Caterina Manco and Matteo Martelli, "Is Gold Yellow? Plant Dyes and Gold-Making in the Ancient Chemical Arts", *Ambix* 71.1 (2024), pp. 10–34.
Martelli 2013: Matteo Martelli (ed. and Engl. transl.), *The* Four Books *of Pseudo-Democritus* (Sources of Alchemy and Chemistry 1; Ambix 60, Supplement 1). Leeds: Maney Publishing, 2013.
Martelli 2016: Matteo Martelli, "Greco-Egyptian and Byzantine Alchemy", in Georgia L. Irby (ed.), *A Companion to Science, Technology, and Medicine in Ancient Greece and Rome* (Blackwell Companions to the Ancient World). Chichester: Wiley Blackwell, 2016, pp. 217–231.
Martelli 2018: Matteo Martelli, "Byzantine Alchemy in Two Recently Discovered Manuscripts in Saint Stephen's (Meteora) and Olympiotissa (Elassona) Monasteries", in Nicolaidis 2018, pp. 99–118.
Martelli 2019a: Matteo Martelli, *L'alchimista antico. Dall'Egitto greco-romano a Bisanzio* (Storie della scienza). Milan: Editrice Bibliografica, 2019.
Martelli 2019b: Matteo Martelli, "Galen in the Late Antique, Byzantine, and Syro-Arabic Alchemical Traditions", in Petros Bouras-Vallianatos and Barbara Zipser (eds), *Brill's Companion to the Reception of Galen* (Brill's Companions to Classical Reception 17). Leiden and Boston, MA: Brill, 2019, pp. 577–593.
Martelli 2022: Matteo Martelli, "Transmuting Tinctures. Water of Sulphur, Quicklime, and 'Washes' in Graeco-Egyptian Alchemy", *Technai* 13 (2022), Special Issue: *Ancient Science and Technology of Colour. Pigments, Dyes, Drugs and Their Experience in Antiquity* (Part 1), pp. 115–142.
Martelli 2024: Matteo Martelli, "Late Byzantine Alchemical Recipe Books: Metallurgy, Pharmacology, and Cuisine", in Petros Bouras-Vallianatos and Dionysios Stathakopoulos (eds), *Drugs in the Medieval Mediterranean: Transmission and Circulation of Pharmacological Knowledge*. Cambridge: Cambridge University Press, 2024, pp. 336–365.
Martelli and Valente 2013: Matteo Martelli and Stefano Valente, "Per una nuova edizione commentata di un lessico alchemico bizantino", *Eikasmos* 24 (2013), pp. 275–296.
Martelli et al. 2022: Matteo Martelli, Sébastien Moureau, and Jennifer M. Rampling, "Theory and Concepts: The Shared Heritage of Byzantine, Arabo-Muslim, and Latin Alchemy", in Charles Burnett and Sébastien Moureau (eds), *A Cultural History of Chemistry*, vol. 2: *A Cultural History of Chemistry in the Middle Ages*. London: Bloomsbury Academic, 2022, pp. 17–34.
Matton 1995: Sylvain Matton, "L'influence de l'humanisme sur la tradition alchimique", *Micrologus* 3 (1995), pp. 279–345.
Mavroudi 2002: Maria Mavroudi, *A Byzantine Book on Dream Interpretation: The* Oneirocriticon of Achmet *and Its Arabic Sources* (The Medieval Mediterranean 36). Leiden: Brill, 2002.
Mavroudi 2006: Maria Mavroudi, "Occult Science and Society in Byzantium: Considerations for Future Research", in Magdalino and Mavroudi 2006, pp. 39–95.
Mavroudi 2013: Maria Mavroudi, "Science, Byzantine", in Roger S. Bagnall et al. (eds), *The Encyclopedia of Ancient History*. Chichester: Wiley-Blackwell, 2013, pp. 6063–6065.
Mavroudi 2015: Maria Mavroudi, "Translations from Greek into Latin and Arabic during the Middle Ages: Searching for the Classical Tradition", *Speculum* 90.1 (2015), pp. 28–59.
Mavroudi 2023: Maria Mavroudi, "Byzantine Translations from Arabic into Greek: Old and New Historiography in Confluence and in Conflict", *Journal of Late Antique, Islamic and Byzantine Studies* 2.1–2 (2023), pp. 215–288.
Merianos 2014: Gerasimos Merianos, "Oil and Wine in Byzantine Alchemical Recipes", in Antonella Pellettieri (ed.), *Identità euromediterranea e paesaggi culturali del vino e dell'olio* (Collana MenSALe, Documenta et monumenta 2). Foggia: Centro Grafico, 2014, pp. 249–259.
Merianos 2017: Gerasimos Merianos, "Alchemy", in Anthony Kaldellis and Niketas Siniossoglou (eds), *The Cambridge Intellectual History of Byzantium*. Cambridge: Cambridge University Press, 2017, pp. 234–251.

Merianos 2022: Gerasimos Merianos, "The Christianity of the Philosopher Christianos. Ethics and Mathematics in Alchemical Methodology", *ARYS* 20 (2022), Special Issue: *Chemical Arts and Religion in Antiquity*, pp. 271–322.

Merianos 2024a: Gerasimos Merianos, "Aspects of the Relationship between Alchemy and Christianity in Byzantium", in Kostas Tampakis and Ronald Numbers (eds), *Orthodox Christianity and the Study of Nature: Histories of Interaction* (Science and Orthodox Christianity 4). Turnhout: Brepols, 2024, pp. 21–54.

Merianos 2024b: Gerasimos Merianos, "Innovation and Byzantine Alchemy in Context: The Constantinian *Solidus* and the Chrysopoetic Goal", *De Medio Aevo* 13.2 (2024), Special Issue: *Scientific Interests and Technological Innovation in Byzantium: Interdisciplinary Perspectives*, pp. 307–319.

Merianos 2024c: Gerasimos Merianos, "Insights on Alchemy, Deception, and Artisanal Knowledge in Byzantium", in Antje Bosselmann-Ruickbie, Zachary Chitwood, Johannes Pahlitzsch, and Martin Marko Vučetić (eds), *Byzanz am Rhein. Festschrift für Günter Prinzing anlässlich seines 80. Geburtstags* (Mainzer Veröffentlichungen zur Byzantinistik 18). Wiesbaden: Harrassowitz, 2024, pp. 145–164.

Merianos 2025: Gerasimos Merianos, "Byproducts of Winemaking in Greek Alchemy: *Pheklē* and *Trygia*", in Steven M. Oberhelman (ed.), *Manuscripts, Plants, Remedies in the Mediterranean Traditions: Studies across Disciplines for Alain Touwaide*, vol. 3: *Remedies: Pharmacy, Drugs, Archaeology, Tradition* (Medical Traditions 6.3). Berlin and Boston, MA: De Gruyter, 2025, pp. 587–610.

Merianos and Sakorrafou 2013: Gerasimos Merianos and Sandy Sakorrafou, "Μαρτυρίες περί αλχημείας στο Βυζάντιο σε μη αλχημικά κείμενα", in Eirini Mergoupi-Savaidou, Gerasimos Merianos, Faidra Papanelopoulou, and Christiana Christopoulou (eds), *Επιστήμη και Τεχνολογία. Ιστορικές και Ιστοριογραφικές Μελέτες* (Εταιρεία Μελέτης και Διάδοσης της Ιστορίας των Επιστημών και της Τεχνολογίας 1). Athens: Ekdotike Athenon, 2013, pp. 45–65.

Mertens 1995: Michèle Mertens (ed. and Fr. transl.), *Zosime de Panopolis, Mémoires authentiques* (Les Alchimistes grecs 4.1). Paris: Belles Lettres, 1995.

Mertens 2006: Michèle Mertens, "Graeco-Egyptian Alchemy in Byzantium", in Magdalino and Mavroudi 2006, pp. 205–230.

Monod 2013: Paul Kléber Monod, *Solomon's Secret Arts: The Occult in the Age of Enlightenment*. New Haven, CT: Yale University Press, 2013.

Moureau 2020: Sébastien Moureau, "*Min al-Kīmiyā' ad Alchimiam*. The Transmission of Alchemy from the Arab-Muslim World to the Latin West in the Middle Ages", *Micrologus* 28 (2020), pp. 87–141.

Newman and Principe 1998: William R. Newman and Lawrence M. Principe, "Alchemy vs. Chemistry: The Etymological Origins of a Historiographic Mistake", *Early Science and Medicine* 3.1 (1998), pp. 32–65.

Nicolaidis 2018: Efthymios Nicolaidis (ed.), *Greek Alchemy from Late Antiquity to Early Modernity* (De Diversis Artibus 104). Turnhout: Brepols, 2018.

Papathanassiou 1990: Maria Papathanassiou, "Stephanus of Alexandria: Pharmaceutical Notions and Cosmology in His Alchemical Work", *Ambix* 37.3 (1990), pp. 121–133.

Papathanassiou 2002: Maria K. Papathanassiou, "Metallurgy and Metalworking Techniques", in Angeliki E. Laiou (ed.-in-chief), *The Economic History of Byzantium: From the Seventh through the Fifteenth Century*, 3 vols (Dumbarton Oaks Studies 39). Washington, DC: Dumbarton Oaks Research Library and Collection, 2002, vol. 1, pp. 121–127.

Papathanassiou 2005: Maria K. Papathanassiou, "L'œuvre alchimique de Stéphanos d'Alexandrie: structure et transformations de la matière, unité et pluralité, l'énigme des philosophes", in Viano 2005a, pp. 113–133.

Papathanassiou 2006: Maria Papathanassiou, "Stephanos of Alexandria: A Famous Byzantine Scholar, Alchemist and Astrologer", in Magdalino and Mavroudi 2006, pp. 163–203.

Papathanassiou 2017: Maria K. Papathanassiou (ed.), *Stephanos von Alexandreia und sein alchemistisches Werk. Die kritische Edition des griechischen Textes eingeschlossen*. Athens: Cosmosware, 2017.

Papathanassiou 2018: Maria K. Papathanassiou, "Stéphanos d'Alexandrie. La tradition patristique dans son œuvre alchimique", in Nicolaidis 2018, pp. 71–97.

Papathanassiou 2020: Maria K. Papathanassiou, "The Occult Sciences in Byzantium", in Stavros Lazaris (ed.), *A Companion to Byzantine Science* (Brill's Companions to the Byzantine World 6). Leiden and Boston, MA: Brill, 2020, pp. 464–495.

Pausillo 2022: Giorgia Pausillo, *Un nuovo catalogo dei manoscritti alchemici greci delle biblioteche italiane*. PhD Dissertation. Alma Mater Studiorum – Università di Bologna, 2022.

Pausillo (Forthcoming): Giorgia Pausillo, *A New Catalogue of Greek Alchemical Manuscripts in Italian Libraries* (Medicine in the Medieval Mediterranean). Berlin and Boston, MA: De Gruyter (Forthcoming).

Pingree and Cutler 1991: David Pingree and Anthony Cutler, "Alchemy", in Alexander P. Kazhdan (ed.-in-chief), *The Oxford Dictionary of Byzantium*, 3 vols. New York and Oxford: Oxford University Press, 1991, vol. 1, pp. 54–56.

Principe 2013: Lawrence M. Principe, *The Secrets of Alchemy* (Synthesis). Chicago, IL: University of Chicago Press, 2013.

Principe and Newman 2001: Lawrence M. Principe and William R. Newman, "Some Problems with the Historiography of Alchemy", in William R. Newman and Anthony Grafton (eds), *Secrets of Nature: Astrology and Alchemy in Early Modern Europe* (Transformations). Cambridge, MA: MIT Press, 2001, pp. 385–431.

Rehm 1939: Albert Rehm, "Zur Überlieferung der griechischen Alchemisten", *Byzantinische Zeitschrift* 39.2 (1939), pp. 394–434.

Roberts 2015: Alexandre M. Roberts, *Matter Redeemed: Alchemy and Exegesis from Antioch to Constantinople, 11th Century*. PhD Dissertation. University of California, Berkeley, 2015.

Roberts 2019: Alexandre M. Roberts, "Framing a Middle Byzantine Alchemical Codex", *Dumbarton Oaks Papers* 73 (2019), pp. 69–102.

Roberts 2020: Alexandre M. Roberts, "A Greek Alchemical Epigram in Its Middle Byzantine Context", *Journal of the Warburg and Courtauld Institutes* 83 (2020), pp. 1–36.

Roberts 2022a: Alexandre M. Roberts, "Byzantine Engagement with Islamicate Alchemy", *Isis* 113.3 (2022), pp. 559–580.

Roberts 2022b: Alexandre M. Roberts, "Hierotechnicians by Name and Their Middle Byzantine Fame", *Journal of Late Antique, Islamic and Byzantine Studies* 1.1–2 (2022), pp. 167–199.

Roberts 2022c: Alexandre M. Roberts, "Islamicate Alchemy in Greek Letters on the First Page of *Marcianus graecus* 299", *Byzantinische Zeitschrift* 115.1 (2022), pp. 341–349.

Roberts 2023: Alexandre M. Roberts, "Thinking about Chemistry in Byzantium and the Islamic World", *Journal of the History of Ideas* 84.4 (2023), pp. 595–619.

Sakorrafou and Merianos 2014: Sandy Sakorrafou and Gerasimos Merianos, "John Kanaboutzes' Commentary on Dionysios of Halikarnassos: A Perception of Alchemy in a Fifteenth-Century Greek Text", in Gianna Katsiampoura (ed.), *Scientific Cosmopolitanism and Local Cultures: Religions, Ideologies, Societies*. Athens: Institute of Historical Research, National Hellenic Research Foundation, 2014, pp. 86–94.

Schreiner and Oltrogge 2011: Peter Schreiner and Doris Oltrogge (ed. and Germ. transl.), *Byzantinische Tinten-, Tuschen- und Farbrezepte* (Österreichische Akademie der Wissenschaften, Philosophisch-Historische Klasse, Denkschriften 419; Veröffentlichungen der Kommission für Schrift- und Buchwesen des Mittelalters, Reihe IV, Monographien 4). Vienna: Verlag der Österreichischen Akademie der Wissenschaften, 2011.

Steiner 2020: Shannon Steiner, *Byzantine Enamel and the Aesthetics of Technological Power, Ninth to Twelfth Centuries*. PhD Dissertation. Bryn Mawr College, 2020.

Steiner 2022: Shannon Steiner, "Nikephoros Blemmydes (1197–c. 1269): Concerning Gold Making", in Foteini Spingou (ed.), *Sources for Byzantine Art History*, vol. 3: *The Visual Culture of Later Byzantium (c. 1081–c. 1350)*. Cambridge: Cambridge University Press, 2022, pp. 432–444.

Stephanides 1927: Michael K. Stephanides, "Μουσικὴ καὶ χρυσοποιία κατὰ τοὺς βυζαντινοὺς χυμευτάς", *Epeteris Hetaireias Byzantinon Spoudon* 4 (1927), pp. 39–45.

Stouraitis 2022: Yannis Stouraitis, "Is Byzantinism an Orientalism? Reflections on Byzantium's Constructed Identities and Debated Ideologies", in Yannis Stouraitis (ed.), *Identities and Ideologies in the Medieval East Roman World* (Edinburgh Byzantine Studies). Edinburgh: Edinburgh University Press, 2022, pp. 19–47.

Tihon 2008: Anne Tihon, "Numeracy and Science", in Elizabeth Jeffreys, John Haldon, and Robin Cormack (eds), *The Oxford Handbook of Byzantine Studies*. Oxford: Oxford University Press, 2008, pp. 803–819.

Tihon 2013: Anne Tihon, "Science in the Byzantine Empire", in David C. Lindberg and Michael H. Shank (eds), *The Cambridge History of Science*, vol. 2: *Medieval Science*. New York: Cambridge University Press, 2013, pp. 190–206.

Viano 1995: Cristina Viano, "Olympiodore l'alchimiste et les présocratiques: une doxographie de l'unité (*De arte sacra*, § 18–27)", in Didier Kahn and Sylvain Matton (eds), *Alchimie: art, histoire et mythes* (Textes et travaux de *Chrysopoeia* 1). Paris: S. É. H. A.; Milan: Archè, 1995, pp. 95–150.

Viano 1996: Cristina Viano, "Aristote et l'alchimie grecque: la transmutation et le modèle aristotélicien entre théorie et pratique", *Revue d'histoire des sciences* 49.2–3 (1996), pp. 189–213.

Viano 2005a: Cristina Viano (ed.), *L'alchimie et ses racines philosophiques. La tradition grecque et la tradition arabe* (Histoire des doctrines de l'Antiquité classique 32). Paris: Librairie Philosophique J. Vrin, 2005.

Viano 2005b: Cristina Viano, "Les alchimistes gréco-alexandrins et le *Timée* de Platon", in Viano 2005a, pp. 91–107.

Viano 2015a: Cristina Viano, "*Mixis* and *Diagnôsis*: Aristotle and the 'Chemistry' of the Sublunary World", *Ambix* 62.3 (2015), pp. 203–214.

Viano 2015b: Cristina Viano, "Une substance, deux natures: les alchimistes grecs et le principe de la transmutation", in Fabienne Jourdan and Anca Vasiliu (eds), *Dualismes. Doctrines religieuses et traditions philosophiques* (= *Chôra* 13, Supplement). Iaşi: Polirom, 2015, pp. 309–325.

Viano 2018a: Cristina Viano, "Byzantine Alchemy, or the Era of Systematization", in Paul T. Keyser and John Scarborough (eds), *The Oxford Handbook of Science and Medicine in the Classical World*. Oxford: Oxford University Press, 2018, pp. 943–964.

Viano 2018b: Cristina Viano, "Olympiodore l'alchimiste et la *taricheia*. La transformation du minerai d'or: technê, nature, histoire et archéologie", in Nicolaidis 2018, pp. 59–69.

Viano 2021: Cristina Viano, "Olympiodorus and Greco-Alexandrian Alchemy", in Albert Joosse (ed.), *Olympiodorus of Alexandria: Exegete, Teacher, Platonic Philosopher* (Philosophia Antiqua 159). Leiden and Boston, MA: Brill, 2021, pp. 14–30.

Watanabe 2023: Masayo Watanabe, *Nature in the Books of Seven Metals – Ǧābirian Corpus in Dialogue with Ancient Greek Philosophy and Byzantine Alchemy*. PhD Dissertation. Alma Mater Studiorum – Università di Bologna, 2023.

Zuretti et al. 1927: Carlo Oreste Zuretti, Otto Lagercrantz, Johan Ludvig Heiberg, Ingeborg Hammer-Jensen, Domenico Bassi, and Emidio Martini, *Catalogue des manuscrits alchimiques grecs*, vol. 2: *Les manuscrits italiens*. Brussels: M. Lamertin, 1927.

Zuretti and Severyns 1928: Carlo Oreste Zuretti and Albert Severyns, *Catalogue des manuscrits alchimiques grecs*, vol. 5.1: *Les manuscrits d'Espagne*, décrits par C. O. Z.; 5.2: *Les manuscrits d'Athènes*, décrits par A. S. Brussels: M. Lamertin, 1928.

Note to the Reader

Following the style guide of the *Medical Traditions* series, the names of ancient Greek and Byzantine authors, as well as the titles of ancient and Byzantine Greek texts, are rendered in their Latin forms when referenced in the footnotes and the bibliographies. Where realistically possible, this practice has also been applied to Arabic work titles. However, some of the alchemical texts discussed in the chapters that follow have not been critically edited or widely studied and therefore lack standardized scholarly titles. In such cases, Latin titles have been created for the sake of consistency, in accordance with traditional scholarly practice.

In line with the series guidelines, ancient and Byzantine Greek personal names are likewise given in their Latinized forms in the text (e.g., "Heraclius" rather than "Herakleios"), though certain conventional exceptions are observed (e.g., "John of Antioch"). Such exceptions apply in the text and not in footnote citations. I have endeavored to maintain homogeneity in these and other matters where consistency of form and usage is desirable throughout the volume. Nevertheless, the reader may notice a few deviations, particularly in cases where individual scholarly preferences have, quite rightly, prevailed (e.g., "Jābir" and "Ǧābir"; "ibn" and "b.").

https://doi.org/10.1515/9783111589619-002

Part I: **Within and Beyond Alchemical Texts: The Byzantine Context**

Michèle Mertens

"I Could not Insert the Complete Chapters": Zosimus of Panopolis' Writings and Byzantine Florilegic Culture

Scattered within the Greek Alchemical Corpus, Zosimus of Panopolis' works raise various and complex issues as to their transmission. The present contribution intends to re-evaluate them in the light of the numerous studies which have recently been devoted to compilation literature, the goal being to better identify to what extent they were tampered with. To do so, I will first give a brief survey of the recent advances in this field, emphasizing what might help understand how our alchemical texts were handled during the Byzantine era. I will next re-examine from a compilatory viewpoint some of Zosimus' writings as handed down by the manuscript *Marcianus graecus* 299 (coll. 584; hereafter: **M**). Lastly, I will provide a preliminary edition, translation, and commentary of one of Zosimus' *Chapters to Eusebeia*, which, if read from a new perspective, could possibly account for a discrepancy between **M**'s table of contents and **M**'s real content.

1 Byzantine "Encyclopedism", "Florilegic Culture", and Compilation Literature

In 1971, Paul Lemerle published a landmark survey of Byzantine learning, the English title of which is *Byzantine Humanism: The First Phase. Notes and Remarks on Education and Culture in Byzantium from its Origins to the 10th Century.*[1] Lemerle claimed, among other ideas, that the tenth century in Byzantium was the golden age of a phenomenon which he qualified as "encyclopedism".[2] Emperor Constantine VII Porphyrogenitus (r. 913–959) supposedly played in this a major role together with the team of scholars he gathered around him, who produced a fair number of works which Lemerle called "encyclopedias". This study was so impressive that, in a manner of speaking, it almost stifled any further research on the subject for about the next two decades.[3]

In 1990, in an article entitled *La cultura della συλλογή: il cosiddetto enciclopedismo bizantino* [...],[4] which also became a milestone in the study of Byzantine literary culture,

1 Lemerle 1971; Engl. transl. Lindsay and Moffatt 1986.
2 After Krumbacher 1897: 258–264, and Dain 1953. For the history of the use of the term "encyclopedia" in Byzantine studies, see Schreiner 2011: 6–8.
3 See Odorico 2017: 23.
4 Odorico 1990.

https://doi.org/10.1515/9783111589619-003

Paolo Odorico challenged the very idea of encyclopedism in Byzantium. He showed that collections of excerpta, condensed libraries, anthologies, florilegia, and so on, to which Lemerle referred as "encyclopedias", do not deserve such a title, but are generally mere textual collections. Instead of the term "encyclopedism", the inappropriateness of which he sharply denounced, Odorico suggested an alternative label: "cultura della συλλογή", "culture du recueil" in French. Paul Magdalino proposed to use "the florilegic habit"[5] in English, and Alexandre Roberts most recently suggested to translate it, following Magdalino, as "the florilegic culture".[6] Even if "the intellectual patterns" that frame all these textual collections are "pretty much the same",[7] they actually differ by the nature of the texts they include, as well as by the way they are compiled. Odorico, therefore, insisted upon the need for making distinctions between them, and called for further studies on compilation literature that would take into account each *syllogē*'s characteristics, goals, functions, and individuality.[8] He later clarified and refined his views in several articles,[9] showing that compilation literature, presented by Lemerle as a distinctive mark of the tenth century in Byzantium, could in fact be traced back at least to Late Antiquity, if not earlier.[10] By then, the "florilegic culture" appeared as well-established and deeply rooted in scholarly practices.

Odorico's call for further research has proved successful indeed. Without any claim of being exhaustive, I have spotted more than thirty-five studies about compilation literature covering various fields of knowledge, all of them published during the last three decades.[11] By going through these studies, I gathered several pieces of information which might be of interest to better understand the context in which **M** was produced in the tenth century and the working methods of those who compiled the Greek Alchemical Corpus.

That the florilegic culture does go back to antiquity is now well established;[12] it nevertheless seems that tenth-century *syllogai* actually have something peculiar which distinguishes them from those produced in the previous centuries. Thus, Lemerle did

5 Magdalino 2011: 143.

6 Roberts 2019: 86n114. Roberts summarizes Odorico's thesis very clearly.

7 To borrow an expression from Marciniak 2013.

8 See Odorico 1990: 20: "[...] siamo costretti per il momento a fermarci, in attesa di studi che esaminino l'individualità di ogni singola silloge e la comune attitudine culturale che sta alla base della 'cultura della συλλογή'. Come si è visto, il problema è ampiamente da ridiscutere e ridefinire".

9 The most significant of which are Odorico 2011; 2014a; 2014b; 2017.

10 See Odorico 2011: 92.

11 In alphabetical order: Bouiron 2012; Cacouros 2011; Ceulemans and Van Deun 2017; Chatzelis 2019; Chatzelis and Harris 2017; Cupane 2019; De Vos et al. 2008; Debié 2015; Demoen 2013; Flusin 2002; 2004; Holmes 2010; Horster and Reitz 2010; Kaldellis 2015; Louyest 2012; Magdalino 2011; 2013; Manafis 2020; McCabe 2007; Németh 2013; 2016; 2017; 2018; Odorico, see note 9 above; Parmentier-Morin 2002; Piccione 1994; 2003; Reydams-Schils 2011; Roberto 2009; Roberts 2019; 2020; Signes Codoñer 2014; Tartaglia 2007; Van Deun and Macé 2011; Zucker 2012.

12 See, e.g., Boehm and Vallat 2020.

not characterize the tenth century as "encyclopedic" for no reason at all. As Magdalino notes, "such a concentration of collecting and compiling projects is not to be found in any other period of Byzantine history".[13] Many of these projects were commissioned by Emperor Constantine VII or by court circles,[14] a phenomenon which Magdalino sees as "an imperially led programme to put the culture of the empire in order",[15] and Ádám Bollók, as a means for Byzantine intellectuals "for expressing their vision of a present rooted in and organically continuing the past", which was a way to distinguish themselves as an elite.[16] Other features unique to the period are that the collections at the same time offer a higher proportion of recent material,[17] they are more antiquarian in character,[18] and that the compilatory movement extended to fields of knowledge not previously included, such as ancient technical literature.[19] These observations led Roberts to try and reconcile Odorico's critique with Lemerle's intuition that the tenth century was indeed special: Roberts suggests that even if most tenth-century compilations are actually no encyclopedias in the modern sense of the term, they nevertheless do reflect what we may call an "*encyclopedic* spirit".[20]

The many studies recently published on compilation literature provide interesting insights into the aims of this literature, the steps of the compiling process, the working methods of the compilers, and the specific vocabulary they used.

As for the aims of compilation literature, it seems that, generally speaking, late antique and Byzantine readers showed a preference for short and easily accessible books. As explained in a passage from the *prooemium* of the *Excerpta Constantiniana*, Emperor Constantine undertook his great enterprise "because of the scarcity of useful books, and because of the fear and dread over the profusion of words".[21] It seems that readers were terrified of the prolixity (πολυλογία) of books. So, the florilegic culture corresponded to a wish to collect a great number of texts in a small digest format. This wish induced working practices which consisted in selecting excerpts and in reorganizing

13 Magdalino 2013: 225, who adds that the label "cultura della *sylloge*" suggested by Odorico "is certainly not inaccurate, though it is also not specific enough to the production of the period". Cf. Roberto 2009: 72; Bollók 2016: 64.
14 See Magdalino 2011: 144–145; Bollók 2016: 64.
15 Magdalino 2011: 153.
16 Bollók 2016: 73–74.
17 See Magdalino 2011: 145.
18 See McCabe 2007: 300; Roberto 2009: 77; Holmes 2010: 61; Magdalino 2011: 145; Bollók 2016: 66–67.
19 See Cupane 2019: 436.
20 See Roberts 2019: 87.
21 Constantinus VII Porphyrogenitus, *Excerpta de virtutibus et vitiis 1* (ed. Büttner-Wobst 1906: 1.18–20): [...] πῇ μὲν σπάνει βίβλων ἐπωφελῶν, πῇ δὲ πρὸς τὴν ἐκτάδην πολυλογίαν δειμαινόντων καὶ κατορρωδούντων [...]; Engl. transl. Roberto 2009: 75. A French translation of the *prooemium* can be found in Lemerle 1971: 281–282.

them to create new types of works. These new compositions were so-to-say reader's digests, cheaper to produce and easier to read than the original books.[22]

In most cases, the aim of these compilations was didactical. But, more commonly, the Byzantines valued all that was useful – ὠφέλεια was very important in their view – including what was useful to the soul (ψυχωφελής). This may explain that they also devoted themselves to scholarly works of compilation.[23]

Several steps can be identified in the compilation process:[24]

- When beginning a compilation work, the first step was to collect sources. As the *prooemium* of the *Excerpta Constantiniana* makes it clear, Constantine "first decided to undertake an active search and to collect from everywhere over the *oikoumenē* books full of diverse and varied erudition".[25] As the florilegic culture was by then already deeply rooted in the scholarly tradition, it cannot be excluded that the collected sources themselves were already the result of a previous compiling process.[26]
- The second step consisted in reading all the collected sources and in selecting interesting texts and passages.[27] This was a tricky task, which required from the excerptor a great soundness of judgment. We are often at a loss to understand which principles of selection were implemented.[28] Moreover, it is very difficult to get a precise idea of to what extent excerpts were selected from the original texts.[29]

---

22 See Odorico 2011: 92: "Cette technique et cette démarche correspondent à la volonté de réunir dans un espace limité un très grand nombre de textes, ce qui procure une remarquable économie de temps et d'argent à ceux qui veulent disposer de connaissances immédiatement accessibles", and p. 103: "recueillir tout le matériel disponible sur des sujets différents, liés par une thématique commune, créer une bibliothèque de référence, transmettre le savoir sans se perdre dans les détails". Cf. Piccione 2003: 47; Zucker 2012: 5; Odorico 2014a: 373–374; Odorico 2017: 38–39.

23 See Holmes 2010: 61–62; Cacouros 2011: 68–69. On the general aims of compilation literature, see Roberto 2009: 76–77. On the importance of ὠφέλεια, see also Cavallo 2006: 81–82; Odorico 2017: 36–37.

24 See Odorico 1990: 9–10; Flusin 2002: 539–544; Piccione 2003: 54; Odorico 2017: 33 and 38–40.

25 Constantinus VII Porphyrogenitus, *Excerpta de virtutibus et vitiis 1* (ed. Büttner-Wobst 1906: 2, 2–4): [...] πρότερον μὲν ζητητικῇ διεγέρσει βίβλους ἄλλοθεν ἄλλας ἐξ ἁπάσης ἑκασταχοῦ οἰκουμένης συλλέξασθαι παντοδαπῆς καὶ πολυειδοῦς ἐπιστήμης ἐγκύμονας [...]; Engl. transl. Roberto 2009: 75.

26 See Magdalino 2011: 145: "in some [...] cases, the existence of an earlier compilation can be inferred, though not always with absolute certainty". Cf. Piccione 2003: 48; Cacouros 2011: 67; Bouiron 2012: 7; Debié 2015: 403.

27 See Piccione 2003: 47: "L'esercizio della libertà di selezione e di rifunzionalizzazione, l'arbitrio del selettore nella costituzione del florilegio [...] è il cuore del processo di elaborazione". Cf. Odorico 1990: 9; Flusin 2002: 541; Roberto 2009: 77.

28 See Brunt 1980: 487: "Epitomators in general seem to have aimed not at producing faithful resumés but at recording, sometimes at length, what they thought of most interest, and their principles of selection are at times impenetrable" (quoted by Zucker 2012: n42).

29 See Flusin 2002: 541: "Il est très difficile de savoir quelle proportion des œuvres exploitées a été recopiée par les excerpteurs".

- Then there was probably an intermediate step consisting in copying the selected texts or passages into temporary notebooks, in order to compile specific files with the textual material ready for further uses.[30] Unfortunately, information about that step is lacking. It is also possible that the selected texts were marked with marginal annotations or diacritic signs.[31]
- In the last step of the process, the selected texts or passages were reorganized in accordance with the general frame of each compilation work,[32] before the final copy was written on a new manuscript.

The final products could vary according to the compilers' working practices. Although no clear typology of compilation literature has been established so far, one may observe that some compilations are made up of a series of excerpts patched together, which Roberts calls "excerpts-collections", while other works consist of entire texts, arranged to form what Odorico calls anthologies, or miniaturized or condensed libraries, and Roberts calls "text-collections".[33] The excerpts-collections can be made of passages quoted *verbatim*, or of passages that have been abridged, summarized, paraphrased.

In order to appreciate how much texts may have been tampered with, it is instructive to go through studies dealing with the *Excerpta Constantiniana*, because in addition to the compiled version, the original sources of the *Excerpta* have sometimes been preserved.[34] This gives us a clearer picture of how the compilers might have worked. A comparison between the original text and the excerpts, when possible, shows that the excerpts are usually rather faithful quotations from the original text. However, it can also occur that the beginning or the end of the excerpt has been slightly altered, that bits of texts considered as unessential have been dropped, that a part of the text has been replaced by an *ad hoc* sentence ensuring the transition between two excerpts, that parts of the text have been summarized, or that some passages have been expanded through digressions.[35]

---

30 See Odorico 2011: 94: "nous pouvons imaginer qu'au fur et à mesure de ses lectures, l'auteur d'une compilation recopiait les textes sur des cahiers à part, avant de les inclure dans sa nouvelle compilation"; Odorico 2014a: 382: "Nous pouvons donc imaginer qu'il y avait des recueils où le matériel à exploiter était d'abord mis ensemble [...] avant de le composer". Cf. Németh 2013: 234n2: [about the *Excerpta Constantiniana*] "there must have been at least one draft copy of each collection, constituting the archetype of each final luxury copy"; and Németh 2017: 279–280 and 288–289.

31 See Flusin 2002: 544: "Une partie au moins des instructions qui leur (*sc.* aux copistes) était nécessaire devait avoir été reportée dans les marges des manuscrits [...]". Cf. Odorico 2010: 212; Németh 2013: 239.

32 See Odorico 2011: 94: "les deux principes qui sont à la base de tout ouvrage de ce type [...]: la *déstructuration* des textes et leur *restructuration* dans un nouveau contexte" [emphasis in the original].

33 See Odorico 2017: 26–27; Odorico 2011: 92; Roberts 2019: 86n116.

34 See Flusin 2004: 121–122. Cf. Kaldellis 2015: 38–39.

35 See Flusin 2004: 124–129; Roberto 2009: 78–84; Signes Codoñer 2014: 68–71.

On the whole, it seems that the essence of the authors' thinking was usually preserved, but one cannot say as much about the phrasing. Juan Signes Codoñer notes that "the freedom with which the compiler operates and summarizes the original text is striking".[36]

It has been noticed that Byzantine compilers used a specific vocabulary to describe their work.[37] I have selected the following terms, which seem particularly significant:[38]

ἀναλέγω	to gather up
ἁρμόζω	to fit together
ἐκλέγω, ἐκλογή	to select, selection
ἐπιτομή	abridgement
ἑρμηνεία, ἑρμηνεύω	interpretation, to interpret
κεφάλαια, κεφαλαιώδης	chapters, in the form of chapters
μετάφρασις, παράφρασις	(stylistic) rewriting, paraphrase
συλλέγω, συλλογή	to bring together, to collect, gathering, collection
συνάγω, συναγωγή	to bring together, bringing together, collection
συναθροίζω, συνάθροισις	to gather together, collecting
σύνταγμα, σύνταξις	arrangement, arranging, collection
συντάττω	to put in order together
συντίθημι	to assemble, to compose
ὑπόμνημα	commentary

2 Zosimus' Writings through the Lens of Compilation Studies

In this section, I will consider how the history of Zosimus' text during the Byzantine times can be evaluated against this intellectual background. More precisely, I will reappraise from a compilatory viewpoint Zosimus' writings as handed down by **M**, the oldest and most beautiful witness of the Greek Alchemical Corpus.[39]

It is worth noting that **M**'s dedicatory epigram claims the belonging of the whole collection to Odorico's "cultura della συλλογή": it says that Theodorus "put together a

36 Signes Codoñer 2014: 71. Cf. Flusin 2004: 129.

37 See Odorico 2010: 210; 2011: 94, 101, 105; Van Deun and Macé 2011b: xvi; Zucker 2012: 2, 4, 8; Demoen 2013: 93; Odorico 2014a: 375–376; Signes Codoñer 2014: 67–86; Debié 2015: 404; Flusin 2015: 37–38; Ceulemans and Van Deun 2017: 362; Odorico 2017: 33, 41; Manafis 2020: 1–10.

38 The exact meaning of some of them is not always clear and would deserve further research. Signes Codoñer (2014: 86) stresses "the importance of a more detailed study of the specific vocabulary used by the Byzantine writers for describing the processes of rewriting".

39 For hypotheses about the constitution of the alchemical corpus, see Mertens 2006: 220–224. See also Touwaide 2018: 53. For a concise and recent survey of the corpus, see Merianos 2017: 235–237.

strange collection, placed it in this book of all-wise thoughts",[40] using precisely the word συλλογή.[41] Furthermore, as **M**'s dating is in line with the wide tenth-century compilation movement, the manuscript has been connected more or less closely to Constantine VII Porphyrogenitus' great enterprises.[42]

M has received the attention of several scholars in recent times. In 1995, Henri Dominique Saffrey showed how easily **M**'s original order can be restored by moving entire quires, and demonstrated that the table of contents at the beginning of **M** is actually that of the manuscript itself.[43] Considering the magnificently decorated *pylē* that opens the manuscript, Saffrey also suggested that it was made for a high-ranking person, perhaps even for the imperial library.[44] In 2002, Jean Letrouit sharply contested Saffrey's reconstruction,[45] but curiously proposed no other solution instead. In 2007, Brigitte Mondrain and Antonio Rigo pointed out that at least four hands played a part in the copying of **M**.[46] Last but not least, in 2019 and 2020, Roberts devoted two major studies to **M** in which he settles many crucial questions. Among others, Roberts confirms that, apart from some minor modifications, Saffrey's reconstruction is well-founded and that the table of contents does reflect **M**'s original order.[47] A comparison with other non-alchemical manuscripts enables him to establish on palaeographical grounds that **M** was likely produced "sometime in the second half of the tenth century".[48] Questioning **M**'s place in Middle Byzantine book culture, Roberts reviews **M**'s contents and notices that **M** is "a collection weaving together texts and excerpts in

---

40 Engl. transl. Roberts 2020: 5. See **M**, f. 5 verso, ll. 26–27: συνῆψεν, ἐντέθεικε συλλογὴν ξένην | ἐν τῇδε βίβλῳ πανσόφων νοημάτων (= Berthelot and Ruelle 1887–1888: 2.4, 8–9). Fr. transl. Saffrey 1995: 8.

41 For other examples of the use of the word συλλογή in dedicatory epigrams, see Roberts 2020: 33–34.

42 See Mertens 2006: 222–223; Marcotte 2017: 38–39; Németh 2018: 94. On the basis of a parallelism in the use of Agatharchides between the anthropozoological collection known as the *Excerpta de natura animalium* (which was assuredly gathered on Constantine VII Porphyrogenitus' order) and the alchemical corpus, Marcotte links the latter to the same emperor's enterprises. Cf. Németh 2018: 94: "Constantine VII patronised the creation of thematic collections beyond the confines of historiography on subjects such as agriculture (*Geoponica*), veterinary medicine (*Hippiatrica*), medicine, natural history and alchemy"; Németh's note 19 refers to the *Marcianus graecus* 299 (coll. 584): "This deluxe codex has never been assigned to Constantine VII's collection. Its structure and the spirit of this alchemical collection is much like the *Hippiatrica*".

43 See Saffrey 1995. I was privileged to benefit from this reconstruction for my edition (Mertens 1995).

44 See Saffrey 1995: 1–2. I mention only for the record Schütt 1996, a published lecture that mainly deals with generalities about the beginnings of alchemy.

45 See Letrouit 2002: 85: "It is easy, however, to establish that: (1) Saffrey's codicological analysis is deficient; (2) Saffrey's hypothesis must be abandoned; the table on fol. 2 is not the old table of M".

46 Unpublished paper read at the international colloquium "Gli alchimisti greci: testi, dottrine, confronti / Les alchimistes grecs: textes, doctrines, comparaisons", held in Venice, 5–7 December 2007. Cf. Martelli 2011: 5n17: "[...] la mano principale avrebbe copiato i ff. 8–57 e 141–196; una seconda mano i ff. 58–140; una terza mano il πίναξ (ff. 2, 5^{v} e 6) ed una quarto mano i ff. 181 e 188".

47 See Roberts 2019: 75–85.

48 Roberts 2020: 25.

a single volume", that the "internal logic of **M** is complex", and that "**M**'s structure is loose and flexible, with a clear overarching structure but many apparent deviations in the details".[49] Re-examining the manuscript's structure closely, he notes that Stephanus of Alexandria's lectures, Cleopatra's texts, and the four iambic poems that are in the initial portion of **M** would have been considered by a learned Middle Byzantine reader as a good introduction to the following works, introducing "the subject matter at hand without jumping immediately into technical details".[50] The organization of the rest of the manuscript is less clear, especially that of the two large clusters of texts revolving around Democritus and Zosimus. Roberts concludes that "the manuscript has few fully integral texts outside of the introductory portion. Most of it is an artful assemblage of digested, distilled, excerpted, and epitomized texts [...] arranged in a loose structure that weaves together these different types of texts".[51] Comparing **M** with the other tenth-century collections, Roberts finds that **M**'s structure "is too loose to be closely associated with the best known set of Byzantine compilations: those produced at the court of Constantine VII Porphyrogenitus".[52] **M** is assuredly a fine manuscript, but "not a luxury manuscript of the highest quality",[53] and was likelier produced for a wealthy patron, rather than for the emperor himself.

Roberts gave a very detailed account of **M**'s introductory texts, but sketched the rest of the manuscript in a rougher manner, acknowledging that "a full account of the manuscript's structure and the logic of its composition would require a comprehensive analysis of all its texts".[54] This is why I am now going to re-analyze Zosimus' two main writings, the *Authentic Memoirs*, and the *Chapters to Eusebeia*, focusing on **M**, and referring to *Parisini graeci* 2325 (hereafter: **B**) and 2327 (hereafter: **A**) when necessary.[55]

---

49 Roberts 2019: 90, 94, and 95 respectively.

50 Roberts 2019: 96.

51 Roberts 2019: 99.

52 Roberts 2020: 3. He thinks that among Middle Byzantine compilations, **M** bears the most resemblance to the *Hippiatrica* (24n67). Németh (2018: 94) reached the same conclusion: see note 42 above.

53 Roberts 2020: 25.

54 Roberts 2019: 95.

55 For an overview of all of Zosimus' writings, see Mertens 1995: XLVII–LXIX (brief summary in English in Mertens 2006: 210–214). I will here leave aside Zosimus' *Chapters to Theodorus*, which Roberts (2019: 90–93) recognizes after myself as "a collection of excerpts (and summaries) of the sources on which it draws". Roberts has established that the Theodorus of the dedicatory epigram cannot be the same as the Theodorus addressed by Stephanus of Alexandria, as the dedicatory epigram was produced for **M** itself, and "not copied into **M** from an earlier compilation" (2020: 11), but the question of whether Zosimus' Theodorus can be identified with either of these remains open (cf. 2020: 7 and n27). As for the fourth group of Zosimus' works, namely the group formed by the *Final Count* and the *Book of Sophe*, it is only handed down by **A**.

The *Authentic Memoirs*

I have already discussed at great length elsewhere the specific problems raised by this group of writings.[56] But I would like here to point to some aspects relating to what I would call "compilatory questions".

Under the title *Authentic Memoirs* (*Hypomnemata authentica*),[57] I edited a collection of thirteen rather heterogeneous texts:

1. *De littera omega* (*On the Letter Omega*)[58]

---

2. (No title – describes a still, and deals with the *phanos*)
3. *De alembico triplici et tubo* (*On the Tribicos and the Pipe*)
4. (No title – deals principally with the making of "waters")
5. *De aqua divina* (*On Divine Water*)
6. (Diagram Ἓν τὸ πᾶν [...])
7. *De organis et caminis* (*On Instruments and Furnaces*)
8. *De evaporatione aquae divinae* (*On the Vaporization of the Divine Water*)

---

9. *De eadem aqua divina* (*On the Same Divine Water*)

---

10. *Zosimi divini De virtute* (*Zosimus the Divine, On Excellence = On the Composition of Waters*)
11. *Lectio secunda* (*Second Lecture*)
12. *Lectio tertia* (*Third Lecture*)
13. *Zosimus dicit* (*De asbesto*) (*Zosimus Says* [*Concerning Quicklime*]).

In *Parisini* **B** and **A**, these texts form a group, while in **M**, the situation is as follows:

56 See Mertens 1995: XLVII–LIV; cf. Mertens 2006: 210–211.

57 On the problems raised by this title, see Mertens 1995: XLVIII; Mertens (Forthcoming): n3 and 4. The adjective "authentic" was probably added by a compiler anxious to make it clear that he was going to copy Zosimus' original text.

58 English translations are given on first mention for clarity and used subsequently for convenience.

Table 1

Folios of M	(detail)	Titles in M	*Authentic Memoirs* (*AM*) (no. and lines of my edition)	*Pinax* no.
92 verso, l. 25–95 verso, l. 24	92 verso, l. 25–95 recto l. 15	*Zosimus the Divine, On Excellence*	= *AM* 10	[29]
	95 recto, l. 16–95 verso, l. 24	*Zosimus Says (Concerning Quicklime)*	= *AM* 13	
186 recto, l. 19–188 verso	186 recto, l. 19–188 recto, l. 11	*Zosimus, On Instruments and Furnaces*	= *AM* 7, 3–14 + *AM* 2, 1–3 + *AM* 3, 21–25 + *AM* 2, 3–20 + *AM* 4, 1–40 + *AM* 3, 1–20 + *AM* 4, 45–93	[39]
	188 recto, ll. 12–25	*On Divine Water*	= *AM* 5	[40]
	(188 verso)	(Diagrams)[59] 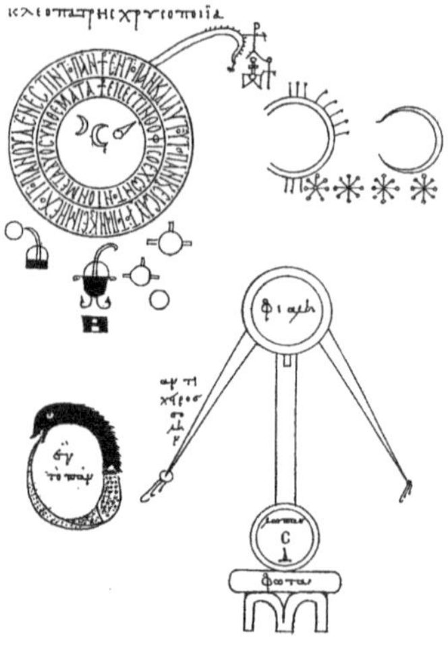	= *AM* 6	
189 recto, l. 1–196 verso	189 recto, l. 1–192 verso, l. 12	*By the Same Zosimus, On Instruments and Furnaces, Authentic Memoirs, On the Letter Omega*	= *AM* 1	[41]

59 Drawings reproduced from Berthelot 1889: 132, 138, 139–140, 146, 148, and 143 respectively.

Table 1 (continued)

Folios of M	**(detail)**	**Titles in M**	***Authentic Memoirs* (*AM*) (no. and lines of my edition)**	***Pinax* no.**
	192 verso, l. 13–193 recto, l. 25	(No title) (Diagrams on f. 193 verso) 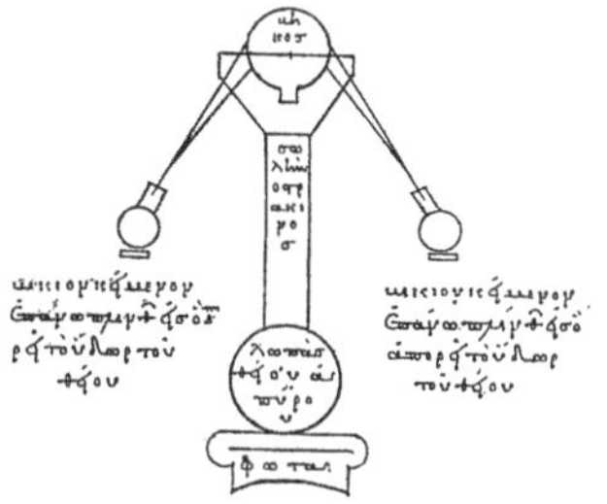 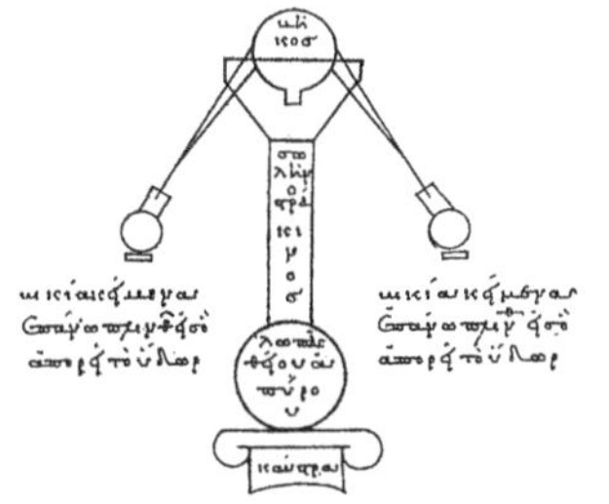	= *AM* 2	
	194 recto, l. 1–194, verso	*On the Tribicos and the Pipe* (Diagrams on f. 194 verso) 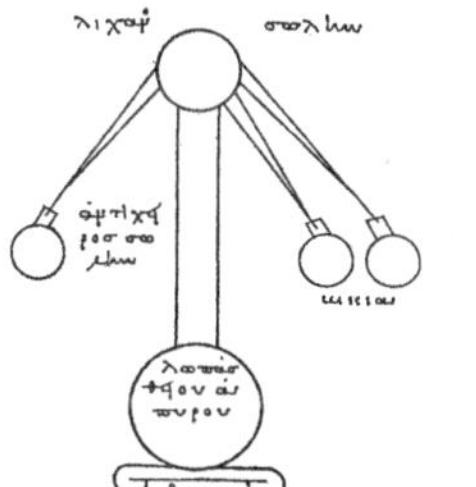 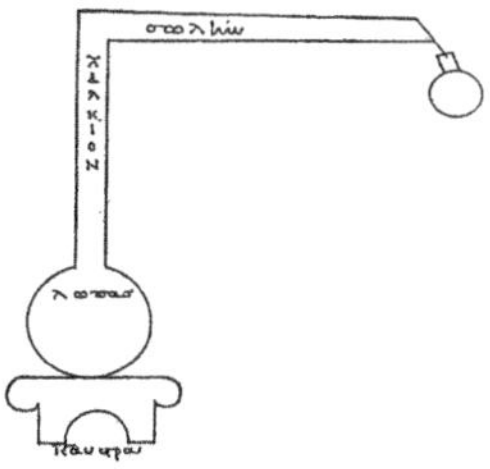	= *AM* 3	

Table 1 (continued)

Folios of M	(detail)	Titles in M	*Authentic Memoirs* (*AM*) (no. and lines of my edition)	*Pinax* no.
	195 recto, ll. 1–21	(No title)	= *AM* 4, 1–44	
	195 recto, l. 22–196 verso	*On Other Furnaces* (Diagram on f. 195 verso [lower part]) 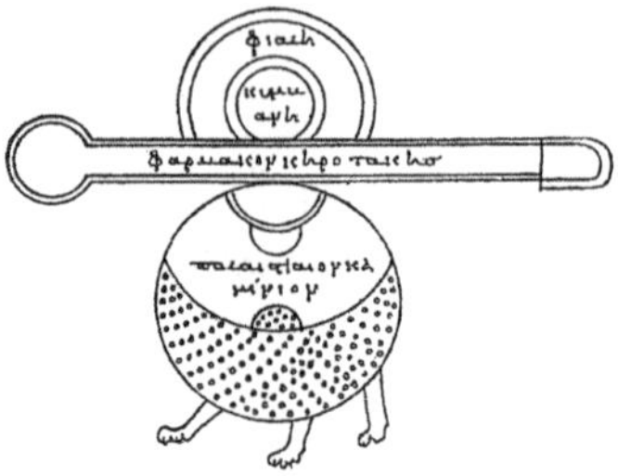 (Diagrams on f. 196 recto [upper part]) 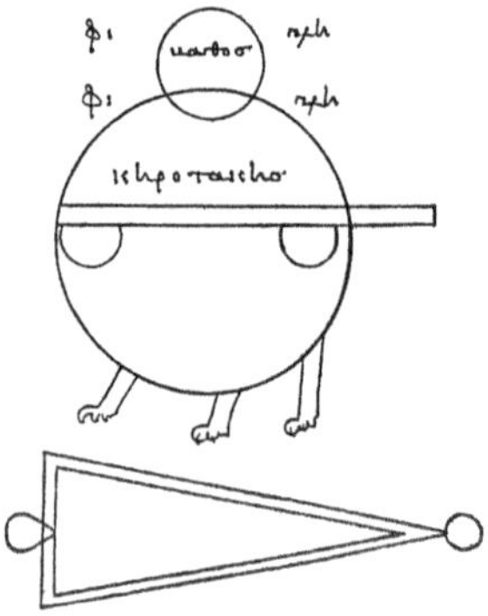 (Diagrams on f. 196 verso) 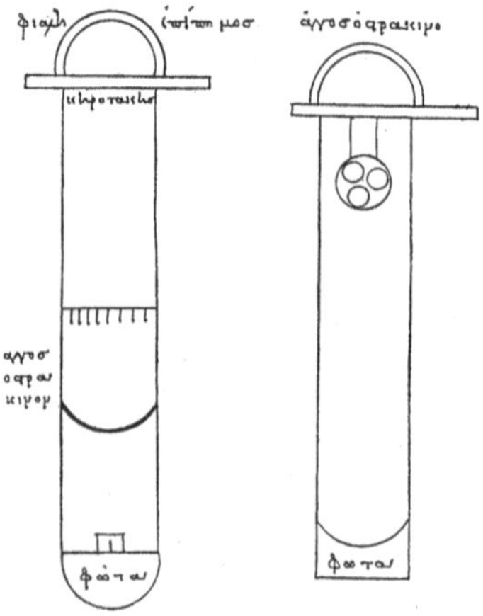	= *AM* 7, 1–7 + 15–32 + 33–42	

Table 1 (continued)

Folios of M	**(detail)**	**Titles in M**	***Authentic Memoirs* (*AM*) (no. and lines of my edition)**	***Pinax* no.**
112 recto, l. 1–115 recto, l. 17	112 recto, ll. 1–12	(No title, continued from previous)[60]	= *AM* 7, 43–52	
	112 recto, l. 13–113 recto, l. 26	*On the Vaporization of the Divine Water*	= *AM* 8	
	113 verso, l. 1–115 recto, l. 4	*On the Same Divine Water*	= *AM* 9	
	115 recto, ll. 5–17	*On the Composition of Waters*	= *AM* 10, 1–16	

It appears that the *Marcianus* actually contains the *Authentic Memoirs* in two distinct versions, but with a different sequence. In my edition, I chose to follow the order of the second version, the one that begins on f. 189, which I named **M'** to differentiate it from the first one. It seems that the compiler of the *Marcianus* had at his disposal two recensions of writings by Zosimus. To the best of my knowledge, such a situation is unique in the alchemical corpus, but not unparalleled in compilation literature.[61] A comparison between **M** and **M'** brings to light that not all the pieces have been taken up in both versions.[62]

Let us now consider each text or each subgroup of texts. The very first text of the series, the treatise *On the Letter Omega*, one of Zosimus' most beautiful texts, has come to us in remarkably good condition, handed down by the sole **M'**. It seems that we are dealing here with a complete text. Had we preserved several pieces of such high quality, we would have had a fine anthology, a pretty miniaturized library!

The same cannot be said of the following texts: *Authentic Memoirs* 2 to 8, on the contrary, are rather badly preserved and sometimes appear as excerpts.[63] For the portions

60 I follow here (as in my edition) Saffrey's reconstruction now confirmed by Roberts (2019: 81), according to which quire 24 (ff. 189–196) is to be followed by quire 14 (ff. 112–118). Why a change of hands took place between f. 196 and f. 112 (cf. note 46 above), that is in the middle of a text, would deserve further investigation.

61 See Flusin 2002: 540: [about the gathering of manuscripts with a view to the composition of the *Excerpta Constantiniana*] "Quant au nombre de manuscrits réunis, il est beaucoup plus difficile de l'estimer. Un même auteur peut en effet avoir été présent au palais en plus d'un exemplaire: c'est ainsi que P. Schreiner, par exemple, est conduit à supposer qu'il y avait deux manuscrits de Théophylacte à la disposition des excerpteurs".

62 See the table in Mertens 1995: LXXII–LXXIII. Are only in **M**: *AM* 4, 45–93; *AM* 5; *AM* 6; *AM* 7, 8–14; *AM* 10, 17–148; *AM* 13. Are only in **M'**: *AM* 1; *AM* 2, 21–39; *AM* 7, 1–3 + 15–52; *AM* 8; *AM* 9. Are common: *AM* 2, 1–20; *AM* 3, 1–27; *AM* 4, 1–44; *AM* 7, 3–7; *AM* 10, 1–16 (for the abbreviation *AM* and the texts corresponding to each number, see Table 1).

63 See Mertens 1995: 128n29: "Les ὅτι successifs trahissent l'intervention d'un compilateur qui a extrait quelques phrases d'un texte sûrement beaucoup plus long".

common to both versions, the succession of the texts is so different that to make **M** correspond to **M'**, the sequence in **M** must be cut up into small sections, and completely reorganized.[64] Moreover, a closer examination shows that the texts common to both versions usually present significant variants.[65] Another remarkable feature of these pieces lies in the great number of scientific illustrations that accompany the texts. In fact, almost all the drawings of technical apparatuses in **M** are related to this section of Zosimus' *Authentic Memoirs*.[66] Interpreting these drawings raises specific difficulties, especially as, very often, the drawing and the text do not correspond very tightly.[67] It should be noted that illustrations never duplicate each other, which may explain why the copyist chose to reproduce both versions. Moreover, there is an important difference between **M** and **M'** in the layout of the illustrations: in **M**, the drawings are put together on one page,[68] whereas in **M'**, they expand on several folios, interspersed with portions of text.

I reluctantly edited the ninth piece, entitled *On the same Divine Water* and dealing with egg distillation, among Zosimus' *Authentic Memoirs*.[69] This text is actually late,[70] and Olivier Dufault argued convincingly that it should be dated "between the age of Heraclius and the compilation of the *Marcianus graecus* 299 (ca. 600–1000 CE)".[71] One must admit that some compiler did not hesitate to interrupt Zosimus' sequence by inserting a recent text dealing with a subject akin to that of the preceding text. As already seen, the incorporation of recent texts, particularly technical recipes, into more ancient works is a characteristic of tenth-century compilations.[72]

The tenth text is in fact the first of Zosimus' so-called *Visions*, in which the alchemical operations are theatricalized and ritualized into symbolic expressions of torture, death, and resurrection, while the alchemical utensils become temples and altars. It is partly duplicated in the *Marcianus*: in **M**, it is entitled *Zosimus the Divine, On Excellence*,

64 Hence the curious sequence *AM* 7, 3–14 + *AM* 2, 1–3 + *AM* 3, 21–25, etc., in the above table.

65 A glance at the critical apparatus of *AM* 2, or at the two-column edition of *AM* 4 (ed. Mertens 1995: 11–12 and 16–18 respectively) will give some idea of the divergences.

66 Except for a strange apparatus on **M**, f. 10 recto; see Berthelot 1889: 141, fig. 18.

67 Sometimes one inevitably comes to the conclusion that some part of the description has been skipped; see Mertens 1995: 121n4: "La description est manifestement incomplète: on ne voit pas l'utilité d'un appareil constitué seulement d'une cucurbite, d'un tuyau posé sur celle-ci et d'un chapiteau; il manque la mention d'un ou de plusieurs tuyaux de décharge, ainsi que des vases de réception correspondants. Ces indications ont sans doute été omises par un épitomateur maladroit, qui aura trouvé que cette deuxième série de tuyaux et de *bicoi* faisait double emploi avec ceux de la phrase précédente".

68 On f. 188 verso. According to Mondrain and Rigo (see note 46 above), the original f. 188 was removed and replaced by the present one, which was copied by a different hand.

69 See Mertens 1995: 201. *AM* 9, as *AM* 8 which it follows, is only in **M'**.

70 See Letrouit 2002: 86: "*On the same divine water*, late anonymous text, with specific title"; Mertens (Forthcoming): n5.

71 Dufault 2017: 84.

72 See note 17 above.

and gives the first Vision in its entirety,[73] whereas in **M'**, it bears the title *On the Composition of Waters*, inspired by the first words of the text itself, and reproduces only the first Vision's first paragraph.[74] Here again, the two versions seem to derive from two separate recensions, for they differ considerably from one another.[75] It is not possible to settle whether the scribe interrupted his copying in **M'** because his model stopped at that point, or because he suddenly realized that he had already copied the text.[76]

As for the eleventh and twelfth texts, which correspond respectively to the second and third Visions, they are found only in **A**.[77] Such a case is of utmost importance to clarify the relations between the main manuscripts, for if **A** contains texts that are not, and never were, in the *Marcianus*,[78] it is clear that, at least for the *Authentic Memoirs*, **A** cannot derive from **M**.[79]

The last text of the *Authentic Memoirs*, simply entitled *Zosimus says*, deals with quicklime and is copied in **M** just after the end of the first Vision. It appears as a patched-up piece whose authenticity is difficult to assess.

To sum up,

- The treatise *On the Letter Omega* appears as a true piece of anthology; at the end of the text, Zosimus says that he is going to deal with apparatuses, so one can infer that the *Letter Omega* and the following texts have been extracted from the same part of Zosimus' works.
- *Authentic Memoirs* 2 to 8 survive in a very poor state of preservation, which is probably partly due to the bad state of the sources used by the compilers,[80] and partly to a deliberate will to abridge.
- The ninth text is not attributable to Zosimus.

---

73 **M**, f. 92 verso, l. 25–95 recto, l. 15, after a piece by the so-called Anonymous Philosopher, and followed by *AM* 13.

74 **M'**, f. 115 recto, ll. 5–17. It comes just after *AM* 9 and is followed by texts that have nothing to do with Zosimus: **M**, f. 115 recto, l. 18 (Περὶ φώτων), and f. 115 recto, l. 20 (Παραινέσεις συστατικαὶ [...]); see Mertens 1995: LIII–LIV.

75 See Zosimus Panopolitanus, *Hypomnemata authentica* (= *AM*) 10 (*De virtute*) (ed. Mertens 1995: 34–35). *Hypomnemata authentica* is cited as *Authentic Memoirs* (*AM*) in Table 1.

76 It seems that both texts were copied by Mondrain and Rigo's second hand (see note 46 above). It is worth noting that **B** and **A** share the same version as **M'**, but while **B** has only the first paragraph (just like **M'**), **A** offers the whole text of the first Vision.

77 It must be noted too that the three Visions are to be found in **A** on ff. 84 verso, l. 26–88 verso, l. 25, that is, in the first section of **A** (see Mertens 1995: XXXVII).

78 Neither in **M** nor in **M'** does the end of the (first paragraph of the) first Vision coincide with the end of a folio, thus the lack of the last two Visions cannot be connected to the loss of a quire.

79 To me, the most likely hypothesis is that **A** and part **M'** of the *Marcianus* have a common ancestor.

80 As stated by Roberto (2009: 83) about the *Excerpta Constantiniana*: "The *excerptores* could not always work on good and complete manuscripts, and as a result the *excerpta* contain many mistakes and misinterpretations".

- The *Visions* also look like pieces of anthology; they are however very different from the texts with apparatuses, and they almost certainly come from another work.

An overview of this group of texts as handed down by **M'** corroborates Roberts' observation about the whole *Marcianus*. It is an assemblage of integral texts and excerpts, arranged in a loose structure.[81] The treatise *On the Letter Omega* plays, in this cluster, the same role as Stephanus' *Lectures* do in the *Marcianus*, that is the role of a pleasant introduction. It looks as if Zosimus' *Authentic Memoirs* are a miniature of the whole collection.

The *Chapters to Eusebeia*

These chapters raise several problems,[82] the first of which concerns the title itself. In the table of contents of **M**, the title is *By Philosopher Zosimus, to Eusebeia on the Sacred and Divine Art, Thirty-Five Chapters.*[83] In **M** itself, the quire containing the title and the beginning of the text is lost, and the first leaf of the following quire (= quire no. 18) begins abruptly in the middle of a chapter. Moreover, only twenty-two chapters out of thirty-five have been preserved, so one may wonder whether the missing chapters were in the lost quire.[84] In manuscripts **B** and **A**, the title is *By Zosimus of Panopolis, Authentic Writing on the Sacred and Divine Art of Making Gold and* [*Quick*]*Silver, According to an Abridgment by Chapters.*[85] It should be noted that **B** and **A** contain only about eight more lines than **M** at the beginning of the first chapter, and not the thirteen chapters that seem to be missing, if we trust the table of contents of **M**. In my opinion, the title in **B** and **A** is a more developed form of the title of manuscript **M**'s table of contents, and must be seen as the main title of the whole work, rather than the title of the very first chapter, as its scope is too large and too general to apply to the first chapter only.[86]

The expression "according to an abridgment by chapters" (κατ' ἐπιτομὴν κεφαλαιώδη) which undoubtedly applies to the whole work,[87] reveals at once that it has

81 As a matter of fact, Odorico's "restructuring" process (2011: 94; see note 32 above) is rather hard to trace in this group of texts.
82 Overview in Mertens 1995: LIV–LX. Edition of Chapters 2, 8, and 16 in Martelli's thesis (2007: 123–139). For connections between Chapters 8 and 9, and Zosimus' Arabic tradition, see Hallum's thesis (2008: 114–132).
83 **M**, f. 2 verso, ll. 9–10: Ζωσίμου φιλοσόφου πρὸς Εὐσέβειαν περὶ τῆς ἱερᾶς καὶ θείας τέχνης κεφάλαια λε.
84 See Saffrey 1995: 5.
85 **B**, f. 118 recto, ll. 1–3 = **A**, f. 112 recto, ll. 22–24: Ζωσίμου τοῦ Πανοπολίτου γνησία γραφὴ περὶ τῆς ἱερᾶς καὶ θείας τέχνης τῆς τοῦ χρυσοῦ καὶ [ὑδρ]αργύρου ποιήσεως κατ' ἐπιτομὴν κεφαλαιώδη (the symbol of quicksilver must have been mixed up with that of silver, which is expected here).
86 If so, the first chapter admittedly does not have its own title. Contrary view in Letrouit 1995: 25n64.
87 As already observed by Hallum (2008: 118n11).

been tampered with. As discussed above, the word ἐπιτομὴ belongs to the particular vocabulary used by Byzantine compilers to describe their work. Arnaud Zucker tried to find out the meaning of ἐπιτομὴ in the context of the zoological *Syllogé* and concluded that the word is rather inconsistent and that the ἐπιτομαὶ do not form a homogeneous category. It seems that the main (and perhaps only) characteristic of an ἐπιτομὴ is to be quantitatively shorter than the original text.[88] To put it simply, an ἐπιτομὴ is "a short version", whose aim is to provide the reader with a new edition, in an easier-to-read and cheaper format.[89] To that end, the epitomizer did not hesitate to make drastic cuts, but the selected parts were usually left unchanged.[90]

So, if we try to make the best of the title in **B** and **A**, it is clear that we are not dealing with integral texts, but with a collection of excerpts, obtained by cutting (ἐπιτομή). The selected extracts were given headings (hence κεφαλαιώδη),[91] which appear to be mostly forged from words of the extracts themselves. As the writings are described as "authentic" (γνησία γραφή), we may perhaps hope that the excerpts in question have not undergone too much rewriting. That the excerpts were obtained by cutting seems confirmed by the not infrequent occurrences in these chapters of the expression καὶ τὰ ἑξῆς (*et caetera*), showing that the remainder of the passage has been dropped. Of course, the length of the lacunas is impossible to determine.

Another problem concerns the last chapters of the series. The sequence of the texts is the same in the three main manuscripts until the end of Chapter 19, but from this point, **M** contains three more chapters, **B** ends, and **A** holds only one more chapter. The extra chapter in **A**, entitled *On the Philosophical Stone*, is certainly not by Zosimus, since it appears as a series of scraps by different late authors. The second and third extra chapters in **M** respectively deal with the making of beer, and a kind of soap, and seem to have little to do with the "divine and sacred art" mentioned in the title of the table of contents. Their authenticity is dubious and should be further investigated before settling the question. As for the first additional chapter in **M**, that is Chapter 20, it will be examined in the next section.

---

88 See Zucker 2012: 5: "l'épitomé ne renvoie pas à une formule particulière de lecture, de rédaction, de composition ou, de manière générale, de récriture, mais strictement à un produit littéraire qui présente pour caractéristique principale et peut-être unique d'être quantitativement *plus courte* que l'oeuvre qu'il signale comme son modèle. Stylistiquement, méthodiquement, les *epitomai* ne constituent pas une catégorie homogène".

89 See Zucker 2012: 5: "En vérité, ce terme ne désigne donc pas une 'opération' mais un 'produit' que l'on peut appeler simplement 'version courte'. L'auteur s'engage uniquement à réduire. L'objectif principal des *epitomai* est, en effet, souvent explicitement, d'offrir la réédition d'un texte dans un format abrégé (économique), constituant une sorte de 'reader's digest'".

90 See Zucker 2012: 5: "L'*epitomé* est une version abrégée obtenue par redécoupage et suppression massive mais conservation de la littéralité originale presque dans tous les cas".

91 See Roberts 2019: 90: "[...] excerpts, usually labeled as 'chapters' or 'headings' (κεφάλαια), which in Byzantine literary culture generally indicated a selection of excerpts arranged under headings", and n135. On Byzantine *kephalaia*, see De Ridder and Levrie 2015.

3 Preliminary Edition, Translation, and Commentary of Chapter 20 to Eusebeia

This chapter has been until now unanimously considered as attributable to Zosimus,[92] but the survey of the florilegic habits we have just carried out should enable us to take a fresh look at it.

> Περὶ ἀφορμῶν συνθέσεως[93]
> Ἡ περὶ ἀφορμῶν σύνθεσις εἰς Θεοσέβειαν τὰς κατὰ μέρος τῶν ἀρχαίων συνθέσεις εἰς ἕνα νοῦν συνῆξεν· ἔτι γε μὴν καὶ ὀνόματα σύνθετα ἐν ταῖς αὐτῶν συντάξεσιν ἀγνοούμενα διὰ τοῦ πράγματος δηλοῖ, ὡς τὴν σποδὸν καὶ τὰ ὁμοιότροπα. Εἰδέναι δὲ δεῖ τίνα κατὰ τοῦ φιλοσόφου ποιεῖ τὸ πυρίμαχον, τὸ δὲ προσπλακὲν ποιεῖ πυρίμαχον, καὶ τὰ ἑξῆς. Ὁ γὰρ σοφὸς ἀφορμὰς λαβὼν πάντως ἀπὸ τῆς ἀρχῆς ἐπὶ τὸ πέρας ἀφίξεται. Ἔνθεν ἐγὼ τὰ τέλεια ἐνθεῖναι οὐκ ἠδυνήθην, ἐπείπερ οὐδὲ παρ' αὐτοῖς εὗρον· οὔτε μὴν τοῦτο προσεθέμην ὅπερ οὐδὲ τοσοῦτος, εἰ μὴ μόνον καθὼς δυνατὸν ὡς εἰκότα τὰ σκορπισθέντα συνάξαι, καὶ τὰ ἀλληγορικὰ εἶναι ἑρμηνεῦσαι. Καὶ ὅσα ἐγχωρεῖ ὑπομνήμασιν γενέσθαι ἐποίησα. Ἔρρωσο.
>
> ---
>
> εἰς Θεοσέβειαν ego : ὡς Θεοσέβειαν **M** : ὦ Θεοσέβεια BeRu

> Compilation "On Principles"
> The compilation "On Principles", addressed to Theosebeia, gathered the different elements of the ancients' compilations into one thought. Moreover, it also discloses, throughout the work, some difficult terms that are not understood in their collections, such as ashes and similar words. It should be known what species, according to the Philosopher,[94] make fire resistance, whereas another species, when mixed, makes <things> fire-resistant, etc. For the wise man, provided he has received principles, starting from the beginning, will, at any rate, reach the goal. As for me, from here onwards, I could not insert the complete <chapters>, as I did not find them by them.[95] And of course, I did not add what such a man[96] had not <added>, except when it was possible to gather the scattered parts suitably, and to interpret allegorical things. And I have done all that is allowed in commentaries. Farewell.

In my opinion, this small chapter was not written by Zosimus, but by a compiler summarizing a text by Zosimus that he was not going to copy in its entirety.

Περὶ ἀφορμῶν συνθέσεως is ambiguous since περὶ is followed by two genitives. This title seems to have been clumsily drawn from the first words of the text, which make it clear that περὶ governs ἀφορμῶν. Σύνθεσις – literally, "putting together" – has a wide array of meanings;[97] the verb συντίθημι is frequently used in compilatory contexts

---

92 Myself included: Mertens 1995: LIX. Cf. Letrouit 1995: 29.

93 **M**, ff. 161 verso, l. 28–162 recto, l. 13 = Berthelot and Ruelle 1887–1888: 2.204, 8–19 (BeRu) = Berthelot and Ruelle 1887–1888: 3.199.

94 i.e., Democritus.

95 i.e., the ancients.

96 i.e., Zosimus?

97 The word is absent from the *Lexikon zur byzantinischen Gräzität*.

with the meaning "to assemble, to compose", especially after συλλέγω "to collect".[98] So, I have translated σύνθεσις by "compilation",[99] that is, "work arising from an assembling of excerpts from previous authors", almost synonymous with σύνταξις, which appears a little below in the text. Ἀφορμαί, with the meaning of "starting-points, principles, directives", occurs several times in Zosimus' *Authentic Memoirs*.[100] Then **M** has ὡς Θεοσέβειαν, which Marcellin Berthelot and Charles-Émile Ruelle turned into a vocative, ὦ Θεοσέβεια; this correction does not fit well, because Zosimus never addresses Theosebeia in the vocative by her name, but always by the words ὦ γύναι. I would rather keep the accusative and correct ὡς in εἰς. I understand that the compiler is dealing here with a compilation by Zosimus entitled *On Principles* and addressed to Theosebeia.[101] The end of the sentence, τὰς κατὰ μέρος τῶν ἀρχαίων συνθέσεις εἰς ἕνα νοῦν συνῆξεν, echoes similar expressions in compilation literature.[102] The compiler seems aware that Zosimus already proceeded, like himself, by compiling previous works.

Then the compiler gives a hint of what Zosimus dealt with in the compilation in question: it seems that the Panopolitan took pains to clarify some difficult terms used by the ancients; the words καὶ τὰ ὁμοιότροπα could indicate that a list of species has been abridged. The next sentence contains a short quotation from Democritus,[103] and ends with καὶ τὰ ἑξῆς, which reveals that the text has been cut away. The sentence beginning with Ὁ γὰρ σοφὸς sounds almost like an excuse for not continuing to copy the text, as the wise man only needs to start from good principles to achieve his goal.

---

98 See Odorico 2011: 101: "L'exposé est fait par le biais d'un recueil des sources (συλλέξαντες), qui ont été assemblées (συνθέντες)". Cf. Demoen 2013: 93.

99 Which I use as an equivalent to "composition" in the sense in which Odorico understands it; see Odorico 2017: 41: "Pendant une longue période, on a produit des 'com-positions', mot plus approprié que 'compilations', où des passages de différents auteurs ont été choisis (ἐκλογή), réunis (ἀναθροίζω, συλλέγω), disposés (ἐκτίθημι), pour former un nouvel ouvrage".

100 Zosimus Panopolitanus, *Hypomnemata authentica* 8.1.6, 8.3.34, 8.5.58 (*De evaporatione aquae divinae*) (ed. Mertens 1995: 26, 27, 28). According to Olympiodorus (ed. Berthelot and Ruelle 1887–1888: 2.102, 17–18), "The Principle" would have been the general title given by Democritus to his four books, but Martelli (2003: 166–167) has shown that this testimony is based upon a misinterpretation by Olympiodorus of his source, Synesius. Incidentally, in the compilation literature, ἀφορμή sometimes denotes the *incipit* of an excerpt to be inserted (see Odorico 1990: 11; 2017: 39; cf. 2010: 211–212), but this meaning does not fit here.

101 About the absence of distinction during the Byzantine era between εἰς and πρὸς in book titles, and the fact that the two prepositions "are used quite indifferently", see Baldwin 1983.

102 See for instance the general preface to the *Geoponica*, Prooem. 7 (ed. Beckh 1895: 2, 16): εἰς ἓν συλλεξάμενος, and the preface to Book 1 (ed. Beckh 1895: 3, 5–8): Τὰ διαφόροις τῶν παλαιῶν [...] εἰρημένα συλλέξας εἰς ἕν, τουτὶ τὸ βιβλίον συντέθεικα (quoted by Odorico 2011: 105–106); cf. the preface to Book 20 (ed. Beckh 1895: 511, 4): [...] καὶ ἐκ διαφόρων τόπων εἰς ἕνα συναγωγήν [...] (quoted by Manafis 2020: 7).

103 [Democritus], *Physica et mystica* 16.173–174 (ed. Martelli 2011: 198 = ed. Martelli 2013: 96): καὶ εἰ τόδε μέν ἐστι πυρίμαχον, τόδε <δὲ> προσπλακὲν πυρίμαχον ποιεῖ ("and whether one species can resist fire, and another, when mixed, can make things fire-resisting"; Engl. transl. Martelli 2013: 97).

The following sentence is the most significant: the compiler confesses that he was unable to insert[104] τὰ τέλεια. The word implied after τὰ τέλεια must be κεφάλαια. I think the compiler means that he could not insert the chapters that would have completed the work, because he was not able to find them among the ancients' writings. He goes on claiming that he acted as a good compiler is supposed to, not adding anything of his own, and doing his best to gather scattered texts. Such assertions are consonant with what is found elsewhere in compilation literature. Particularly striking is the resemblance with the preface to a *synaxarion* studied by Odorico: οὐκ ἀφ' ἑαυτοῦ τι προστιθεὶς ("having added nothing by himself"), [...] ἀλλ' συνάγων εὐλαβῶς τὰ ἐσκορπισμένα ("[...] but cautiously bringing together the scattered pieces").[105] The words ἑρμηνεῦσαι and ὑπομνήμασιν in the next sentences also belong to the compilatory vocabulary. The text closes with Ἔρρωσο, which usually marks the end of a text or a collection.[106]

So, it seems that the compiler who composed this little text was commissioned to copy more chapters than those he had at his disposal. But how did he notice that some chapters were missing? We can imagine that he had received instructions to copy more chapters, presumably thirty-five, and presumably from the same source from which the drafter of the table of contents of **M** drew the number "thirty-five" mentioned in the title of this collection of chapters. Although we can take for granted that the table of contents was meant to be that of **M** itself,[107] other inconsistencies in the table suggest that its drafter[108] might have simply copied the *pinax* of his source without checking the actual contents of the manuscript.[109] Unfortunately, information is lacking about how tables of contents were worked out.[110] So we can just imagine that there was some mistake at some point in the counting of Zosimus' chapters.[111]

---

104 Note the verb ἐντίθημι, which appears in the dedicatory epigram of **M** (cf. note 40 above).

105 See Odorico 2011: 105. Full text in *Synaxarium Constantinopolitanum*, Prolegomena (ed. Delehaye 1902: col. ɪx, 23–25; Engl. transl. mine). Cf. George Cedrenus' preface (*Historiarum compendium*, Prooem. 3 [ed. Tartaglia 2016: 1.72, 25–26]): ἡμεῖς [...] τὰ εἰκότα συνελέξαμεν (quoted by Odorico 2014a: 375); "we have brought together the most appropriate passages"; Engl. transl. mine.

106 Cf. Zosimus Panopolitanus, *Hypomnemata authentica* 7.2.10 (*De organis et caminis*) (ed. Mertens 1995: 23), and note *ad loc.* (Mertens 1995: 189n10).

107 See Roberts 2019: 76–83.

108 Whose hand is different from the other hands playing a part in the copying of **M**; cf. note 46 above.

109 For instance, is it not curious that the chapter by Agathodaimon, which is only one line and a half long (**M**, f. 95 verso, ll. 26–27), deserves one full line in the table of contents (no. 30), that is as much as the *Chapters to Eusebeia* (no. 32), which occupy more than twenty folios in the manuscript? Another example is that the table announces fifteen *Chapters to Theodorus* (no. 34), whereas the title in **M** mentions no number, and whereas **M** actually contains sixteen chapters (cf. Mertens 1995: ʟxɪ), so the drafter of the table must have had another source of information.

110 It seems that they did not necessarily intend to reflect the manuscripts' exact contents; see Odorico 2014a: 381n16: "Même le pinax, qui parfois se trouve au début d'un manuscrit, ne sert à rien d'autre qu'à donner une idée du contenu, et nullement à créer une sorte d'aide au repérage des passages".

111 Did, for instance, some compiler erroneously add the sixteen *Chapters to Theodorus* to the extant nineteen *Chapters to Eusebeia*? This is purely hypothetical.

Be that as it may, the fact that the compiler took pains to note that he could find the chapters nowhere seems to account for the discrepancy between the number of chapters announced in the table of **M**, and the real content of the manuscript. If my interpretation of this small text is correct, the missing (or supposedly missing) chapters were already to be found nowhere when **M** was copied, and probably even earlier.[112]

May this little mix of compilation and Zosimean studies be useful to both domains!

Bibliography

Sources

Collected Sources

Berthelot and Ruelle 1887–1888: Marcellin Berthelot and Charles-Émile Ruelle (ed. and Fr. transl.), *Collection des anciens alchimistes grecs*, 3 vols. Paris: G. Steinheil, 1887–1888 (reprint: Osnabrück: O. Zeller, 1967).

Synaxarium Constantinopolitanum: Hippolyte Delehaye (ed.), *Propylaeum ad Acta Sanctorum Novembris. Synaxarium Ecclesiae Constantinopolitanae e codice Sirmondiano nunc Berolinensi adiectis synaxariis selectis*. Brussels: Société des Bollandistes, 1902.

Authors and Works

Constantinus VII Porphyrogenitus, *Excerpta de virtutibus et vitiis 1*: Theodor Büttner-Wobst (ed.), *Excerpta de virtutibus et vitiis, pars 1* (Excerpta historica iussu imp. Constantini Porphyrogeniti confecta 2.1). Berlin: Weidmann, 1906.

[Democritus], *Physica et mystica*: Matteo Martelli (ed. and It. transl.), *Pseudo-Democrito, Scritti alchemici, con il commentario di Sinesio* (Textes et travaux de *Chrysopoeia* 12). Paris: S. É. H. A.; Milan: Archè, 2011, pp. 180–205.

Matteo Martelli (ed. and Engl. transl.), *The* Four Books *of Pseudo-Democritus* (Sources of Alchemy and Chemistry 1; Ambix 60, Supplement 1). Leeds: Maney Publishing, 2013, pp. 79–103.

Geoponica: Heinrich Beckh (ed.), *Geoponica sive Cassiani Bassi scholastici De re rustica eclogae*. Leipzig: Teubner, 1895 (reprint: 1994).

Georgius Cedrenus, *Historiarum compendium*: Luigi Tartaglia (ed.), *Georgii Cedreni Historiarum compendium*, 2 vols (Bollettino dei Classici, Supplemento 30). Rome: Bardi, 2016.

Zosimus Panopolitanus, *Hypomnemata authentica*: Michèle Mertens (ed. and Fr. transl.), *Zosime de Panopolis, Mémoires authentiques* (Les Alchimistes grecs 4.1). Paris: Belles Lettres, 1995.

112 So, they were not in the lost quires of **M**. As for the situation in **B** and **A**, which have only eight more lines than **M** at the beginning, if we assume that they do not derive from **M**, it must be supposed that, provided these missing chapters are not fictional, they were lost before **B** and **A**'s ancestors parted from that of **M**.

Literature

Baldwin 1983: Barry Baldwin, "Book Titles in the *Suda*", *Journal of Hellenic Studies* 103 (1983), pp. 136–137.

Berthelot 1889: Marcellin Berthelot, *Introduction à l'étude de la chimie des anciens et du Moyen Âge*. Paris: G. Steinheil, 1889.

Boehm and Vallat 2020: Isabelle Boehm and Daniel Vallat (eds), *Epitome: abréger les textes antiques* (Littérature & Linguistique 2). Lyon: Maison de l'Orient et de la Méditerranée, 2020.

Bollók 2016: Ádám Bollók, "Rediscovering Antiquities and Reviving the Past: Byzantium during the 9th and 10th Centuries", *Convivium* 3 (2016), pp. 60–74.

Bouiron 2012: Marc Bouiron, "Du texte d'origine à l'*Épitomé* des *Ethnika*: les différentes phases de réduction et la transmission du lexique géographique de Stéphane de Byzance", *Rursus* 8 (2012), https://doi.org/10.4000/rursus.1027.

Brunt 1980: Peter A. Brunt, "On Historical Fragments and Epitomes", *The Classical Quarterly* 30.2 (1980), pp. 477–494.

Cacouros 2011: Michel Cacouros, "Littérature philosophique à Byzance et sa postérité à l'époque moderne", *Annuaires de l'École pratique des hautes études, Section des sciences historiques et philologiques* 142 (2011), pp. 63–70.

Cavallo 2006: Guglielmo Cavallo, *Lire à Byzance*, transl. Paolo Odorico and Alain Segonds (Séminaires byzantins 1). Paris: Belles Lettres, 2006.

Ceulemans and Van Deun 2017: Reinhart Ceulemans and Peter Van Deun, "Réflexions sur la littérature anthologique de Constantin V à Constantin VII", in Bernard Flusin and Jean-Claude Cheynet (eds), *Autour du* Premier humanisme byzantin *et des* Cinq études sur le XIe siècle, *quarante ans après Paul Lemerle* (= *Travaux et mémoires* 21.2). Paris: Association des Amis du Centre d'histoire et civilisation de Byzance, 2017, pp. 361–388.

Chatzelis 2019: Georgios Chatzelis, *Byzantine Military Manuals as Literary Works and Practical Handbooks: The Case of the Tenth-Century* Sylloge Tacticorum. London: Routledge, 2019.

Chatzelis and Harris 2017: Georgios Chatzelis and Jonathan Harris, *A Tenth-Century Byzantine Military Manual: The* Sylloge Tacticorum (Birmingham Byzantine and Ottoman Studies 22). London: Routledge, 2017.

Cupane 2019: Carolina Cupane, "The Age of Collecting and the Culture of the Sylloge (9th–10th Centuries)", in Falko Daim (ed.), *History and Culture of Byzantium* (Brill's New Pauly Supplements 10). Leiden and Boston, MA: Brill, 2019, pp. 434–436.

Dain 1953: Alphonse Dain, "L'encyclopédisme de Constantin Porphyrogénète", *Lettres d'Humanité* 12 (1953), pp. 64–81.

Debié 2015: Muriel Debié, "L'historiographie tardo-antique: une littérature en extraits", in Sébastien Morlet (ed.), *Lire en extraits: lecture et production des textes, de l'Antiquité à la fin du Moyen Âge* (Cultures et civilisations médiévales 63). Paris: Presses de l'Université Paris-Sorbonne, 2015, pp. 393–413.

Demoen 2013: Kristoffel Demoen, "La poésie de la συλλογή: les paratextes métriques des manuscrits byzantins et le (vocabulaire du) recueil", in Christian Gastgeber, Charis Messis, Dan Ioan Mureşan, and Filippo Ronconi (eds), *Pour l'amour de Byzance: Hommage à Paolo Odorico* (Eastern and Central European Studies 3). Frankfurt am Main: P. Lang, 2013, pp. 89–98.

De Ridder and Levrie 2015: Eva De Ridder and Katrien Levrie, "*Capita* Literature in Byzantium", in Reinhart Ceulemans and Pieter De Leemans (eds), *On Good Authority: Tradition, Compilation and the Construction of Authority in Literature from Antiquity to the Renaissance* (Lectio 3). Turnhout: Brepols, 2015, pp. 123–137.

De Vos et al. 2008: Ilse De Vos, Erika Gielen, Caroline Macé, and Peter Van Deun, "L'art de compiler à Byzance: la lettre Γ du Florilège Coislin", *Byzantion* 78 (2008), pp. 159–223.

Dufault 2017: Olivier Dufault, "Transmutation Theory and the Dating of the Alchemical Recipe *On the Same Divine Water*", in Andrea Le Moli and Lela Alexidze (eds), *Prote Hyle: Notions of Matter in the Platonic and Aristotelian Traditions* (Filosofie). Palermo: Palermo University Press, 2017, pp. 67–84.

Flusin 2002: Bernard Flusin, "Les *Excerpta* constantiniens: logique d'une anti-histoire", in Sylvie Pittia (ed.), *Fragments d'historiens grecs: autour de Denys d'Halicarnasse* (Collection de l'École française de Rome 298). Rome: École française de Rome, 2002, pp. 537–559.

Flusin 2004: Bernard Flusin, "Les *Excerpta* constantiniens et la *Chronographie* de Malalas", in Joëlle Beaucamp et al. (eds), *Recherches sur la Chronique de Jean Malalas*, vol. 1 (Monographies 15). Paris: Centre de recherche d'histoire et civilisation de Byzance, 2004, pp. 119–136.

Flusin 2015: Bernard Flusin, "Un auteur impérial: Constantin Porphyrogénète", in Reinhart Ceulemans and Pieter De Leemans (eds), *On Good Authority: Tradition, Compilation and the Construction of Authority in Literature from Antiquity to the Renaissance* (Lectio 3). Turnhout: Brepols, 2015, pp. 23–41.

Hallum 2008: Bink Hallum, *Zosimus Arabus: The Reception of Zosimos of Panopolis in the Arabic/Islamic World*. PhD Dissertation. Warburg Institute, 2008.

Holmes 2010: Catherine Holmes, "Byzantine Political Culture and Compilation Literature in the Tenth and Eleventh Centuries: Some Preliminary Enquiries", *Dumbarton Oaks Papers* 64 (2010), pp. 55–80.

Horster and Reitz 2010: Marietta Horster and Christiane Reitz (eds), *Condensing Texts – Condensed Texts* (Palingenesia 98). Stuttgart: Franz Steiner, 2010.

Kaldellis 2015: Anthony Kaldellis, *Byzantine Readings of Ancient Historians: Texts in Translation, with Introductions and Notes* (Routledge Classical Translations). London: Routledge, 2015.

Krumbacher 1897: Karl Krumbacher, *Geschichte der byzantinischen Litteratur von Justinian bis zum Ende des oströmischen Reiches (527–1453)* (Handbuch der klassischen Altertumswissenschaft 9.1). Second edition. Munich: C. H. Beck, 1897.

Lemerle 1971: Paul Lemerle, *Le premier humanisme byzantin. Notes et remarques sur enseignement et culture à Byzance, des origines au Xe siècle* (Bibliothèque byzantine, Études 6). Paris: Presses universitaires de France, 1971.
Engl. transl.: *Byzantine Humanism: The First Phase. Notes and Remarks on Education and Culture in Byzantium from its Origins to the 10th Century*, transl. by Helen Lindsay and Ann Moffatt (Byzantina Australiensia 3). Canberra: Australian Association for Byzantine Studies, 1986.

Letrouit 1995: Jean Letrouit, "Chronologie des alchimistes grecs", in Didier Kahn and Sylvain Matton (eds), *Alchimie: art, histoire et mythes* (Textes et travaux de *Chrysopoeia* 1). Paris: S. É. H. A.; Milan: Archè, 1995, pp. 11–93.

Letrouit 2002: Jean Letrouit, "Hermétisme et alchimie: contribution à l'étude du *Marcianus graecus* 299 (= M)", in Carlos Gilly and Cis van Heertum (eds), *Magia, alchimia, scienza dal '400 al '700. L'influsso di Ermete Trismegisto*, 2 vols. Florence: Centro Di, 2002, vol. 1, pp. 85–109.

Lexikon zur byzantinischen Gräzität: Erich Trapp et al., *Lexikon zur byzantinischen Gräzität, besonders des 9.–12. Jahrhunderts* (Österreichische Akademie der Wissenschaften, Philosophisch-Historische Klasse, Denkschriften 238, 250, 276, 293, 326, 352, 417, 494; Veröffentlichungen der Kommission für Byzantinistik 6.1–8). Vienna: Verlag der Österreichischen Akademie der Wissenschaften, 1994–2017.

Louyest 2012: Benoît Louyest, "L'épitomé du *Banquet des Sophistes* d'Athénée: vers une étude comparative entre l'épitomé et le *Marcianus*", *Rursus* 8 (2012), https://doi.org/10.4000/rursus.1045.

Magdalino 2011: Paul Magdalino, "Orthodoxy and History in Tenth-Century Byzantine 'Encyclopedism'", in Van Deun and Macé 2011a, pp. 143–159.

Magdalino 2013: Paul Magdalino, "Byzantine Encyclopaedism of the Ninth and Tenth Centuries", in Jason König and Greg Woolf (eds), *Encyclopaedism from Antiquity to the Renaissance*. Cambridge: Cambridge University Press, 2013, pp. 219–231.

Manafis 2020: Panagiotis Manafis, *(Re)writing History in Byzantium: A Critical Study of Collections of Historical Excerpts*. London: Routledge, 2020.

Marciniak 2013: Przemyslaw Marciniak, "Book review of Peter Van Deun and Caroline Macé (eds), *Encyclopedic Trends in Byzantium?* Leuven: Peeters, 2011", *Bryn Mawr Classical Review*, 2013.02.11, https://bmcr.brynmawr.edu/2013/2013.02.11/.

Marcotte 2017: Didier Marcotte, "Les mines d'or des Ptolémées: d'Agatharchide aux archives de Photios, suivi d'une Note sur les extraits constantiniens d'Agatharchide", *Journal des Savants* 2017 (no. 1), pp. 3–49.

Martelli 2003: Matteo Martelli, "L'opera alchemica dello Pseudo-Democrito: un riesame del testo", *Eikasmos* 14 (2003), pp. 161–184.

Martelli 2007: Matteo Martelli, *Zosimo di Panopoli e Sinesio: due commentatori dell'opera alchemica pseudo-democritea*. PhD Dissertation. Alma Mater Studiorum – Università di Bologna, 2007.

Martelli 2011: Matteo Martelli, *Pseudo-Democrito, Scritti alchemici, con il commentario di Sinesio* (Textes et travaux de *Chrysopoeia* 12). Paris: S. É. H. A.; Milan: Archè, 2011.

Martelli 2013: Matteo Martelli, *The* Four Books *of Pseudo-Democritus* (Sources of Alchemy and Chemistry 1; Ambix 60, Supplement 1). Leeds: Maney Publishing, 2013.

McCabe 2007: Anne McCabe, *A Byzantine Encyclopaedia of Horse Medicine: The Sources, Compilation, and Transmission of the* Hippiatrica. Oxford: Oxford University Press, 2007.

Merianos 2017: Gerasimos Merianos, "Alchemy", in Anthony Kaldellis and Niketas Siniossoglou (eds), *The Cambridge Intellectual History of Byzantium*. Cambridge: Cambridge University Press, 2017, pp. 234–251.

Mertens 1995: Michèle Mertens, *Zosime de Panopolis, Mémoires authentiques* (Les Alchimistes grecs 4.1). Paris: Belles Lettres, 1995.

Mertens 2006: Michèle Mertens, "Graeco-Egyptian Alchemy in Byzantium", in Paul Magdalino and Maria Mavroudi (eds), *The Occult Sciences in Byzantium*. Geneva: La Pomme d'or, 2006, pp. 205–230.

Mertens (Forthcoming): Michèle Mertens, "Zosime de Panopolis: état de la question à l'aube du XXIe siècle" (Forthcoming).

Németh 2013: András Németh, "The Imperial Systematisation of the Past in Constantinople: Constantine VII and His *Historical Excerpts*", in Jason König and Greg Woolf (eds), *Encyclopaedism from Antiquity to the Renaissance*. Cambridge: Cambridge University Press, 2013, pp. 232–258.

Németh 2016: András Németh, "Excerpts versus Fragments: Deconstructions and Reconstitutions of the *Excerpta Constantiniana*", in Anthony Grafton and Glenn W. Most (eds), *Canonical Texts and Scholarly Practices: A Global Comparative Approach*. Cambridge: Cambridge University Press, 2016, pp. 253–274.

Németh 2017: András Németh, "Compilation Methods of the *Excerpta Constantiniana* Revisited: From one Compilator to the Three-Stage Model of Teamwork", *Byzantinoslavica* 75 (2017), pp. 265–290.

Németh 2018: András Németh, *The* Excerpta Constantiniana *and the Byzantine Appropriation of the Past*. Cambridge: Cambridge University Press, 2018.

Odorico 1990: Paolo Odorico, "La cultura della συλλογή. 1) Il cosiddetto enciclopedismo bizantino. 2) Le tavole del sapere di Giovanni Damasceno", *Byzantinische Zeitschrift* 83 (1990), pp. 1–21.

Odorico 2010: Paolo Odorico, "'Parce que je suis ignorant': *Imitatio/Variatio* dans la chronique de Georges le Moine", in Andreas Rhoby and Elisabeth Schiffer (eds), *Imitatio – Aemulatio – Variatio* (Österreichische Akademie der Wissenschaften, Philosophisch-Historische Klasse, Denkschriften 402; Veröffentlichungen zur Byzanzforschung 21). Vienna: Verlag der Österreichischen Akademie der Wissenschaften, 2010, pp. 209–216.

Odorico 2011: Paolo Odorico, "Cadre d'exposition / Cadre de pensée: la culture du recueil", in Van Deun and Macé 2011a, pp. 89–107.

Odorico 2014a: Paolo Odorico, "Dans le cahier des chroniqueurs: le cas d'Eustathe d'Antioche", in Juan Signes Codoñer and Inmaculada Pérez Martín (eds), *Textual Transmission in Byzantium: Between Textual Criticism and Quellenforschung* (Lectio 2). Turnhout: Brepols, 2014, pp. 373–389.

Odorico 2014b: Paolo Odorico, "Prolégomènes à la littérature byzantine du IXe et du Xe siècle", in Dean Sakel (ed.), *Byzantine Culture: Papers from the Conference "Byzantine Days of Istanbul"* (AKDTYK, Türk Tarih Kurumu Yayınları, Series VIII, 12). Ankara: Türk Tarih Kurumu, 2014, pp. 93–108.

Odorico 2017: Paolo Odorico, "Du premier humanisme à l'encyclopédisme: une construction à revoir", in Bernard Flusin and Jean-Claude Cheynet (eds), *Autour du* Premier humanisme byzantin *et des* Cinq études sur le XIe siècle, *quarante ans après Paul Lemerle* (= *Travaux et mémoires* 21.2). Paris: Association des Amis du Centre d'histoire et civilisation de Byzance, 2017, pp. 23–42.

Parmentier-Morin 2002: Édith Parmentier-Morin, "Les fragments de Denys d'Halicarnasse attribués à Nicolas de Damas: recherches sur la composition des *Excerpta* constantiniens", in Sylvie Pittia (ed.), *Fragments d'historiens grecs: autour de Denys d'Halicarnasse* (Collection de l'École française de Rome 298). Rome: École française de Rome, 2002, pp. 461–479.

Piccione 1994: Rosa Maria Piccione, "Sulle fonti e le metodologie compilative di Stobeo", *Eikasmos* 5 (1994), pp. 281–317.

Piccione 2003: Rosa Maria Piccione, "Scegliere, raccogliere e ordinare: letteratura di raccolta e trasmissione del sapere", in Enrico V. Maltese (ed.), *Bisanzio tra storia e letteratura* (= *Humanitas* 58.1). Brescia: Morcelliana, 2003, pp. 44–63.

Reydams-Schils 2011: Gretchen Reydams-Schils (ed.), *Thinking through Excerpts: Studies on Stobaeus* (Monothéismes et Philosophie 14). Turnhout: Brepols, 2011.

Roberto 2009: Umberto Roberto, "Byzantine Collections of Late Antique Authors: Some Remarks on the *Excerpta historica Constantiniana*", in Martin Wallraff and Laura Mecella (eds), *Die Kestoi des Julius Africanus und ihre Überlieferung* (Texte und Untersuchungen zur Geschichte der altchristlichen Literatur 165). Berlin: De Gruyter, 2009, pp. 71–84.

Roberts 2019: Alexandre M. Roberts, "Framing a Middle Byzantine Alchemical Codex", *Dumbarton Oaks Papers* 73 (2019), pp. 69–102.

Roberts 2020: Alexandre M. Roberts, "A Greek Alchemical Epigram in its Middle Byzantine Context", *Journal of the Warburg and Courtauld Institutes* 83 (2020), pp. 1–36.

Saffrey 1995: Henri Dominique Saffrey, "Historique et description du manuscrit alchimique de Venise *Marcianus Graecus* 299", in Didier Kahn and Sylvain Matton (eds), *Alchimie: art, histoire et mythes* (Textes et travaux de *Chrysopoeia* 1). Paris: S. É. H. A.; Milan: Archè, 1995, pp. 1–10.

Schreiner 2011: Peter Schreiner, "Die enzyklopädische Idee in Byzanz", in Van Deun and Macé 2011a, pp. 3–25.

Schütt 1996: Hans-Werner Schütt, *Ein verborgener Schatz Venedigs: Der alchemische Kodex 299 aus der Biblioteca Marciana* (Quaderni 50). Venice: Centro Tedesco di Studi Veneziani, 1996.

Signes Codoñer 2014: Juan Signes Codoñer, "Towards a Vocabulary for Rewriting in Byzantium", in Juan Signes Codoñer and Inmaculada Pérez Martín (eds), *Textual Transmission in Byzantium: Between Textual Criticism and Quellenforschung* (Lectio 2). Turnhout: Brepols, 2014, pp. 61–90.

Tartaglia 2007: Luigi Tartaglia, "Meccanismi di compilazione nella *Cronaca* di Giorgio Cedreno", in Fabrizio Conca and Gianfranco Fiaccadori (eds), *Bisanzio nell'età dei Macedoni: forme della produzione letteraria e artistica* (Quaderni di Acme 87). Milan: Cisalpino, 2007, pp. 239–255.

Touwaide 2018: Alain Touwaide, "The Alchemical Manuscript Tradition: An Overview", in Efthymios Nicolaidis (ed.), *Greek Alchemy from Late Antiquity to Early Modernity* (De Diversis Artibus 104). Turnhout: Brepols, 2018, pp. 45–58.

Van Deun and Macé 2011a: Peter Van Deun and Caroline Macé (eds), *Encyclopedic Trends in Byzantium?* (Orientalia Lovaniensia Analecta 212). Leuven: Peeters, 2011.

Van Deun and Macé 2011b: Peter Van Deun and Caroline Macé, "Encyclopedic Trends in Byzantium? An Introduction", in Van Deun and Macé 2011a, pp. XIII–XIX.

Zucker 2012: Arnaud Zucker, "Qu'est-ce qu'épitomiser? Étude des pratiques dans la *Syllogé* zoologique byzantine", *Rursus* 7 (2012), http://rursus.revues.org/961.

Maria K. Papathanassiou

The Poet-Alchemists and Their Relation to Stephanus of Alexandria

Stephanus of Alexandria (late sixth–early seventh century) was a well-known and eminent scholar in Alexandria, even before his invitation by Emperor Heraclius (r. 610–641) to settle in Constantinople and teach the *quadrivium*. His lectures, comprising *On the Great and Sacred Art of Making Gold*,[1] constitute a commentary on earlier alchemical texts and influenced the poet-alchemists, as is evident from several passages of their poems. A close study of these poems reveals their dependence on Stephanus' lectures and the differing interests of their authors. Here, we focus on those topics that deserve special attention.

1 Heliodorus

Heliodorus addressed his poem to "Theodosius, the great emperor";[2] no other reference to this name appears in the verses. In Byzantine history, there were only three emperors with this name: Theodosius I the Great (r. 379–395), Theodosius II (r. 408–450), and Theodosius III (r. 715–717).[3] Whether any of these emperors were interested in the great and sacred art of chrysopoeia is doubtful. Theodosius III reigned for only two years, and we know little about him; therefore, he cannot be excluded as a possible addressee of the poem. On the other hand, we know that Heraclius had an interest in occultism, especially astrology and alchemy. He asked Stephanus to cast his horoscope and attended at least one of Stephanus' lectures on chrysopoeia.[4]

Heliodorus encourages "the emperor", without referring to him by name,[5] to introduce wise men initiated into the subject of his interest as sensible teachers, to listen attentively to their lectures and riddles, and to welcome their hopeful explanations of these matters.[6] The poem ends with a prayer to Christ, in which the poet exhorts the adepts to prove themselves wise men, caring for nothing except the worship and

---

1 Stephanus Alexandrinus, *De magna et sacra arte* (ed. Papathanassiou 2017: 157–218).

2 [Heliodorus], *De philosophorum arte mystica* (ed. Goldschmidt 1923: 26–34). Unless otherwise noted, translations are by the author.

3 Marc Lauxtermann (2019: 205–207) argues that Theodosius II is the most likely dedicatee, although he entertains the possibility that the reference to an Emperor Theodosius may be a hoax. He nevertheless dates the poems to the fifth or early sixth century on the basis of metrical analysis.

4 Stephanus Alexandrinus, *De magna et sacra arte* 9 (Title) (ed. Papathanassiou 2017: 213).

5 [Heliodorus], *De philosophorum arte mystica*, vv. 18, 32 (ed. Goldschmidt 1923).

6 [Heliodorus], *De philosophorum arte mystica*, vv. 32–41.

https://doi.org/10.1515/9783111589619-004

praise of God.[7] In these final verses, we find a significant reference to the "faithful and very pious emperors who, as worshipers of Christ, are entrusted with the state and the dominion on earth".[8] We know that Heraclius, crowned in Constantinople on 5 October 610, had, from 22 January 613, his son Heraclius Constantine (b. 3 May 612) as co-emperor. The existence of one (Heraclius) or two emperors (Heraclius and his son) could explain the references to the "emperor"/"emperors".

In Heliodorus' poem, we find specific words referring to the wise men that are not present in Stephanus' text, such as: "sophist(s)",[9] "sophistic",[10] "orators",[11] "education",[12] "instructor",[13] "mortal",[14] "initiation",[15] "initiated",[16] "those who perform a mystic rite",[17] "of the art of the chief of initiated",[18] "artistry",[19] and "rendering service".[20]

Heliodorus uses the word "science" once,[21] while Stephanus uses it twice, as well as its derivatives "scientist" (ἐπιστήμων) and "scientifically" (ἐπιστημονικῶς, ἐνεπίστημος).[22] The same applies to the word "lurking in",[23] which appears frequently in Stephanus' text.[24]

Heliodorus' poem unfolds the story of the red galactic stone. The various phases of its chemical treatment, culminating in its final transformation, are likened to the stages of an embryo's development in its mother's womb until birth.[25] The narrative includes

7 [Heliodorus], *De philosophorum arte mystica*, vv. 227–268.
8 [Heliodorus], *De philosophorum arte mystica*, vv. 257–260: [...] εὐσεβεστάτοις / ἄναξι πιστοῖς [...] / τοῖς τὸ κράτος καὶ σκῆπτρα γῆς πιστουμένοις, / ὡς χριστολάτραις [...].
9 [Heliodorus], *De philosophorum arte mystica*, v. 45: σοφισταί; v. 52: σοφιστά.
10 [Heliodorus], *De philosophorum arte mystica*, v. 8: σοφιστικῶν; vv. 15–16: σοφιστικῆς ἐννοίας.
11 [Heliodorus], *De philosophorum arte mystica*, v. 16: ῥητόρων.
12 [Heliodorus], *De philosophorum arte mystica*, vv. 5, 23, 33: παιδεία.
13 [Heliodorus], *De philosophorum arte mystica*, v. 35: παιδευτής.
14 [Heliodorus], *De philosophorum arte mystica*, v. 26: βροτοῖς; v. 228: βροτοῦ.
15 [Heliodorus], *De philosophorum arte mystica*, vv. 17, 36, 47: μύησιν.
16 [Heliodorus], *De philosophorum arte mystica*, v. 45: μύσται.
17 [Heliodorus], *De philosophorum arte mystica*, v. 2: μυστιπόλοι.
18 [Heliodorus], *De philosophorum arte mystica*, v. 59: μυσταρχικῆς τέχνης; v. 55: μυσταρχικῶς. For the translation of the term μυσταρχικῆς, see Liddell–Scott–Jones, s.v. "μυστάρχης" ("chief of μύσται"); Liddell–Scott–Jones, Supplement, s.v. "μυσταρχικός" ("of or for a μυστάρχης"). Cf. *Lexikon zur byzantinischen Gräzität*, s.vv. "μυσταρχία" (*mystische Kraft*), "μυσταρχικός" (*mystisch*).
19 [Heliodorus], *De philosophorum arte mystica*, v. 60: τεχνουργίας.
20 [Heliodorus], *De philosophorum arte mystica*, v. 91 ὑπουργός.
21 [Heliodorus], *De philosophorum arte mystica*, v. 20: ἐπιστήμην.
22 Stephanus Alexandrinus, *De magna et sacra arte* 2.134; Theod. 54; 4.65; 6.25, 85 (ed. Papathanassiou 2017: 165; 169; 175; 187; 190).
23 [Heliodorus], *De philosophorum arte mystica*, v. 207: ἐμφωλεῦον.
24 Stephanus Alexandrinus, 4.56, 70, 166, 170, 182, 186; 8.15, 17–18, 33, 64, 68, 83 (ed. Papathanassiou 2017: 174–175, 178–179; 207–210).
25 [Heliodorus], *De philosophorum arte mystica*, beginning at v. 72.

a very confusing reference to the first and second "key",[26] marking two phases in the accomplishment of the work. On the other hand, Stephanus refers to and elaborates on three "keys", which correspond to the relationships among the four primary elements (fire, air, water, earth) and the four qualities (dry, wet, warm, cold), with the four humors of the body (blood, yellow bile, phlegm, black bile) and the components of the alchemical work (quicksilver, copper, copper soul, cinnabar/rust).[27]

2 Theophrastus

Beginning his poem,[28] Theophrastus praises the wise men using the words "sophist", "orator", and "teacher", and characterizes them as "mortal".[29] He uses the word "initiation", refers to the "initiated" men, and the "very wise artifice".[30] Notably, these words do not exist in the text of Stephanus' lectures. We find only the word διδάσκαλος as his epithet in the titles of his lectures.

Theophrastus continues his narration, praising the astronomical knowledge of these men:

> ἡμεῖς μὲν ἄστρων τάξιν ὥσπερ καὶ θέσιν φύσιν τε καὶ ὕψωσιν ὡς διάστασιν, μέτρων ὁρισμοὺς ἀκριβῶς καὶ ἀνατολὰς κρύψεις τε δείξει προλέγοντες καὶ δύσεις οὐ σφαλλόμεθα, μακρὰν ὄντων τὴν θέαν τούτων ἀφ' ἡμῶν, καὶ νοὸς γνώσει μάλα ὁρῶντες, ὥσπερ εἰσὶν αἰσθήσει φρενῶν, γινώσκομεν τἀληθὲς ὡς εἶναι ἐκεῖ ἐν οὐρανῷ καὶ μηδὲν ἀγνοεῖν ὅλως τελουμένων ἐκεῖσε· [...].
>
> We do not fail in predicting with high accuracy the order of the stars, their place and nature, and their altitude as a dimension; also, their risings, settings, and periods of invisibility. Although the stars are seen far away from us, we see them through our intellectual knowledge, as if we did it through our senses, and we truly know all that is in the sky and do not ignore all that happens there.[31]

This astronomical knowledge, acquired through geometrical models, allows the prediction of the motions and positions of the stars. Ptolemaic astronomy provides kinematic geometrical models for these astronomical predictions regarding fixed stars and the

---

26 [Heliodorus], *De philosophorum arte* mystica, vv. 201, 211: κλείς.
27 Stephanus Alexandrinus, *De magna et sacra arte* 5.27–59 (ed. Papathanassiou 2017: 181–183).
28 [Theophrastus], *De arte divina* (ed. Goldschmidt 1923: 34–42).
29 [Theophrastus], *De arte divina*, v. 1 (ed. Goldschmidt 1923): Οἱ τῶν σοφιστῶν ἄνδρες ὥσπερ ῥήτορες; v. 10: βροτοὺς σοφιστάς; v. 16: σοφοὶ διδάσκαλοι; v. 31: ἐχέφροσι βροτοῖς; v. 69: σοφιστικῆς ἐννοίας; v. 90: σοφιστικῆς τέχνης.
30 [Theophrastus], *De arte divina*, vv. 18, 37, 68: μύησιν; v. 11: μεμυημένους; vv. 239–240: πανσόφου τεχνουργίας.
31 [Theophrastus], *De arte divina*, vv. 21–31.

planets. Thus, wise men who know theoretical astronomy can see through their intellect all that happens in the sky without the aid of their senses.

Further, Theophrastus refers to the "medical art and science" (ἰατρικῆς τέχνης τε καὶ ἐπιστήμης).[32] By contrast, Stephanus refers to these only in his *Letter to Theodorus*, where he mentions the "medical scientist" (ἰατρικῆς ἐπιστήμονος), the "medical treatment" (ἰατρείας), and what is done "in medical terms" (ἰατρικῶς).[33] Theophrastus' view is that astronomical knowledge is helpful in astrological predictions, especially those referring to medical astrology (*iatromathematica*). This becomes clear from the verses that refer to an initiation into the medical art and science, enabling the prediction of bodily injuries and diseases, the outcome of the disease and its remedy, and even a possible death due to the disease.[34] Both Ptolemy and Hephaestion of Thebes lay out the method of investigation regarding injuries and diseases. However, Hephaestion included a chapter on patients, explaining the influence of the moon's phases on the evolution of the disease.[35]

The wisdom of the wise men extends to all species on the earth through the principle of cosmic sympathy.[36] Their writings discuss the qualities, mixtures, and taste of dried and fresh flowers and the juices of plants. They also include the colors of stones and metals and their benefits and harms. These ideas are also applied to terrestrial animals.

Thereafter, Theophrastus' poem closely relates to Stephanus' work. Theophrastus begins by blaming the uninitiated men who lack sophistication and accuse the wise men of being liars, saying that it is possible to make gold. These ignorant men ask how it is possible to change the nature of something that God created. In answer, we show them that the result of this art is straightforward to apprehend and valuable, and that it needs only one nature that brings forth all.[37]

According to Stephanus, the sacred art has four turning points (two solstices and two equinoxes), each one having three towers (signs of the zodiac) and revolves around them. Consequently, the art has "twelve towers" (δωδεκάπυργος) and "twelve signs" (δωδεκάζῳδος). He further explains that the philosophers' mystery is accomplished by the seven planets and the twelve zodiacal signs; in fact, the color of the compound of the four bodies (τετρασωμία) changes according to the shift in the colors of the seven

32 [Theophrastus], *De arte divina*, v. 35.

33 Stephanus Alexandrinus, Theod. 39, 54, 76 (ed. Papathanassiou 2017: 168–169).

34 [Theophrastus], *De arte divina*, vv. 33–42.

35 Ptolemaeus, *Apotelesmatica* (*Tetrabiblos*) 3.12 (ed. Robbins 1940); Hephaestion, *Apotelesmatica* 2.13; 3.31 (ed. Pingree 1973). See Papathanassiou 1999: 367–370.

36 [Theophrastus], *De arte divina*, vv. 43–66.

37 [Theophrastus], *De arte divina*, vv. 67–87. Cf. Stephanus Alexandrinus, *De magna et sacra arte* 7.26–48, 60–67 (ed. Papathanassiou 2017: 199–200, 201); see the comments in Papathanassiou 2017: 95.

planets as they pass from one sign to the next.[38] Consequently, making the philosophers' stone relates to the seasons and the planetary aspects.

Theophrastus and Stephanus describe physical phenomena related to each of the four turning points of the year (seasons), each having three towers (months). Thus, the sun, passing through these four turning points, circles the twelve towers of the whole year and shines its bright light, warming all that is on the earth.[39] Likewise, the moon receives its light from the sun, as do the planets and the fixed stars.[40] Remarkably, Theophrastus does not distinguish between the light emitted by the fixed stars as fiery objects and the sun's light reflected by the planets. He also states that the work composed of four elements is completed in four phases, and the seven colors that it displays during its making correspond to the nature and color of the seven planets.[41]

After its whitening and yellowing, the moon can dye any "body" (object) and transform it into gold, which is a miracle.[42] This stone is three-subsistent and made of metal. It is well-known, but all ignore it and seek it everywhere. It has no value but carries a hidden treasure, the divinity, as a mystery.[43] This stone, macerated in horse dung for seven days, gives birth to the dragon that, in forty days, bites his tail. This *ouroboros* dragon looks white, and his hide is spotted.[44]

Stephanus describes quite briefly the spotted serpent-dragon of the chemical operations for the making of the philosophers' stone in his *Letter to Theodorus*:

> Τί γὰρ ἢ δράκοντα ἔχει ὁ ἀγρὸς καὶ διὰ τοῦ φυσήματος ἀναξηραίνει τὸν τόπον [...]. Ἐγὼ δὲ αὐτὸν εἶδον καὶ τὰς ποικίλας τοῦ σώματος αὐτοῦ φολίδας· καὶ ἡ μὲν ἀρχὴ τῆς οὐρᾶς αὐτοῦ λευκὴ ὡς γάλα, ἡ δὲ κοιλία καὶ ὁ νῶτος κροκοειδής, καὶ ἡ κεφαλὴ μελάγχλωρος.
>
> For the field has a serpent, and he dries up the place with his breath [...]. I saw him and the many-colored scales of his body. The beginning of his tail was white as milk, but his belly and back were saffron-colored, and his head was greenish-black.[45]

On the other hand, Theophrastus extensively describes the passion of the *ouroboros* dragon, which corresponds to the successive chemical operations for making the

38 Stephanus Alexandrinus, *De magna et sacra arte* 5.78–89 (ed. Papathanassiou 2017: 183–184); see the comments in Papathanassiou 2017: 92–93.
39 [Theophrastus], *De arte divina*, vv. 88–113.
40 [Theophrastus], *De arte divina*, vv. 114–116.
41 [Theophrastus], *De arte divina*, vv. 88–126.
42 [Theophrastus], *De arte divina*, vv. 118–132.
43 [Theophrastus], *De arte divina*, vv. 133–140. Cf. Stephanus Alexandrinus, *De magna et sacra arte* 4.40–42 (ed. Papathanassiou 2017: 174); see the comments in Papathanassiou 2017: 90.
44 [Theophrastus], *De arte divina*, vv. 141–145.
45 Stephanus Alexandrinus, *De magna et sacra arte*, Theod. 5–9 (ed. Papathanassiou 2017: 167); see the comments in Papathanassiou 2017: 85. Illustration in *Parisinus graecus* 2327, f. 279 recto (copied in 1478).

stone.[46] Moreover, Zosimus, in his *Authentic Memoirs*, describes his apocalyptic dreams, which reveal the passion and sacrifice of the *ouroboros* dragon on the bowl-shaped altar and those of the copper and lead man.[47] It is therefore very likely that Zosimus' work was Theophrastus' source of inspiration on this subject.

Theophrastus ends his poem praising this divine work of "very wise artistry" (πάνσοφος τεχνουργία) with rhetorical expressions,[48] as well as a prayer. He praises God, the benefactor and Lord, who shows providence and mercy towards those who live piously.[49]

3 Hierotheus

Hierotheus,[50] following Stephanus, refers to the enigmatic teachings of the ancient wise men,[51] who disclose their secrets to piously living men in a state of grace.[52] He refers to the "divine wise man called the first leader, initiated, and teacher",[53] and he ends his poem with a prayer to Christ the Lord.[54]

The most crucial issue in his poem is the extensive description of an astronomical phenomenon,[55] which was recorded by Stephanus. Notably, Hierotheus is the only poet-alchemist who refers to it. He too describes it at length, but his reference to the planetary positions is unclear and lacks the accuracy of Stephanus'. Hence, his numerous rhetorical expressions and allegories regarding the marriage of the Sun and the Moon to make the goldstone are at times confusing; in his allegories, Hierotheus refers to Rome (Roman father), Persia (Persian mother), and Egypt (Egyptians). The verses related to this phenomenon are as follows:

> Λευχειμονούσης ἐκ σελήνης Περσίδος / λαχοῦσα τῆς ἑῴας ἡ Ἀφροδίτη / τοῦ ἡλίου ἔκλαμψιν αὐγῆς προσμένει, / Ἑρμῆς δὲ ὡς ἀλλοῖος ἄλλως δείκνυται / πρὸ ἡλίου βαίνων τε οὐ φαίνει ὅλως, / καλύπτεται αὐτὸς δὲ τοῦ Κρόνου μένει [...].

46 [Theophrastus], *De arte divina*, vv. 147–212.

47 Zosimus Panopolitanus, *Hypomnemata authentica* 10.2, 5 (*De virtute*) (ed. Mertens 1995: 35–36, 39–40); 11.2 (*Lectio secunda*) (ed. Mertens 1995: 44–45); 12 (*Lectio tertia*) (ed. Mertens 1995: 46–47).

48 [Theophrastus], *De arte divina*, vv. 213–238. Cf. Stephanus Alexandrinus, *De magna et sacra arte* 1.83 (ed. Papathanassiou 2017: 159); see the comments in Papathanassiou 2017: 84.

49 [Theophrastus], *De arte divina*, vv. 239–265. Cf. Papathanassiou 2018.

50 [Hierotheus], *De arte divina et sacra* (ed. Goldschmidt 1923: 42–49).

51 [Hierotheus], *De arte divina et sacra*, vv. 7–12 (ed. Goldschmidt 1923).

52 [Hierotheus], *De arte divina et sacra*, vv. 190–199.

53 [Hierotheus], *De arte divina et sacra*, vv. 34–35: [...] θεῖον σοφόν τε ἄνδρα τὸν κεκλημένον / ἀρχηγέτην μύστην τε καὶ διδάσκαλον [...].

54 [Hierotheus], *De arte divina et sacra*, vv. 202–242.

55 [Hierotheus], *De arte divina et sacra*, vv. 94–136.

The Moon [is] dressed in white garments, and Venus (Aphrodite), being at the Persian dawn, is waiting for the sunshine of the daybreak. Mercury (Hermes) appears in different ways as it precedes the Sun and does not shine being fully covered [by the sunrays]; while [the planet of] Saturn (Cronus) stays.[56]

τῆς Περσαίας πρόεισι καὶ Ἀφροδίτη / [...] / τὸν ἄρρενα πρόσζευξον ἐκ Ῥώμης τόπων / ἥκοντα πορφυροῦντα [...].

Venus comes from the Persian area / [...] let couple [to her] the male in purple who comes from Roman countries.[57]

Τὴν νυμφικῶς πρὸς αὐτὸν ἐκ τῆς Περσίδος / ἑῴας ὡς τάχιστα ἤκουσαν, ὅπως / τούτῳ συναφθῇ πρὸς γάμων κοινωνίαν.

The bride swiftly comes to him from the Persian dawn to join him in matrimony.[58]

A comparison between Stephanus' and Hierotheus' accounts of this phenomenon shows that Stephanus' strict and accurate description occupies only a short passage of his lecture, in contrast to Hierotheus' description in verses, which forms half his poem.

The great importance of Stephanus' laconic and accurate account of this astronomical phenomenon lies in the possibility that we can use it to date his alchemical work. Taking into account the known correspondence between planets and metals (Sun–gold, Moon–silver, Mercury–quicksilver, Venus–copper, Mars–iron, Jupiter–tin, Saturn–lead), Stephanus relates this astronomical phenomenon to the making of the "philosophers' stone" as follows:

Πάλιν ὁ τῆς Ἀφροδίτης τὴν περσικὴν λαχὼν ἑῴαν προηγεῖται τὰς τοῦ ἡλίου αὐγάς· πάλιν ὁ τοῦ Ἑρμοῦ ὑπὸ τὰς τοῦ ἡλίου αὐγὰς ἐπὶ τὰ ἑπόμενα εὑρίσκεται· πάλιν ὁ τοῦ Κρόνου διὰ τὴν τοῦ ὕψους βαθύτητα ἀμυδρῶς προσφαίνεται· πάλιν ὁ τοῦ Ἄρεως τὴν πυρώδη τομὴν ἀπεργάζεται· ἐν οἷς ἡ νυμφικῶς διεσκευασμένη προέρχεται σελήνη, τὰς ἐννέα τῶν τμημάτων ἀναλαμβάνει ὁλκάδας, δι' ἧς τὸ συγκιρνώμενον τελειοῦται κρᾶμα.

Again, [the planet of] Venus attained the Persian dawn and now precedes the rays of the Sun; again, [the planet of] Mercury, under the rays of the Sun, is found in the following [zodiacal signs]; again [the planet of] Saturn is faintly discernible due to the steepness of its height; again [the planet of] Mars is preparing the burning cut; towards these [planets] comes the Moon dressed as a bride, [and] takes up the towed ships of the nine parts. By means [of the Moon], the alloy, mixing itself, does so to perfection.[59]

56 [Hierotheus], *De arte divina et sacra*, vv. 94–99.

57 [Hierotheus], *De arte divina et sacra*, vv. 112, 114–115.

58 [Hierotheus], *De arte divina et sacra*, vv. 129–131.

59 Stephanus Alexandrinus, *De magna et sacra arte* 6.65–71 (ed. Papathanassiou 2017: 189).

In astronomical terms, this passage means the following: "At dawn", the Sun is under the horizon. "Mercury, under the rays of the Sun, is found in the following [zodiacal signs]" means that Mercury is also under the horizon and is therefore invisible. "Venus attained the Persian dawn and now precedes the rays of the Sun" means that Venus is visible as the "morning star" near the eastern horizon at dawn. The phrase "Saturn is faintly discernible due to the steepness of its height" refers to Saturn's great distance from the Earth, which is also true according to ancient cosmological models. "Mars is preparing the burning cut" means that Mars, which relates to violent activities, weapons, cuts, burns, and iron, is preparing to leave Pisces (the last zodiacal sign and a watery one) and enter Aries (the first zodiacal sign and a fiery one). Finally, "the Moon [comes] dressed as a bride" towards them means that the Moon is about to come in conjunction with the Sun (new moon), a phenomenon allegorically described as "sacred marriage" (θεογαμία). Consequently, after the full moon phase, the Moon moves towards these planets and the Sun without being in conjunction with any of them.

Stephanus' description of this particular planetary phenomenon implies that the order of the planetary positions from east to west is as follows: Mercury, Sun (both invisible under the horizon), Venus, Saturn, Mars, and Moon (all four visible above the horizon). The lack of reference to Jupiter in the text means it did not participate in the assembly and was not visible.

Based on our calculations made with the Voyager computer program, we have been able to identify a great assembly of the Sun, the Moon, Mercury, Venus, Mars, and Saturn ordered according to Stephanus' description and visible in the eastern sky at dawn during the reign of Heraclius (5 October 610–11 January 641), at Constantinople on 7 June 617.

If we were at Constantinople on that date and Stephanus invited us to admire the splendid view of the starry sky, he would first show us Mercury, visible in the twilight as an evening star low in the west. The next day, early at dawn (4:05 a.m. local time, 24 minutes before sunrise), and in increasing height from the eastern horizon, Stephanus would show us the planet Venus as a morning star very low in the east but visible thanks to its excellent brightness. A little higher than Venus, Saturn would be in conjunction with the crescent of the waning Moon, and finally, red Mars would appear high in the sky. The position of Mars in 1°39' Aries, the first zodiacal sign and a fiery one, explains why "Mars is preparing the burning cut". Stephanus must have been observing the planets for many days while this exceptional planetary phenomenon gradually evolved. Mars was moving straight forward (towards the following zodiac sign) through the last degrees of Pisces before entering Aries on 4 June. Meanwhile, the Moon, after the full moon of 26 May, would come successively into conjunction with Mars (3 June), Saturn (7 June), and Venus (8 June), reaching its next conjunction with the Sun (new moon) on 9 June 617. Stephanus does not mention the 3 June conjunction of the Moon and Mars in Pisces, possibly because he wrote this lecture sometime after 26 May but before 3 June 617.[60]

The question is whether Hierotheus could have referred to an astronomical phenomenon if he had not observed it with his own eyes. I believe that Hierotheus – or, more likely, the single author of all four poems (see below on the question of authorship) – was one of Stephanus' close disciples and attended his lectures on the *quadrivium* of mathematics. He accompanied his master early at dawn to observe this astronomical phenomenon, which inspired him to describe it freely in verses.

Hierotheus ends his poem with a prayer encouraging the initiated men to praise the mysteries of Christ the Lord.[61]

4 Archelaus

Archelaus,[62] influenced by Stephanus, also refers to the knowledge bestowed by God's grace to pious men who retire and live in purity, fasting, and praying; moreover, by denying their flesh, they mortify their body and shed many tears to be illuminated by God.[63] He focuses on the famous alchemical principle, "Nature rejoices in Nature, and Nature masters Nature, and Nature conquers Nature",[64] and refers to the four primary elements and their transformations into one another through the four qualities.[65]

Beginning his poem, Archelaus states that one should think of the work as an "initiation to knowledge, science"[66] that should be accomplished "with art and wisdom".[67] The author ends his poem by affirming to the adept that he did not conceal anything. Instead, he disclosed the "very sage experience and practice of the art",[68] and he advises the adept "to examine all texts in the books of the wise men".[69]

---

60 Papathanassiou 2006: 180–188.

61 [Hierotheus], *De arte divina et sacra*, vv. 190–229.

62 [Archelaus], *De arte sacra* (ed. Goldschmidt 1923: 50–59).

63 [Archelaus], *De arte sacra*, vv. 40–48, 314–322 (ed. Goldschmidt 1923).

64 [Archelaus], *De arte sacra*, vv. 62–90.

65 [Archelaus], *De arte sacra*, vv. 97–121; Cf. Stephanus Alexandrinus, *De magna et sacra arte* 5.27–59 (ed. Papathanassiou 2017: 181–183).

66 [Archelaus], *De arte sacra*, v. 14: [...] μύησιν γνώσεως, ἐπιστήμην [...].

67 [Archelaus], *De arte sacra*, v. 2: τεχνουργικῶς; v. 3: σοφουργικήν.

68 [Archelaus], *De arte sacra*, vv. 298–299: [...] πάνσοφον τέχνης τε τὴν ἐμπειρίαν / καὶ πρᾶξιν [...].

69 [Archelaus], *De arte sacra*, vv. 309–310: [...] καὶ ἐξεταστὴς ὧν ἔφη λόγων γένῃ / τῶν κειμένων πάντων μὲν ἐν βίβλοις σοφῶν [...].

Also, Archelaus encourages the adept "to mortify his body" and "grow wings of the soul";[70] moreover, "to wash the shame of the body by shedding tears",[71] and "to succeed in seeing the unshaded and eternal light".[72] All this reminds us of related passages in Stephanus' text.[73]

Chemical writers used allegories regarding chemical operations as the passion and death of the compound of the "philosophers' stone", considered a "living stone" that has to pass through various phases of treatment before attaining perfection – namely, to pass from death to life. In Archelaus' poem, we find verses that remind us of those from the Service and the famous *Epitaphios thrēnos* (Ἐπιτάφιος θρῆνος; Lamentations before the Holy Sepulchre) of Good Friday:

- ἰδοῦσα κείμενον δὲ ταύτης ὄργανον / ἄφωνον ἄψυχόν τε μὴ κινούμενον / ὡς προσδοκοῦν δὲ τοῦ τάφου ἀνάστασιν.
  Having seen her organ / voiceless and lifeless, motionless / as being waiting for the resurrection from the tomb.[74]

- Ἔξελθε βαῖνον ἐκ σκότους ὡς ἐξ ᾄδου / [...] / ῥῖψον φθορᾶς χιτῶνα καὶ πάλιν, ἔνδυμα / πρῴην ὅπερ σε εἶχεν ὡς κατάκριτον [...].
  Go out, coming from the darkness as from Hades / [...] / throw away the tunic of corruption and former garment that had thee as condemned.[75]

- νεκρὸς γὰρ ἦσθα τὸ πρὶν εἰς φθορὰν πέλων / τάφῳ κατῴκησάς τε νεκρός, ἄψυχος, / ἄμορφος, ἄπνους, μηδόλως κάλλος φέρων [...].
  Previously thou were dead, being in corruption / and thou dwelt in the tomb a lifeless corpse / without breath, without form, without any beauty in thee.[76]

- [...] καλύπτεταί σου κάλλος ἔκλαμπρον τάφῳ, / ταῖς κειρίαις δὲ νεκρὸς ἄπνους, ἄψυχος, / εἱλίσσῃ μὲν πρόσωπον ἐκφαίνων μόνον. / ἐρριμμένος γυμνὸς δὲ τοῖς ὁρῶσί σε / ἄφωνος οὕτως φαίνῃ τὸ ζῆν προσδοκῶν / ἔργοις λέγων· ποῦ ἐστιν ἡ ζῶσα ψυχή;

---

70 [Archelaus], *De arte sacra*, vv. 314–315: νέκρωσον οὖν τὸ σῶμα [...] / ψυχὴν πτέρωσον [...].

71 [Archelaus], *De arte sacra*, vv. 321–322: [...] τοῦ σώματος τὸ αἶσχος ἐκπλύνων ῥοαῖς / ἐξ ὀμμάτων σου δακρύων [...].

72 [Archelaus], *De arte sacra*, vv. 325–326: οὕτως βιῶν γὰρ ἀξιωθήσῃ βλέπειν / τὸ φῶς τὸ ἄσκιόν τε καὶ ἀΐδιον [...].

73 Cf. Papathanassiou 2017: 141–142.

74 [Archelaus], *De arte sacra*, vv. 207–209. Cf. Stephanus Alexandrinus, *De magna et sacra arte* 6.131–134 (ed. Papathanassiou 2017: 192): ὅτε γὰρ ἐπὶ τὸ φθοροποιὸν νεύσει, ἄφωνον τοῖς ὁρῶσιν εὑρίσκεται· κείμενον γὰρ θνητὸν φωνὴν ὅλως οὐκ ἐκπέμπει, ἐπειδὰν δὲ εἰς ἄφθαρτον μετέλθῃ σῶμα, τότε ἔμψυχον καὶ τὸ σῶμα γίνεται.

75 [Archelaus], *De arte sacra*, vv. 214–217.

76 [Archelaus], *De arte sacra*, vv. 224–226.

Thy splendid beauty is covered in the tomb, / wound in grave-clothes, a lifeless corpse without breath, / thou reveal only thy face to those seeing thee lying down naked, / voiceless, thou appear as being expecting to live, / asking by actions, where is the living soul?[77]

The following verses from the *Vespers* (*Friday Afternoon*) and *Matins* (*Friday Evening*) *of Holy Saturday* contain words or expressions similar to those found in Archelaus' poem (emphasis added in the excerpts below highlights relevant words and expressions):

- Σὲ τὸν ἀναβαλλόμενον τὸ φῶς, ὥσπερ ἱμάτιον, καθελὼν Ἰωσὴφ ἀπὸ τοῦ ξύλου σὺν Νικοδήμῳ, καὶ θεωρήσας *νεκρόν*, *γυμνόν*, *ἄταφον*, εὐσυμπάθητον θρῆνον ἀναλαβών, ὀδυρόμενος ἔλεγεν· [...].
  You, who cover Yourself with light as with a garment, were taken down from the Cross by Joseph, with the help of Nicodemus. When he saw You *dead, naked, and unburied*, he took up a moving lamentation; and stricken with grief he said [...].[78]

- Ὁ εὐσχήμων Ἰωσὴφ ἀπὸ τοῦ ξύλου καθελὼν τὸ ἄχραντόν σου *σῶμα*, σινδόνι καθαρᾷ *εἰλήσας* καὶ ἀρώμασιν, ἐν μνήματι καινῷ κηδεύσας ἀπέθετο.
  When he took down Your immaculate *Body* from the Cross, the honorable Joseph *wrapped* it in a clean linen shroud with spices and laid it for burial in a new tomb.[79]

- Ἄφραστον θαῦμα! ὁ ἐν καμίνῳ ρυσάμενος τοὺς ὁσίους παῖδας ἐκ φλογός, *ἐν τάφῳ νεκρός, ἄπνους* κατατίθεται εἰς σωτηρίαν ἡμῶν τῶν μελῳδούντων· Λυτρωτά, ὁ Θεὸς εὐλογητός εἶ.
  Ineffable wonder! He who delivered the pious youths in the furnace from the raging fire is laid out today, *dead and breathless, in a tomb*, for the salvation of us who sing, "O our God and Redeemer, You are blessed".[80]

- Πέπαυται τόλμα μαθητῶν, Ἀριμαθαῖος δὲ ἀριστεύει Ἰωσήφ· *νεκρὸν γὰρ καὶ γυμνὸν θεώμενος τὸν ἐπὶ πάντων Θεόν*, αἰτεῖται καὶ κηδεύει κραυγάζων· Οἱ παῖδες εὐλογεῖτε, ἱερεῖς ἀνυμνεῖτε, λαὸς ὑπερυψοῦτε εἰς πάντας τοὺς αἰῶνας.
  Though the Disciples' courage failed, yet that of Joseph of Arimathea has excelled. For *when he sees that the God over all is hanging dead and naked*, he asks for Him and buries Him, crying, "Bless the Lord, O you Servants; sing a hymn, Priests; and People, exalt Him beyond measure unto all the ages".[81]

---

77 [Archelaus], *De arte sacra*, vv. 237–242.
78 *Synecdemus Christiani Orthodoxi*: 1159; Engl. transl. *Vespers of Holy Saturday*: 14.
79 *Synecdemus Christiani Orthodoxi*: 1160; Engl. transl. *Vespers of Holy Saturday*: 15.
80 *Synecdemus Christiani Orthodoxi*: 1167 (Ode 7, Heirmos); Engl. transl. *Graveside Lamentation*: 26.
81 *Synecdemus Christiani Orthodoxi*: 1168 (Ode 8, Troparion); Engl. transl. *Graveside Lamentation*: 27.

– Ἐπὶ τῷ ξένῳ σου τόκῳ τὰς ὠδῖνας φυγοῦσα, ὑπερφυῶς ἐμακαρίσθην, ἄναρχε Υἱέ· *νῦν δέ σέ*, Θεέ μου, *ἄπνουν ὁρῶσα νεκρόν*, τῇ ρομφαίᾳ τῆς λύπης σπαράττομαι δεινῶς· ἀλλ' ἀνάστηθι, ὅπως μεγαλυνθήσωμαι.
I was blest in escaping labor, in a manner past nature, at Your unusual birth, O unoriginate Son. *But now, as I see You*, my God, *not breathing and dead*, I am painfully torn by the sword of my sorrow. So arise, so that I may be magnified.[82]

With regard to the *Lamentations* (Τὰ ἐγκώμια), the following verses are also noteworthy:

– *Ὁ ὡραῖος κάλλει παρὰ πάντας βροτοὺς ὡς ἀνείδεος νεκρὸς καταφαίνεται*, ὁ τὴν φύσιν ὡραΐσας τοῦ παντός.
Fairer in His beauty than all creatures on earth, He is seen now lying lifeless, his beauty gone, yet all beauty in creation springs from Him.[83]

– *Κάλλος, Λόγε, πρίν, οὐδὲ εἶδος ἐν τῷ πάσχειν ἔσχες*, ἀλλ' ἐξαναστὰς ὑπερέλαμψας, καλλωπίσας τοὺς βροτοὺς θείαις αὐγαῖς.
Suffering in pain, neither form, hadst Thou, O Word, nor beauty, but by Thine arising, Thy beauty shines, and Thy holy rays adorn all those on earth.[84]

– Ἔφριξεν ἰδὼν τὸ ἀόρατον φῶς σε, Χριστέ μου, *μνήματι κρυπτόμενον ἄπνουν τε*, καὶ ἐσκότασεν ὁ ἥλιος τὸ φῶς.
Struck with fear, the sun saw Thy light invisible as Thou lay *lifeless and concealed in the grave*, my Christ, and it shuddered and relinquished its own light.[85]

– *Ὡς νεκρὸν τὸν ζῶντα* σὺν μυροφόροις πάντες μυρίσωμεν ἐμφρόνως.
As women bearing myrrh did, let us in our awareness anoint *as dead the Living*.[86]

– *Ὕπτιον ὁρῶσα* ἡ πάναγνός σε, Λόγε, μητροπρεπῶς ἐθρήνει.
The All-pure Virgin wept with a mother's grief, O Word, *when she saw Thee lying dead*.[87]

– Ὢ γλυκύ μου ἔαρ, γλυκύτατόν μου τέκνον, *ποῦ ἔδυ σου τὸ κάλλος;*
Thou art my sweetest Springtime, My sweetest Son, I ask Thee, *where has Thy beauty faded?*[88]

---

82 *Synecdemus Christiani Orthodoxi*: 1169 (Ode 9, Troparion); Engl. transl. *Graveside Lamentation*: 28–29.
83 *Synecdemus Christiani Orthodoxi*: 1170 (Stasis 1); Engl. transl. Takis 2003: 2.
84 *Synecdemus Christiani Orthodoxi*: 1179 (Stasis 2); Engl. transl. Takis 2003: 5.
85 *Synecdemus Christiani Orthodoxi*: 1181; Engl. transl. Takis 2003: 5.
86 *Synecdemus Christiani Orthodoxi*: 1182 (Stasis 3); Engl. transl. Takis 2003: 7.
87 *Synecdemus Christiani Orthodoxi*: 1183; Engl. transl. Drillock and Erickson 2005: 50.
88 *Synecdemus Christiani Orthodoxi*: 1183; Engl. transl. Takis 2003: 7.

- *Ἀνάστηθι, οἰκτίρμον*, ἡμᾶς ἐκ τῶν βαράθρων, ἐξανιστῶν τοῦ Ἅιδου.
  Arise, O Lord of Mercy, and with Thee, also raise us who linger deep in Hades.[89]

- *Ἀνάστα*, ζωοδότα, ἡ σὲ τεκοῦσα Μήτηρ, δακρυρροοῦσα λέγει.
  "*Arise*, Thou Who bestows Life!" the Mother who didst bear Thee through flowing tears entreats Thee.[90]

Modern scholars who have focused on the chronology of the *Epitaphios thrēnos* have not agreed on its dating. However, we know that the *Epitaphios thrēnos* has inspired Byzantine iconography from the eleventh century onwards,[91] as well as the Lamentation scene embroidered on the liturgical cloths named *epitaphioi*, the earliest surviving examples of which appeared around the year 1300.[92] Thus, we could venture a guess, namely that of an early composition of such verses related to Christ's passion and their later incorporation into Byzantine hymnography. In this case, the specific words or expressions appearing in Archelaus' poem are evidence of the early chronology of some verses of the *Epitaphios thrēnos*.

Why did these verses inspire Archelaus? I think that the mourning atmosphere of Holy Week in the Orthodox Easter, especially that of the evening of Good Friday, would have significantly influenced Archelaus, who was – very likely – composing his verses during the period preceding (or following) Easter Sunday of 617, celebrated on 3 April.[93]

Had Archelaus and other disciples, together with Stephanus, observed the astronomical phenomenon unfolding between 26 May and 3 June 617? I think that Archelaus – or whoever the sole author of the poems was – also participated in his master's astronomical activity. So, the mystical symbolism of this astronomical phenomenon and its correspondence to the phases of making the philosophers' "living stone", which, like a man, suffers its passion unto death to achieve its perfection – namely, its resurrection – inspired him to write his poem.

5 On the Authorship of the Poems

There is a question regarding the authorship of these poems: is there one author who wrote the four poems in iambic verses, or are there four authors?

---

89 *Synecdemus Christiani Orthodoxi*: 1184; Engl. transl. Takis 2003: 7.
90 *Synecdemus Christiani Orthodoxi*: 1184; Engl. transl. Takis 2003: 8.
91 Maguire 1981: 102; also Myslivec 1968.
92 Gonosová 1991; Ćurčić 1991: 252; Woodfin 2004: 296–297; Antonaras 2004; Deriziotis 2004; Petrinski 2004.
93 Parise 1982: 323.

It is difficult to answer this question because the oldest chemical manuscript, *Marcianus graecus* 299 (coll. 584), contains these four poems as works of four different persons. Even so, it is possible to point out some issues:

1) The name of "Theodosius, the emperor" is mentioned only in the title of Heliodorus' poem; this could be a later addition by a copyist. As we have already mentioned, the anonymous references to the "emperor" and to the "very pious emperors" who are "worshipers of Christ" could refer to Heraclius and his son Heraclius Constantine. In this case, Heliodorus would have written his poem during the reign of Heraclius.

2) All poems contain specific words and expressions that are the same, especially those not found in Stephanus' lectures. Therefore, if it is not a case of coincidence or of technical terminology in use at that time, it could indicate a single author.

3) Each poem focuses on a different subject. Heliodorus unfolds the phases of life of the "living stone". Theophrastus explains the passion of the *ouroboros* dragon. Hierotheus draws specific attention to the rare planetary phenomenon in 617, while Archelaus reproduces the mystical atmosphere of Easter, which is related to the view of this planetary phenomenon in the same year. We could then ask whether this distinction of subjects is due to the personal interests of four authors or if it indicates the authorship of only one person who unfolds his ideas in four lectures written in iambic verses.

The manuscript tradition attributes each of the four poems to a different author; the names of these authors are otherwise unknown. According to Goldschmidt, only one author wrote the four poems.[94] His conclusion seems the most reasonable explanation for the aforementioned issues, especially the gradually increasing mystical excitement in the verses of Hierotheus and Archelaus. In this case, whoever the author may have been – likely one among the four poets – belonged to Stephanus' intellectual environment in Constantinople and was one of his close disciples. He participated in Stephanus' teaching activity in 617 and attended his master's lectures on chrysopoeia. Inspired by them, he reproduced the most inspiring parts in his verses in the same year, 617.

Note

The four alchemical poems have recently been re-edited, with a Spanish translation, by María Belén Boned Fernández in her PhD dissertation *Los poemas alquímicos atribuidos a Heliodoro, Teofrasto, Hieróteo y Arquelao: edición, traducción y comentario*. Universidad Complutense de Madrid, 2025. At the time of writing this paper, however, the dissertation was not yet complete.

94 Goldschmidt 1923: 11–15.

Bibliography

Sources

Collected Sources

Goldschmidt 1923: Günther Goldschmidt (ed.), *Heliodori carmina quattuor ad fidem codicis Casselani* (Religionsgeschichtliche Versuche und Vorarbeiten 19.2). Giessen: A. Töpelmann, 1923.

Synecdemus Christiani Orthodoxi: *Μέγας καὶ Ἱερὸς Συνέκδημος Ὀρθοδόξου Χριστιανοῦ*. Thirteenth edition. Athens: Papadimitriou, 2001.

Engl. transl.: ***Graveside Lamentation***: *Holy Week 2025 – Graveside Lamentation on Friday Evening: Matins of Holy Saturday. Texts in Greek and English* (Digital Chant Stand of the Greek Orthodox Archdiocese of America), https://dcs.goarch.org/goa/dcs/p/s/2025/04/18/em/gr-en/se.m04.d19.em.pdf.

Engl. transl.: ***Lamentations***: Nancy Chalker Takis, *Τὰ ἐγκώμια – The Lamentations before the Holy Sepulchre*. Williamston, MI: New Byzantium Publications, 2003; David Drillock and John H. Erickson, *Holy Week*, vol. 3: *The Services of Holy Saturday*. Crestwood, NY: St Vladimir's Seminary Press, 2005.

Engl. transl.: ***Vespers of Holy Saturday***: *Holy Week 2025 – Removal from the Cross on Friday Afternoon: Vespers of Holy Saturday. Text in English* (Digital Chant Stand of the Greek Orthodox Archdiocese of America), https://dcs.goarch.org/goa/dcs/p/s/2025/04/18/ve/en/se.m04.d19.ve.pdf.

Authors and Works

[Archelaus], *De arte sacra*: ed. Goldschmidt 1923, pp. 50–59.

[Heliodorus], *De philosophorum arte mystica*: ed. Goldschmidt 1923, pp. 26–34.

Hephaestion, *Apotelesmatica*: David Pingree (ed.), *Hephaestionis Thebani Apotelesmaticorum libri tres*, vol. 1 (Bibliotheca Scriptorum Graecorum et Romanorum Teubneriana). Leipzig: Teubner, 1973.

[Hierotheus], *De arte divina et sacra*: ed. Goldschmidt 1923, pp. 42–49.

Ptolemaeus, *Apotelesmatica (Tetrabiblos)*: Frank E. Robbins (ed. and Engl. transl.), Ptolemy, *Tetrabiblos* (Loeb Classical Library 435). Cambridge, MA: Harvard University Press; London: W. Heinemann, 1940 (reprint 1964).

Stephanus Alexandrinus, *De magna et sacra arte*: Maria K. Papathanassiou (ed.), *Stephanos von Alexandreia und sein alchemistisches Werk. Die kritische Edition des griechischen Textes eingeschlossen*. Athens: Cosmosware, 2017.

[Theophrastus], *De arte divina*: ed. Goldschmidt 1923, pp. 34–42.

Zosimus Panopolitanus, *Hypomnemata authentica*: Michèle Mertens (ed. and Fr. transl.), *Zosime de Panopolis, Mémoires authentiques* (Les Alchimistes grecs 4.1). Paris: Belles Lettres, 1995.

Literature

Antonaras 2004: Anastasios Antonaras, "Epitaphios", in Evans 2004, pp. 312–313, cat. no. 187A.

Ćurčić 1991: Slobodan Ćurčić, "Late Byzantine *Loca Sancta*? Some Questions Regarding the Form and Function of Epitaphioi", in Slobodan Ćurčić and Doula Mouriki (eds), *The Twilight of Byzantium: Aspects of Cultural and Religious History in the Late Byzantine Empire*. Princeton, NJ: Department of Art and Archaeology, Princeton University, 1991, pp. 251–261.

Deriziotis 2004: Lazaros Deriziotis, "Epitaphios", in Evans 2004, pp. 313–314, cat. no. 187B.

Evans 2004: Helen C. Evans (ed.), *Byzantium, Faith and Power (1261–1557)*. New York: The Metropolitan Museum of Art, 2004.

Goldschmidt 1923: Günther Goldschmidt, *Heliodori carmina quattuor ad fidem codicis Casselani* (Religionsgeschichtliche Versuche und Vorarbeiten 19.2). Giessen: A. Töpelmann, 1923.

Gonosová 1991: Anna Gonosová, "Epitaphios. Liturgical Cloth", in Alexander P. Kazhdan (ed.-in-chief), *The Oxford Dictionary of Byzantium*, 3 vols. New York and Oxford: Oxford University Press, 1991, vol. 1, pp. 720–721.

Lauxtermann 2019: Marc D. Lauxtermann, *Byzantine Poetry from Pisides to Geometres. Texts and Contexts*, vol. 2 (Wiener Byzantinistische Studien 24.2). Vienna: Austrian Academy of Sciences Press, 2019.

Lexikon zur byzantinischen Gräzität: Erich Trapp et al., *Lexikon zur byzantinischen Gräzität, besonders des 9.–12. Jahrhunderts* (Österreichische Akademie der Wissenschaften, Philosophisch-Historische Klasse, Denkschriften 238, 250, 276, 293, 326, 352, 417, 494; Veröffentlichungen der Kommission für Byzantinistik 6.1–8). Vienna: Verlag der Österreichischen Akademie der Wissenschaften, 1994–2017.

Liddell–Scott–Jones: Henry George Liddell, Robert Scott, and Henry Stuart Jones, *A Greek-English Lexicon. Ninth Edition with a Revised Supplement*. Oxford: Clarendon Press, 1996.

Maguire 1981: Henry Maguire, *Art and Eloquence in Byzantium*. Princeton, NJ: Princeton University Press, 1981.

Myslivec 1968: Josef Myslivec, "Epitaphios", in Engelbert Kirschbaum and Wolfgang Braunfels (eds), *Lexikon der christlichen Ikonographie*, 8 vols. Freiburg im Breisgau: Herder, 1968–1976, vol. 1, pp. 654–655.

Papathanassiou 1999: Maria Papathanassiou, "Iatromathematica (Medical Astrology) in Late Antiquity and the Byzantine Period", *Medicina nei Secoli* 11.2 (1999), pp. 357–376.

Papathanassiou 2006: Maria Papathanassiou, "Stephanos of Alexandria: A Famous Byzantine Scholar, Alchemist and Astrologer", in Paul Magdalino and Maria Mavroudi (eds), *The Occult Sciences in Byzantium*. Geneva: La Pomme d'or, 2006, pp. 163–203.

Papathanassiou 2017: Maria K. Papathanassiou, *Stephanos von Alexandreia und sein alchemistisches Werk. Die kritische Edition des griechischen Textes eingeschlossen*. Athens: Cosmosware, 2017.

Papathanassiou 2018: Maria K. Papathanassiou, "Stéphanos d'Alexandrie. La tradition patristique dans son œuvre alchimique", in Efthymios Nicolaidis (ed.), *Greek Alchemy from Late Antiquity to Early Modernity* (De Diversis Artibus 104). Turnhout: Brepols, 2018, pp. 71–97.

Parise 1982: Frank Parise (ed.), *The Book of Calendars*. New York: Facts On File, 1982.

Petrinski 2004: Ivan Petrinski, "Epitaphios of the Shepherd of the Bulgarians", in Evans 2004, pp. 314–315, cat. no. 188.

Woodfin 2004: Warren Woodfin, "Liturgical Textiles", in Evans 2004, pp. 295–299.

Gerasimos Merianos

Alchemy in Byzantine Hagiography: Menia the Wise and Heliodorus the Sorcerer

Byzantine literature offers valuable glimpses into how alchemy was conceived within the broader culture. Explicit references to alchemy – or rather, to terms later identified with what we now call "alchemy" – in non-alchemical texts are rare. Nevertheless, such references, including more implicit ones, are increasingly identified thanks to our present tendency to more critically scrutinize texts of various genres. Careful reading thus brings to light significant testimonies in non-specialist contexts, which not only enrich our understanding of the perception of alchemy, but also challenge the assumption that it was marginal or isolated, revealing instead the subtle and varied ways the "sacred art" intersected with different strands of Byzantine culture.

1 Menia the Wise: Divine Metallurgical Knowledge

A case that vividly illustrates this intersection is the *Life of Saint Pancratius of Taormina* (*BHG* 1410). This hagiographical work contains passages that, although superficially connected with alchemy in the past,[1] have yet to be systematically and comprehensively examined through the lens of the history of alchemy. The analysis of these references offers an opportunity to explore how "alchemical" themes and terminology could surface in unexpected literary settings.

This *Life* recounts the mission and martyrdom of Saint Pancratius, a disciple of Saint Peter. Sent by Peter to evangelize Taormina in Sicily, Pancratius subsequently became its first bishop. The *Life* was purportedly written by Evagrius, Pancratius' assistant and successor.[2] Cynthia Stallman-Pacitti, the editor of the text, suggests that it was probably composed in Sicily at the end of the seventh or the beginning of the eighth century.[3] Its primary aim was to enhance the prestige of the diocese of Taormina by asserting apostolic origins, promoting its patron saint, and justifying its jurisdictional rights. The

1 See, e.g., Acconcia Longo 1989: 57; 2001a: 10; Brandes 2005: 194–195; Merianos and Sakorrafou 2013: 47–48.

2 On the figure of Evagrius, see Lampadaridi 2020.

3 See Stallman-Pacitti 2018: 11–19; on pp. 11–18 she examines a range of possible *termini ante quos* for the creation of the text. On its dating, cf. Alexakis 2019, who proposes "a date before and very close to the outbreak of Iconoclasm for its composition"; Lampadaridi 2022: 98, who argues for "une datation haute, avant 730"; also Acconcia Longo 2001b. On the *Life's* recensions, translations, and its use in later literature, see Stallman-Pacitti 2018: 23–35.

https://doi.org/10.1515/9783111589619-005

work's wide diffusion is indicated by its popularity as evidenced by the copies owned by Theodore the Studite (759–826) and Patriarch Nicephorus I (806–815).[4]

The *Life* also engages with the history of the city. To this end, it incorporates "the story of Taurus and Menia", an etiological myth explaining the origin of the city's name, Taormina (Greek: *Tauromenion* or *Tauromenia*).[5] In this tale, Menia, a Macedonian descendant of Alexander the Great,[6] is joined with Taurus, a Canaanite descendant of Nimrod.[7] Alexander Kazhdan has noted that, beyond constituting a "romance", this tale may also be regarded as "the first Byzantine chivalrous story".[8] The text further exhibits distinct Eastern coloring, especially Syrian. Pancratius and Evagrius are presented as Syrians, natives of Antioch. Moreover, the romance of Taurus and Menia connects Sicily to the East – through Taurus to Syria and Menia to Macedonia – and alludes to ancient dynastic alliances, since Taurus is depicted as descending from Nimrod.[9] This dimension is further contextualized by the strong Syrian presence in seventh- and eighth-century Italy, particularly in the south and in Rome.[10]

The myth begins with two kings, Rhemaldus and Aquilinus, ruling different parts of ancient Calabria. Aquilinus coveted Rhemaldus' territories and eighteen-year-old wife, Menia, whom the author calls "wise". Taurus, at the age of fifteen, had been taken captive with his mother in Syria and eventually sold to Rhemaldus. He grew up to be a man of great stature, like a giant – an allusion to his descent, since Nimrod is regarded as a giant in many traditions.[11] After Rhemaldus was killed in a battle against Aquilinus, Taurus assumed power and married the deceased king's widow, Menia. When Aquilinus

4 See Re 2011: 248; also Acconcia Longo 1989: 55; 2001b; Efthymiadis 2017: 378. Cf. Stallman-Pacitti 2018: 30–31.

5 On etiological thinking and reasoning, see, e.g., Wessels and Klooster 2022.

6 On Alexander in Byzantine tradition, see, e.g., Jouanno 2002; Kaldellis 2022 (with references to earlier scholarship). On the figure of Menia, see Stallman-Pacitti 2018: 498.

7 The fictive descent of Taurus from Nimrod may be linked to another place name. Not far from Taormina lies the Nebrodi mountain range, which Strabo, *Geographica* 6.2.9 (ed. Meineke 1877: 1.377, 20–21) describes as follows: Ἀνταίρει δὲ τῇ Αἴτνῃ τὰ Νεβρώδη ὄρη ταπεινότερα μὲν πλάτει δὲ πολὺ παραλλάττοντα; Engl. transl. Jones 1917–1932: 3.91: "Over against Aetna rise the Nebrodes Mountains, which, though lower than Aetna, exceed it considerably in breadth". It is perhaps not coincidental that the name of Nimrod in Greek is Νεβρώδ, a form that could suggest a connection between the Nebrodi range and the biblical figure. See also Stallman-Pacitti 2018: 425n258. On the figure of Taurus, see Stallman-Pacitti 2018: 499. According to Noyé (2014: 339–342; 2021: 437), Taurus was largely inspired by the figure of Emperor Constans II (r. 641–668).

8 Kazhdan et al. 1999: 305–307.

9 Van Esbroeck and Zanetti 1988: 163–164; Lourié 2021: 177–178.

10 On the phenomenon of the migration of Syrians and Palestinians in the seventh century, which also extended to Sicily and Rome and even influenced papal affairs, see Theodoropoulos 2020; also Lourié 2021: 177. Cf. Stallman-Pacitti 2018: 424n256.

11 See, e.g., Tuval 2016: 44, 54, 56; Rapp 2014.

attacked again, Taurus and Menia, together with their followers, fled to the site that would become Tauromenion, where they founded the homonymous city.[12]

It is at this point in the narrative that we reach the episode most relevant to our discussion. Menia, guided by one of her seers, discovered a plot of land rich in metal ores and produced abundant quantities of pure gold, silver, bronze, and iron.

> ἡ οὖν πάνσοφος Μενία ἡ Μακεδόνισσα ἐνέγκασα τὸν μαντεῶνα αὐτῆς φησίν· “ὦ μάντεών μου, πῶς τὰ μεταλλικὰ εἴδη εὕροιμεν;” ὁ δὲ ἀπελθὼν εὗρεν πάντα, ἃ καὶ αὐτῇ μόνῃ ὑπέδειξεν. ἐπικαλεσαμένη δὲ τὸν τῆς ἄνω προνοίας θεὸν ἐποίησεν χρυσίου καθαροῦ πλῆθος πολὺ καὶ ἀργύριον εἰς πλησμονὴν καὶ χαλκὸν καὶ σίδηρον, μόλιβδόν τε καὶ κασσίτερον μόνον οὐχ εὗρεν ποιῆσαι. οὕτως λοιπὸν πάνυ πεπλούτηκεν ἅπαντας, ὥστε καὶ τὴν ἀφορμὴν τῆς χυμεύσεως ὑποδεῖξαι πᾶσιν.
> ἡνίκα δὲ ἅπαντα τὰ κατὰ τὴν πόλιν πεπλήρωντο χρείας, ἤκουσεν Ἀκυλῖνος ὅτι ὁ Ταῦρος πεπέρακε ποιήσας πόλιν ἰσχυράν, καὶ τὰ ἐγχρῄζοντα ἅπαντα καὶ ἄνδρας πολεμιστὰς ἐκτήσατο βοηθεῖν αὐτῷ, καὶ πλοῦτος πολὺς ἐν τῇ πόλει εὑρέθη· τὸ γὰρ χυμευτικὸν γῄδιον τοῦ μεταλλικοῦ εὑρέθη ἐν τῷ τόπῳ αὐτοῦ καὶ ἐκ τούτου πεποίηκε πλοῦτον πολύν.
>
> Then the all-wise Menia the Macedonian took one of her seers and said, “O seer, how can we find metal ores?” He went off and found them all and showed them to her alone. She called upon the god of the Higher Providence and manufactured a great quantity of pure gold and silver in ample quantity and bronze and iron. Only lead and tin she did not find to manufacture. So she then made everyone extremely rich, since she revealed to everyone the means for *chymeusis* (τὴν ἀφορμὴν τῆς χυμεύσεως).
> When everything the city needed was completed, Aquilinus heard that Taurus had crossed over and made the city strong, and possessed everything he needed and fighting men to help him, and enormous wealth had been found in the city. For a tract of land with metal for *chymeusis* (τὸ γὰρ χυμευτικὸν γῄδιον τοῦ μεταλλικοῦ) had been found in his territory and from this he had created enormous wealth.[13]

Moreover, at an earlier point in the *Life of Pancratius*, Menia is presented in the following terms:

---

12 *Vita Pancratii Tauromenii* 282–296 (ed. Stallman-Pacitti 2018: 418–437). In her preface, dated February 1992, Stallman-Pacitti (2018: IX) cites the principal studies up to that time that considered the text as a whole (e.g., Veselovskiy 1886). It should be added that the story of Taurus and Menia in particular was also edited in Angiò 1994, which Stallman-Pacitti could not have taken into account due to her untimely passing.

13 *Vita Pancratii Tauromenii* 297–298 (ed. and Engl. transl. [adapted from] Stallman-Pacitti 2018: 436, 10–20 [text], 437 [transl.]). For an Italian translation of the story of Taurus and Menia, see Angiò 2024. Stallman-Pacitti (2018: 437n266), drawing on Veselovskiy 1886, offers interpretive hypotheses concerning the figure of Menia and her abilities. Wolfram Brandes (2005) has shown that the image of the gold-making Menia traveled across time and cultures, her name reappearing in the Eddic poem *Grottasǫngr*, the “Song of [the Mill] Grotti”, where Menja, together with her sister Fenja, is depicted as a giantess endowed with the power to grind out gold and prosperity.

> ἐμφέρεται γὰρ ἐν τοῖς ἱστορικοῖς, ὅτι αὕτη ἡ Μενία, Μακεδόνισσα ἐκ γένους Ἀλεξάνδρου τοῦ βασιλέως τῶν Μακεδόνων, ἥτις ἁπάσης μεταλλικῆς τέχνης ἔμπειρος ἦν καὶ τὰ μαζικὰ στίλβουσα, ἀόκνως τὴν πόλιν αὐτῆς ἔπλησεν χρυσίου καὶ ἀργυρίου, ὡς μὴ εὑρίσκεσθαι ἐν τῇ πόλει ἐν ἐνδείᾳ τινά. πολλοὶ δὲ τῶν πέριξ ἐπαρχιῶν χάριν πλούτου ἤρχοντο καὶ ἐνεπιμπλοῦντο.
>
> For it is related in historical works that Menia, a Macedonian descended from Alexander, king of the Macedonians, who was skilled in every form of metallurgical art and made the lumps/bun ingots(?)[14] of metal gleam, unstintingly filled her city with gold and silver, so that you could not find anyone in the city in need. Many people from the surrounding areas came thanks to this wealth and were richly rewarded.[15]

One might argue that these excerpts pertain not to alchemy but rather to mining and metal-working. At first glance, Menia's role could be understood simply as supplying her people with metals through her divinely granted metallurgical skills. Although this is a hagiographical text rather than historical – one that admits myth and exaggeration more readily – we may still wonder: if nothing alchemical was involved in Menia's actions, then it would have sufficed for her merely to identify the ore deposits and leave the tasks of extraction and smelting to specialized artisans. Yet, the narrative leaves no room for such speculation: Menia herself produced abundant quantities of both precious and base metals from the ores in a short time. A crucial element implied in the account is the pressure of time. With Aquilinus threatening the newly-founded city, Menia had to enrich it rapidly so that it would possess the resources to confront the enemy. Instead of describing a straightforward miracle, the hagiographer presents a two-stage solution to this urgent problem: first, at Menia's request, a seer identifies the location of the ores; and second, she invokes the god of Higher Providence and is divinely endowed with the knowledge to manipulate the metal ores herself. In Greek chrysopoetic literature, transmutation is sometimes described as a process that may occur in a relatively short time, once the necessary agent has been obtained.[16] At the

14 See *Lexikon zur byzantinischen Gräzität*: s.v. μαζικόν (*Metallklumpen*). Cf. Acconcia Longo 2001a: 10n57: "[...] καὶ τὰ μαζικὰ στίλβουσα (che faceva risplendere – στίλβω –, cioè diventare oro o argento, gli impasti, le leghe di metalli, oppure τὰ μαζικὰ indica oggetti d'uso comune?) [...]". It is noteworthy that Liddell–Scott–Jones: s.v. μᾶζα II, defines this term generally as "lump, mass, ball", but in an alchemical context specifically as "amalgam", on the basis of references in the Leiden X and Stockholm chemical papyri. This is not an implausible explanation for our passage, since an amalgam coating makes (base) metals gleam – if, indeed, that is what the text intends to convey. Techniques of amalgam gilding and silvering are attested in the Greek alchemical tradition (Merianos 2017: 247–248). Be that as it may, Halleux (1981: 168 [p. 85n6]) explains that μᾶζα designates a barley cake, and stresses that, behind the name in Greek alchemy, "il y a deux idées: la forme de la galette des lingots de métal et d'autre part l'idée d'une pâte qui lève, rejoignant les idées sur la croissance des minéraux". On the term μᾶζα, see also the comments of Stephanides 1909: 27 and n4, 28.

15 *Vita Pancratii Tauromenii* 126 (ed. and Engl. transl. [adapted from] Stallman-Pacitti 2018: 230, 13–17 [text], 231 [transl.]).

16 See, e.g., Nicephorus Blemmydes, *De confectione auri* (ed. Berthelot and Ruelle 1887–1888: 2.457, 11–17).

same time, the depiction of Menia enriching her city through metallurgical knowledge brings to mind an occasionally attested aphorism: that alchemy is the remedy for the incurable disease of poverty,[17] provided one can grasp its secrets.

Menia's invocation of the deity of "Higher Providence" to manufacture metals also lends alchemical connotations to the story. While the invocation of God to accomplish an extraordinary deed is commonplace in a Christian context, in Menia's case it also recalls late antique narratives in which women acquire alchemical knowledge through revelation. In the late second- or early third-century text known as *The Letter of Isis to Horus*, the angel Amnael initiates the goddess Isis into the secret of gold- and silver-making, though he seeks to reveal this knowledge in exchange for sexual favors.[18] This text is usually studied in parallel with a well-known passage of Zosimus of Panopolis (late third–fourth century), preserved by George Syncellus (d. after 810), which refers to the myth – derived from 1 Enoch – that attributes the first teaching of the alchemical art to angels, who, desiring women, revealed their knowledge to them. Because of this, and because they taught humankind every wickedness, they were cast out of heaven.[19] Naturally, the link between sexual desire and the revelation of esoteric knowledge could not have been incorporated into a hagiographical narrative. Yet, in Menia's case too, alchemical knowledge is portrayed as divinely bestowed upon a woman. This aspect is also consistent with the appellations "sacred" and "divine art", long used to describe what modern scholarship refers to as "alchemy".[20]

This notion of divinely-granted alchemical wisdom persisted as alchemy evolved within a Christian framework. It is therefore no coincidence that Byzantine alchemical authors – such as Christianus (late sixth–seventh century) and Stephanus of Alexandria (seventh century), among others – regarded participation in divine knowledge, of which alchemy formed a part, as a gift of God. Comprehending matter required illumination bestowed by divine grace upon a worthy individual, enabling the unraveling of the lessons the universe imparted about its divine design.[21]

Notably, the god of Higher Providence, whom Menia invokes, also appears in the *Alexander Romance* attributed to Pseudo-Callisthenes.[22] Menia's story in the *Life of Pancratius* – where, as already noted, she is presented as a descendant of Alexander the Great – may therefore have drawn inspiration from the *Alexander Romance*.[23] More crucial, however, is that she is depicted as believing in the god of Higher Providence,

17 Colinet 2010: 93n64, 117n4.

18 [Isis Prophetissa], *Filio suo Horo* (ed. Mertens 1983–1984). See Festugière 1944: 253–260; Idel 1986; Mertens 1988; 1989a; Blanco Cesteros 2022; Lopes da Silveira 2022. See also the next footnote.

19 Georgius Syncellus, *Ecloga chronographica* (ed. Mosshammer 1984: 14, 2–14). See Mertens 1989b; Fraser 2004; Martelli 2014; Martelli and Rumor 2014; Reed 2014; Bull 2018. See also the previous footnote.

20 On these terms see, e.g., Merianos 2024a: 23, 27.

21 Merianos 2022; 2024a: 35–39. See also Maria Mavroudi's contribution to this volume.

22 Stallman-Pacitti 2018: 437n267.

23 Stallman-Pacitti 2018: 498. On the Greek *Alexander Romance*, see, e.g., Jouanno 2002; Djurslev 2024. Particularly on the *Alexander Romance* and Byzantine chronicles, see Jouanno 2018.

a name evocative of the Christian concept of Divine Providence. Although a pre-Christian figure, Menia is framed in such a way as to bring her religiously close to Christian audiences – an aspect of particular importance for the positive portrayal of Taormina's co-founder.[24]

Terminological Insights from a Middle Byzantine Alchemical Fragment

Concerning the specific words and phrases in the excerpts from the *Life of Pancratius of Taormina* as discussed above, it must be acknowledged that Byzantine terminology is seldom clear-cut. Particular attention, however, should be given to the context-sensitive terms χύμευσις and χυμευτικόν, which in the texts of the so-called Greek Alchemical Corpus usually denote the alchemical process. Indeed, χύ(εί)μευσις, which may be translated as "fusion" and "casting",[25] or "alloying" and "enameling",[26] is among the terms that, in chemical writings, designate a domain largely overlapping with alchemy. It is also important to keep in mind that words such as χυμευτός / χειμευτός ("fused", "cast", or "enameled")[27] refer to enamel-decorated objects and occur in texts of various genres. These and related terms (e.g., χύμευσις, χυμευτικός) suggest an interrelation between the luxury-goods industry and alchemy, a connection likewise reflected in the vocabulary.[28]

Furthermore, in the excerpts, Menia is explicitly associated with μεταλλικὴ τέχνη (literally, "metallurgical art"), which has been rendered as "metal-working"[29] or, in a more traditional sense, as "mining".[30] It is true that χύμευσις and μεταλλικὴ τέχνη are

24 Cf. Brandes 2005: 195–196.

25 Sophocles 1914: s.v. χύμευσις.

26 *Lexikon zur byzantinischen Gräzität*: s.v. χύμευσις (*Metallegierung*; *Emailarbeit*).

27 Sophocles 1914: s.v. χυμευτός; *Lexikon zur byzantinischen Gräzität*: s.v. χυμευτός (*emailliert*, *aus Email*).

28 Merianos 2017: 241–242. On the relationship between enameling and alchemy, see Steiner 2020.

29 Μεταλλικὴ τέχνη is translated in this way in *Vita Pancratii Tauromenii* 104; 126; 151 (ed. and Engl. transl. Stallman-Pacitti 2018: 199; 231; 267).

30 See, for instance, *Vita et miracula Theocleti Lacedaemonii* 3.2 (ed. and Engl. transl. Kaldellis and Polemis 2019: 174 [text], 175 [transl.]): Καὶ πονοῦσι μὲν ὡς τὰ πολλὰ οἱ τὴν μεταλλικὴν μετερχόμενοι τέχνην, ὀρύγματά τινα ὀρύσσοντες ὑπόγαια καὶ βαθύτατα τῇ ἀβύσσῳ πλησιάζοντα, [...] ("Miners work hard to open up very deep channels underground, reaching to the very depths of the earth [...]"). Indeed, according to Liddell–Scott–Jones (s.v. μεταλλικός), μεταλλική (sc. τέχνη) denotes the "art of mining"; μεταλλικός II, however, is defined as "possessing a knowledge of metals". It should be noted that, in a medico-pharmaceutical context, the adjective may assume specialized meanings. For instance, μεταλλικὰ φάρμακα designates "mineral drugs"; see, e.g., Galenus, *De compositione medicamentorum per genera libri VII* (ed. Kühn 1827: 705, 18).

different terms with distinct senses; yet, in alchemical discourse, there are instances where they are closely linked.

In particular, this close association can be evidenced in a Middle Byzantine fragment of considerable relevance to our discussion, preserved in several manuscripts – *Parisinus graecus* 2327 (= **A**; 1478), *Laurentianus Pluteus* 86.16 (= **L**; 1492), *Meteora, Mone Hagiou Stephanou* 97 (= **Met**; 1503/1504), and *Elassona, Mone Panagias Olympiotisses* 197 (= **El**; copied by three different copyists, partly in 1507, partly by the priest Margarites in 1741).[31] Although Marcellin Berthelot and Charles-Émile Ruelle published it erroneously in their edition within the section devoted to the writings of Zosimus of Panopolis,[32] elsewhere they suggest that it may in fact represent the title, or preface, of a Byzantine handbook of practical chemistry dating between the eighth and tenth centuries.[33] In this short text, the adjective μεταλλικὴ is set alongside the adjective χυμευτικὴ in reference to a particular book, and the passage concludes with a mention of τέχνη μεταλλική.

Ὅτι τὸ παρὸν βιβλίον ὀνομάζεται βίβλος μεταλλικὴ <καὶ> χυμευτική, περὶ χρυσοποιίας, ἀργυροποιίας, ὑδραργύρου πήξεως, ἔχον αἰθάλας, βαφάς, φούρμας ἀπὸ βροντησίου, ὡσαύτως καὶ λίθων πρασίνων, καὶ λυχνιτῶν, καὶ ἑτέρων πάντων χρωμάτων, καὶ μαργάρων, καὶ δερμάτων ἐρυθροδανώσεως βασιλικῶν. Ταῦτα δὲ πάντα γίνονται ὑπὸ ὑδάτων, ᾠῶν, διὰ τέχνης μεταλλικῆς.

1 Ὅτι … ὀνομάζεται om **MetEl** ‖ <καὶ> scripsi : καὶ **MetEl** : <καὶ> BeRu Martelli 2011 : καὶ Martelli 2018 ‖ χυμευτικῇ **ALMetEl** ‖ ☉ποιίας [i.e. χρυσοποιίας] **ALMetEl** ‖ 2 ☾ποιίας [i.e. ἀργυροποιίας] **ALMetEl** ‖ ☽ [i.e. ὑδραργύρου] **ALMetEl** ‖ ἔχον scripsi : ἔχων **ALMetEl** : ἔχουσα corr BeRu Martelli 2011 et 2018 ‖ θ˥θ˥ [i.e. αἰθάλας] **MetEl** ‖ βαφὰς καὶ **MetEl** ‖ φούρμουσαι **AL** BeRu : φρούμας **MetEl** : φούρμας corr Martelli 2011 et 2018 ‖ βροντησίου scripsi : βροτισίων **AL** : βροντησίου **MetEl** : βροντησίων corr Martelli 2011 : βροντησίου Martelli 2018 ‖ 3 λυχνίτων **AL** : λυχνιτὸν **MetEl** : λυχνίτων Martelli 2018 ‖ παντίων **MetEl** ‖ 4 ἐρυθρωδανώσεις **AL** : ἐρυθροδανόσις **MetEl** : ἐρυθροδανώσεως corr Martelli 2011 et 2018 ‖ γίνονται scripsi : γίνωνται **AL** : γύνοντε **Met** : γύνονται **El** : γίγνονται BeRu Martelli 2011 : γίνονται Martelli 2018 ‖ ὑδάτων scripsi : ≈ **ALMetEl** : ὑδάτων θαλασσίων BeRu Martelli 2011 et 2018 ‖ ό͞ό [i.e. ᾠῶν] **ALMetEl**.

The present book is called the book on metallurgy <and> *chymeusis*,[34] concerning chrysopoeia, argyropoeia, the solidification of mercury – it [also] contains vapors, tinctures, and [hollow] molds

31 A, f. 240 recto–verso; **L**, f. 219 verso; **Met**, f. 1 recto–verso; **El**, f. 1 recto; Berthelot and Ruelle 1887–1888: 2.220, 11–16 (= BeRu); Martelli 2011: 133; 2018: 108 (Martelli 2011 edits from **A**; Martelli 2018 edits from **Met** and **L**). See also Halleux 2021: 9n11. On **Met**, see Sofianos 1986: 305–309; on **El**, see Skouvaras 1967: 375–378 (a reproduction of f. 1 recto occupies p. 376). In the following edition, the text of **A** is used as the basis.

32 Berthelot and Ruelle 1887–1888: 2.220, 11–16.

33 Berthelot and Ruelle 1887–1888: 3.360: "On voit qu'il s'agit d'un manuel byzantin de Chimie. La composition même de l'ouvrage remonte à une époque ancienne, telle que le VIIIe ou le Xe siècle". See also Martelli 2013: 62–63 and n366; 2018: 109.

34 Cf. Olympiodorus the alchemist (sixth century), who mentions "a book on *chymeusis*" (βίβλον […] χημευτικήν) in *De arte sacra* 18 (ed. Berthelot and Ruelle 1887–1888: 2.80, 13). See Viano 2022: 145.

of bronze[35] – as well as green stones, red stones, and all the other colors, and pearls, together with the red dyeing[36] of imperial leathers. All these are produced with waters,[37] [and] eggs,[38] through the metallurgical art.[39]

---

35 The reference to *phourmai*, hollow bronze molds for counterfeiting coinage, is particularly intriguing. First, the text suggests that the book on metallurgy and *chymeusis* placed special emphasis on the counterfeiting of coinage, presumably gold. Second, since other Greek chemical texts also contain references to the making of *phourmai*, we may speculate that this technical procedure was considered significant, and that the lost book to which the fragment alludes may have incorporated the extant – or similar – recipes for producing counterfeit gold coinage. On the term *phourma* (its standard numismatic rendering, occurrences in chemical texts, and connections with state workshops, mint workers, and techniques), see Merianos 2024b. For the interpretation of the phrase φούρμας ἀπὸ βροντησίου, attention should be given to the text entitled Εἰ θέλεις ποιῆσαι φούρμας καὶ τύλους ἀπὸ βροντησίου, ποίει οὕτως (If You Want to Make Bronze Molds and Casts, Do as Follows), found in *Marcianus graecus* 299 (coll. 584), ff. 128 verso – 130 recto; see the recent edition and commentary in Halleux 2021: 129–138 (no. V). For an overview of the minting techniques, see Casoli and Wahl 2019: 27–32.

36 The term ἐρυθροδάνωσις, rendered here as "red dyeing", is derived from the verb ἐρυθροδανόω ("to dye with madder, to dye red"), itself formed from the noun ἐρυθρόδανον, i.e., madder. See Liddell–Scott–Jones: s.vv. ἐρυθρόδανον, ἐρυθροδανόω, and ἐρυθροδάνωσις ("dyeing scarlet"); *Lexikon zur byzantinischen Gräzität*: s.vv. ἐρυθρόδανον and ἐρυθροδανόω (*mit Krapp [rot] färben*). The roots of madder (*Rubia* species) have been used since ancient times to extract red colorants for textile dyeing and pigment production; see, e.g., Daniels et al. 2014. On madder and related dye plants of the Rubiaceae family, see Cardon 2007: Chapter 4.

37 I render the sign ≈ here as ὑδάτων ("waters"), not as ὑδάτων θαλασσίων ("sea waters") as in Berthelot and Ruelle (1887–1888: 2.220, 16), followed by Martelli (2011: 133; 2018: 108; cf. Martelli 2013: 62: "sea water"), and Halleux (2021: 9n11). In **A** (f. 240 verso) and **L** (f. 219 verso), the sign carries a *daseia* (rough breathing), an *oxeia* (acute accent), and the abbreviation "-ων" above it, whereas in **Met** and **El** it appears without any diacritics. The earlier choices to render the sign as ὑδάτων θαλασσίων are not arbitrary. Several Greek manuscripts preserve lists of alchemical signs with their interpretations, intended to guide the reader. In most of these, the sign ≈ is indeed glossed as θαλάσσια ὕδατα (Zuretti 1932: no. 80; cf. nos. 216, 848, 968, 1308), while other signs are reserved for ὕδωρ (Zuretti 1932: no. 82; cf. nos. 218, 327, 477, 675, 768, 850, 1063). Yet, these lists do not present the signs in a fixed or uniform manner across all manuscripts. It is particularly striking that in the list of **L** – one of the very manuscripts that transmits the text under discussion – the sign for ὕδωρ is represented by a three-wave line (f. 5 recto; Zuretti 1932: no. 1180; cf. nos. 861, 960). This form is very close to the two-wave line traditionally denoting θαλάσσια ὕδατα, although in **L** θαλάσσια ὕδατα are still represented in the conventional way (f. 3 recto; Zuretti 1932: no. 848; cf. nos. 861, 960). More important, however, is that the sign I render as ὑδάτων in the text must denote the chemical solutions known specifically as ὕδατα ("waters"). In alchemical discourse, the term refers to solutions capable of dyeing metals, with θεῖον ὕδωρ ("divine water") being the most renowned case (see Martelli 2009; 2022). Within the Greek Alchemical Corpus, two of the quintessential methods of the "art" are precisely the use of "waters" and of (distilled) eggs (see next note). In this light, the reference to "waters" complements the subsequent mention of eggs, forming a readily recognizable pair of techniques, whereas the reading "sea waters" proves semantically puzzling.

38 There are several texts in the Greek Alchemical Corpus that refer to the processing of eggs for the production of the transmutational agent. The underlying concept is that the matter is quadripartite and corresponds to the four parts of an egg (shell, membrane, white, yolk). Since ancient times, the egg was regarded as an image of the world, whose four parts correspond to the four elements. For Greek alchemical texts on egg distillation, see Colinet 2000: 171; Dufault 2017; Merianos 2022.

39 Cf. the translation in Martelli 2013: 62.

This text raises several noteworthy points that warrant discussion. The mentioned book is referred to as βίβλος μεταλλικὴ καὶ χυμευτική. This reading is preserved in the later witnesses to the text (**Met** and **El**), whereas in the older manuscripts (**A** and **L**) it appears as βίβλος μεταλλικὴ χυμευτική. However, as Günter Prinzing demonstrates in this volume – in the case of the recipe collection on the art of goldsmithing preserved in **A** and **Met** – the older manuscript does not always offer the best reading. Returning to the paratactic syntax μεταλλικὴ καὶ χυμευτική, the conjunction links the two terms, a connection reinforced by the final sentence, which stresses that all the aforementioned products are achieved through metallurgical art. It should be noted in this context that the paratactic καὶ does not necessarily imply separateness, but may also function as the conjunction in a hendiadys – that is, two terms expressing a single, broader concept.

With regard to the adjective χυμευτική, the *Suda* lexicon defines χεί(ύ)μευσις as a synonym of χει(η)μεία.[40] In our passage it denotes the broader spectrum of alchemical aims and techniques, more inclusive than the terms chrysopoeia and argyropoeia, which are oriented exclusively toward metallic transmutation. The text itself makes this clear: gold-making and silver-making are presented as only two among the pursuits encompassed by the book on metallurgy and *chymeusis*. Indeed, the fourfold goals described in the book – gold-making, silver-making, the production of precious stones and pearls, and purple dye – reflect the wider conception of alchemy rather than the narrower focus on gold- and silver-making alone. Both tendencies are, in fact, attested in the Greek alchemical tradition.[41]

As for the expression τέχνη μεταλλική ("metallurgical art"), it seems to function in the closing sentence as an umbrella term for techniques, encompassing the fourfold aims together with others, such as the making of bronze molds. This is significant because it not only indicates an interest that extended beyond the exclusive production of precious metals, but also an emphasis on the technical side of the art – on distinct methods and procedures applicable, to varying degrees, across all four objectives. One might, however, interpret the phrase μεταλλικὴ καὶ χυμευτικὴ at the beginning of the text as denoting two subsets of objectives, with μεταλλικὴ referring to metals and χυμευτικὴ to precious stones, pearls, and purple dye. Although this reading is possible, the mention of "metallurgical art" in the concluding sentence seems to confer upon it a central focus, indicating that metallurgy is treated as the operative framework of the techniques described. If this observation holds, the phrase functions as an inclusive term, an implication likewise suggested by the paratactic syntax of μεταλλικὴ καὶ χυμευτικὴ at the beginning of the fragment. From a broader historiographical perspective, such fluidity of terminology illustrates precisely the difficulty of defining "alchemy" as

40 *Suda lexicon*, s.v. Χειμεία (X 227) (ed. Adler 1928–1938: 4.800): Χειμεία: ἡ χείμευσις. ζήτει ἐν τῷ χημεία; see also s.v. Χημεία (X 280) (ed. Adler 1928–1938: 4.804). On these entries in the *Suda*, see Merianos 2017: 238, 239, 241, and 242, as well as Alexandre Roberts' contribution to this volume.
41 Martelli 2013: 58–63.

a stable category.[42] The evidence discussed here, in which metallurgical and chemical terminology overlap, exemplifies this conceptual blending and underscores the need to consider these arts as part of a continuous field of transformative knowledge rather than as sharply divided disciplines. Moreover, this once again confirms that there was not a single, fixed designation or concept in Byzantium for what we now refer to as "alchemy".[43]

A similar, though somewhat inverted, point emerges in the tale of Menia. In the last excerpt of the *Life of Pancratius* cited above, she is praised for her metallurgical expertise – "she was skilled in every form of metallurgical art" (ἁπάσης μεταλλικῆς τέχνης ἔμπειρος ἦν).[44] Later we are told that "the city of Taurus and Menia was the richest in gold and silver and celebrated for the other types of metallurgical art".[45] Menia's expertise and technical mastery appear to have been transmitted by her to the city. Yet, the foundation of this divinely bestowed knowledge, I would argue, was (al)chemical, conceived as a kind of overarching framework that combined *know-why* with *know-how*.[46] In the story of Taurus and Menia, unlike in the anonymous description of the book on metallurgy and *chymeusis*, priority is given – albeit *implicitly* – to "chymeutic" knowledge. This difference reflects perspective and emphasis rather than contradiction. What emerges in both cases is the perception that "chymeutic" knowledge could be conceptualized as closely related, even kindred, to metallurgical art. From this perspective, and within specific contexts, "metallurgical art" may thus be understood as being intertwined with *chymeusis*. In any case, the difficulty of drawing a clear line between artisanal crafts and the "sacred art", owing to their mutual exchange, is explicitly acknowledged.[47]

It should be added that the term "metallurgical art" is rare in the Greek Alchemical Corpus. It is therefore noteworthy that **Met** and **El** preserve a pair of passages serving as "introductory texts", in both of which the phrase appears. Each manuscript transmits two concise pieces that seem to offer readers a preliminary overview of the (al)chemical writings copied within. The first is likely unique to **Met** and **El**, as it does not appear in any other known alchemical manuscript; the second is the brief text concerning the book on metallurgy and *chymeusis* already discussed.

The first passage contains an intriguing statement: Ἐπαινετὴ δὲ ἡ ἑλικοειδὴς γραφὴ τῆς μεταλλικῆς τέχνης, ἐπεὶ ἡ ἀρχὴ τοῦ συγγράμματος τοῖς ἑλικτοῖς ἔοικεν ὄφεσιν [...] ("Praiseworthy, too, is the spiral writing of the metallurgical art, since the beginning of

42 See Roberts 2022: 169–170; 2023: 598–599.
43 Merianos 2024a: 23.
44 *Vita Pancratii Tauromenii* 126 (ed. Stallman-Pacitti 2018: 230, 14–15); Engl. transl. mine.
45 *Vita Pancratii Tauromenii* 151 (ed. and Engl. transl. [adapted from] Stallman-Pacitti 2018: 266, 22–24 [text], 267 [transl.]).
46 Cf. Halleux 2021: 9–10.
47 Halleux 2021: 9; Merianos 2024b: 158.

the work resembles coiled serpents").[48] As Matteo Martelli has observed, the latter part of this sentence, together with what follows, echoes Michael Psellus (1018–late 1070s).[49] Here, the reference to "metallurgical art" clearly serves to characterize the contents of the texts copied in the Meteora and Elassona manuscripts. The "spiral-like writing", on the other hand, points to the traditional challenge, recognized since antiquity, of making sense of the notoriously perplexing alchemical writings.[50] Such difficulty necessitated close engagement with both alchemical and philosophical literature, together with a meticulous "cover-to-cover" reading of the manuscripts themselves.

Taken together, the two introductory texts in **Met** and **El** – the first alluding to "spiral writing" and the second concerning the book on metallurgy and *chymeusis* – invoke the notion of metallurgical art in an (al)chemical context. Since the second text is probably older, I would suggest that the anonymous author of the first may have derived its reference to the metallurgical art from it. In any case, the two passages appear complementary: the first offers a reflection on style and structure; the second, a description of practical content; and together they provide a kind of introductory framework for the texts copied in these volumes.

To return to Menia, the discussion of the terms *chymeusis* and "metallurgical art" indicates that, in the *Life of Pancratius*, they may be understood as closely associated with alchemy.

2 Heliodorus the Sorcerer: Diabolical Metallurgical Deception

Modern scholarship has examined the analogies and differences between the *Life of Saint Pancratius of Taormina* and the *Life of Saint Leo of Catania*, including their references to alchemical and magical metal-making.[51] It is perhaps not accidental, as Stephanos Efthymiadis has noted, that both *Lives* (and their saints) are associated with southern Italy and with the imaginary.[52] The *Life of Leo* survives in several versions[53] and was composed after that of Pancratius. It has long been recognized that the text displays a striking imbalance: the saint appears merely as a secondary protagonist, while the bulk of the narrative is devoted to his archenemy, the sorcerer Heliodorus, who is credited with extraordinary feats. Augusta Acconcia Longo, editor of the short prose

48 **Met**, f. 1 recto; **El**, f. 1 recto; Martelli 2018: 107. Engl. transl. mine.
49 Michael Psellus, *De Heliodoro et Achille Tatio judicium* (ed. Dyck 1986: 92, 24–28). See Martelli 2018: 108.
50 Cf., e.g., Merianos 2022: 294.
51 Acconcia Longo 1989: 55–61; 2001a; Kazhdan et al. 1999: 295–308; Brandes 2005: 195.
52 Efthymiadis 2024: 11; cf. 2017: 375–379.
53 For versions, editions, and discussions of the *Vita*, see Acconcia Longo 2007: 7–30; Alexakis 2011: 9–37.

version of the Greek *Life of Saint Leo of Catania* (*BHG* 981), considered it to display iconoclastic sympathies, pointed to southern Italy or Sicily as its place of composition, and dated it to the Second Iconoclasm (815–843). Alexander Alexakis, editor of the longer prose version of the *Vita* (*BHG* 981b), has argued that it is a ninth-century, anti-iconoclastic text, probably written in Constantinople or its environs.[54]

Whereas Menia's deeds are portrayed as enriching and protective of her city, Heliodorus' powers are cast in a destructive light. Having made a pact with the devil and receiving an assistant demon, he performs sinister marvels. Most pertinent to our discussion is his alleged power to transmute stones into gold and wood into silver – a feat with devastating economic consequences. These transformed products (coins and/or other objects?) circulated in the market of Catania. But once transactions were completed, they were found to have reverted to their original state and appearance. As a result, citizens refrained from buying and selling, and the city fell into acute poverty.

> Ἡ μέντοι δεινότης τῆς διαβολικῆς ἐκείνου δολιότητος καὶ δημοσίᾳ προέβη, μέχρι τῶν πωλούντων λέγω καὶ τῶν ἀγοραζόντων· μεταλλάττων γὰρ τῷ δοκεῖν ἦν τὰς προκειμένας ὕλας καὶ δεικνύων λίθον μὲν ὡς χρυσίον, ξύλον δὲ ὡς ἀργύριον καί τι τυχὸν ἄλλο τῶν εὐώνων καὶ οὐδαμινῶν ὡς πολύτιμόν τι χρῆμα καὶ μέγα, καὶ τῶν ἀχρείων καὶ ἀ<νω>φελῶν ὡς ἀναγκαῖον καὶ χρήσιμον, ὅπερ μετὰ τὸ ἀπεμποληθῆναι εἰς τὴν οἰκείαν αὖθις εὑρίσκετο μένον ποιότητα καὶ θεωρίαν, οὐ τὴν τυχοῦσαν παρέχων ζημίαν καὶ φιλονεικίαν τοῖς τε πιπράσκουσιν ἅμα καὶ τοῖς ἀγοράζουσιν. Τὴν γοῦν τοιαύτην ἐναργῶς καταμανθάνοντες δυσπραγίαν, διὰ δεινότητα γινομένην Ἡλιοδώρου, τοῦ πωλεῖν καὶ ἀγοράζειν πάντες ἀπέσχοντο, πρὸς ἐσχάτην ἐντεῦθεν πενίαν καὶ ἀπορίαν καταντήσαντες.
>
> Indeed, the shrewdness of his diabolical deceit also reached the public domain, including the buyers and merchants, I mean. For he transformed (μεταλλάττων) the appearance of the materials at hand, making stone [appear] as gold, and wood as silver. What was cheap and worthless, [he transformed into] something great and costly, and what was obsolete and useless, into [something] important and useful. After the same [item] was sold, it was found to have returned to its original state and appearance, which caused no small loss for and was a source of contention between the merchants and buyers. When they clearly realized that their bad luck happened because of Heliodorus' wickedness, everyone refrained from selling and buying, which resulted in the most extreme poverty and destitution.[55]

---

54 Acconcia Longo 1989: 36–55; Alexakis 2011: 73–74. For a concise overview, see Efthymiadis 2023: 39n19, who also cites the bibliography of the debate between Acconcia Longo and Auzépy, initiated by an article of Auzépy (1992: 62–67) and her interpretation of the *Life*. This debate is not considered here, since its connection with the topic is indirect. See also D'Aiuto 2025 for an overview of modern scholarship on the Greek hagiographical dossier of Saint Leo of Catania, together with a focused discussion of philological observations on the metrical *Vita* (*BHG* 981c).

55 *Vita Leonis Cataniae* (longior) 15.1–12 (ed. and Engl. transl. Alexakis and Wessel 2011: 158 [text], 159 [transl.]). Among the sources he proposes as possible influences on Heliodorus' story, Alexakis (2011: 109–110) refers to a short tale that comes nearest to Heliodorus' transformation of materials into gold and their circulation in the market of Catania. This is the story of Pases and his hemiobol coin, attributed to Apion, the Alexandrian grammarian (fl. early first century CE). An entry in the *Suda* lexicon on Pases mentions in connection with his hemiobol: "Pases had also a half-obol coin made by means

As the narrative progresses and the sorcerer is taken to prison,

> [...] διαλαθὼν τοὺς ἕλκοντας αὐτὸν στρατιώτας καὶ γῆθεν λίθον ἀνελόμενος, διὰ τῆς συνήθους μαγγανείας τοῦτον ἐχρύσωσεν καί φησι πρὸς αὐτούς· "Ἐὰν ἄνετον ἀφῆτέ με πορευθῆναι, δίδωμι ὑμῖν ποσότητα χρυσίου λιτρῶν τριῶν". Οἱ δὲ τὸ ὑπέρογκον τοῦ χρήματος κερδᾶναι γλιχόμενοι, "Δίδου ἡμῖν", φασί, "τὸ ποσόν, καθάπερ ὑπέσχου, καὶ χαίρων ὅπου βούλει πορεύθητι". Ὁ δὲ δίδωσιν αὐτοῖς τὸν χρυσοειδῆ λίθον ἐκεῖνον· [...]. Οἱ δὲ τὸν νομιζόμενον χρυσὸν διαμερίζειν βουληθέντες, λίθον εὗρον μηδαμῶς τῆς οἰκείας μεταβληθέντα ποιότητος, ὃν καὶ ζυγοστατήσαντες εὗρον ὁλκὴν ἔχοντα λιτρῶν τριῶν, ὡς ἐκεῖνος ἔχειν προείρηκεν.
>
> [...] he picked up a stone from the ground without the soldiers who were dragging him noticing. By his usual sorcery (μαγγανείας) he turned the stone into gold and said to the soldiers, "If you let me go free, I'll give you the sum of three pounds of gold". Eager for this great sum of money, they said: "Give us the sum you promised and go freely, to wherever you wish". So Heliodorus gave them that golden-looking stone. [...] When the soldiers wished to divide the purported gold, they found that it [was a stone] that had in no way been altered from its original state. They weighed it and found it to be three pounds, as Heliodorus had earlier said that it was.[56]

The verb μεταλλάττω in the first excerpt – one of whose meanings is "to transmute" – is employed to denote transformation. The same verb also occurs several times in the Greek Alchemical Corpus, for instance, in the so-called *Work of the Four Elements* (possibly twelfth century): "This is the nature that conquers nature, that transmutes (μεταλλάττουσα) natures [...]". The phrase presents a shortened and adapted form of the well-known, pseudo-Democritean formula.[57] The transformations performed by Heliodorus may be seen as alluding to transmutation, not constituting, of course, true transmutation. However, there are two notable dissimilarities. First, alchemists are expected to transmute base metals, not wood and stones, into precious ones. Second,

---

of his incantations, which he used to give to the sellers from whom he wished to buy something and then, if he wished so, the coin was found again [in his hands]"; see *Suda lexicon*, s.v. Πάσης (Π 752) (ed. Adler 1928–1938: 4.65, 8–10); Engl. transl. Alexakis 2011: 110. On Pases and the *Suda* entry, see Rives 2009.

56 *Vita Leonis Cataniae* (longior) 31.1–18 (ed. and Engl. transl. [adapted from] Alexakis and Wessel 2011: 178 [text], 179 [transl.]).

57 *De quattuor elementis* (ed. Berthelot and Ruelle 1887–1888: 2.338, 18–19): αὕτη ἐστὶν ἡ φύσις ἡ νικῶσα τὴν φύσιν, ἡ *μεταλλάττουσα* τὰς φύσεις, [...] (emphasis in all instances is mine). Cf. [Democritus], *Physica et mystica* 3.61–63 (ed. and Engl. transl. Martelli 2013: 84 [text], 85 [transl.]): ἡ φύσις τῇ φύσει τέρπεται, καὶ ἡ φύσις τὴν φύσιν νικᾷ, καὶ ἡ φύσις τὴν φύσιν κρατεῖ ("Nature delights in Nature, nature conquers nature, nature masters nature"); see p. 85, *apparatus fontium*, on alchemical texts quoting the formula. For other instances of the verb μεταλλάττω, see *De quattuor elementis* (ed. Berthelot and Ruelle 1887–1888: 2.339, 7–8): καὶ εἰ μὲν λευκαίνεται τὸ πέταλον, ἁπαλύνεται καὶ *μεταλλάττεται* ἐκ τῆς οὐσίας αὐτοῦ, καὶ γίνεται τέλειον, χρυσοῦ κάλλιον; *Lexicon chrysopoeiae*, s.v. Ψωμοί (ed. Berthelot and Ruelle 1887–1888: 2.17, 6): Ψωμοί εἰσι τὰ *μεταλλάσσοντα* τῷ εἴδει; *Serpens Ouroboros* (ed. Berthelot and Ruelle 1887–1888: 2.21, 21–22): [...] τουτέστι συμφαγώνεται καὶ συγχωνεύεται, λειώνεται καὶ *μεταλλάττεται* τὸ σύνθεμα ἐν τῇ σήψει. On the verb μεταλλάττω, see Liddell–Scott–Jones: s.v. μεταλλάσσω III, and Alexakis 2011: 232, who cites only the first of these passages. On *De quattuor elementis*, see Colinet 2000.

Heliodorus' achievement pertains to illusion, since its effect is short-lived.[58] It has been noted that, in fact, the temporary duration of the magical effect is a standard feature distinguishing magic from true miracles.[59]

At the same time, we should not expect a non-specialized text to be precise about materials and processes – especially when its purpose is to exaggerate the powers of Heliodorus – or to draw a clear distinction between alchemy and magic. After all, esoteric knowledge was often thought to pertain to "magic", and alchemy could easily be regarded as such in the public imagination. Indeed, in Byzantium and beyond, "magic" may be considered an overarching category that, depending on the perspective, encompassed divination and other esoteric fields of knowledge while also blurring the boundaries with miracles. Richard Kieckhefer's characterization of magic as an "aggregating term" is particularly apt here as well.[60]

Michael Psellus offers an illuminating indication of this in the eleventh century, emphatically stressing in *On How to Make Gold* that the transmutation of matter occurs through natural alteration, not by spells, wonder-working, or other unspeakable deeds. He thus accentuates the rational character of the "art of change" (τέχνη τῆς μεταβολῆς), insisting that it operates within the limits of nature, and should therefore not be regarded with wonder.[61] Earlier in the same work, he notes that those unfamiliar with its principles might marvel at it and associate it with suspect operative arts.[62] This is hardly surprising, partly because alchemy employs a cryptic technical language replete with code names and symbols. Moreover, the uninitiated could readily perceive the process of metallic transmutation – if not in practice, then at least in concept – as "miraculous".[63]

The concern over association with magic, in fact, had loomed over alchemical practice since its early days. Olivier Dufault has contextualized Zosimus of Panopolis within his intellectual and social milieu, offering an insightful analysis of the ways in which scholarly patronage influenced the development of alchemy. He explores scholarly patronage from the late first to the fourth centuries CE, with particular attention to

58 Alexakis 2011: 232.

59 Van Pelt 2021: 73.

60 Kieckhefer 2019: 15–16. For Byzantine concepts and practices related to magic, see Greenfield 2017. For the historical development and shifting connotations of the terms "superstition", "magic", and "occult", see Hanegraaff 2012: 156–191. For the term "esoteric", see Hanegraaff 2012: 334–335; Geller and Geus 2014.

61 Michael Psellus, *Epistula de chrysopoeia* 5 (ed. Bidez 1928: 30, 10–19). For an Italian translation of this work, see Albini 1988. On Psellus and alchemy, see Merianos 2017: 243–246, 250; 2024a: 24, 40–43, 47; Roberts 2023: 611–615.

62 Michael Psellus, *Epistula de chrysopoeia* 1 (ed. Bidez 1928: 26, 8–11). Notably, alchemy's identification as, or association with, a form of magic occasionally appears in Islamicate thought; see Moureau 2020: 93, 95n33, 96.

63 Merianos 2024a: 43.

scholars engaged with *mageia* – an ambiguous term that carried both positive connotations, as an intellectually appealing "Persian" or "Eastern philosophy", and negative ones, as the practice of illegitimate rituals. This ambiguity enabled scholars to equate the legitimate form of *mageia* with witchcraft in order to discredit their peers who were interested in *mageia*. Dufault analyzes the character-type of the "learned sorcerer", the polemical use of which reflects tensions among client scholars competing for patronage. He is portrayed as a Greek-educated scholar of, for example, Egyptian, Jewish, or Samaritan origin, whose activities at aristocratic banquets included *paignia* – entertaining tricks that displayed knowledge of natural philosophy and stimulated discussion. Dufault demonstrates through case studies how client scholars could appeal to patrons by presenting themselves as masters of Eastern wisdom.[64]

In this context, for Zosimus of Panopolis, the correct practice of alchemy relied on Greek scholarship, especially the pseudo-Democritean tradition. By contrast, the "daimonic" alchemy of his rivals is portrayed as associated with *mageia* in its negative sense, that is, with illegitimate rituals, whereas Zosimus strives to distance himself from it.[65] Thus, his efforts to dissociate his "true alchemy" from *mageia* remind us that such concerns are of ancient pedigree.

What's in a Name? Heliodorus between Alchemy, Magic, and Literature

It may be helpful to understand the persona of Heliodorus and its alchemical resonances by revisiting an older discussion concerning his very name and its possible connections with his most famous namesake: Heliodorus of Emesa (third or fourth century CE), the author of *Aethiopica*, a late antique Greek romance set in Egypt.[66] Acconcia Longo, and more recently Laura Mecella, have argued that Heliodorus' novel – admired for its sophistication but noted for its erotic themes and references to magic – placed its author in a precarious position and gave rise to a legendary tradition that, over time, transformed his image.[67] Beginning in the fifth century, he was depicted as a bishop of

64 Dufault 2019: 9–92.

65 Dufault 2019: 122–141. Dufault (2019: 139) rightly stresses that: "Alchemical literature itself does not suggest a link with *mageia* broadly considered but with *mageia* conceived as the wisdom of the Persian *magoi*. Democritus, who was considered one of the most important figures of Greek alchemy, was also commonly thought to have studied with *magoi*".

66 For Heliodorus of Emesa, see concisely Jeffreys and Jeffreys 1991. For close readings and new approaches to key passages and themes of the *Aethiopica*, see Repath and Whitmarsh 2022. For the figure of Calasiris, a priest of Isis, and for Egyptian magic in the novel, see, e.g., Dufault 2019: 75–84 and Escolano-Poveda 2020.

67 For the discussion that follows, see Acconcia Longo 1989: 13–15; Mecella 2014: 634–637.

Tricca in Thessaly.[68] Subsequently, he was subject to a gradual process of distortion that increasingly tied the author of *Aethiopica* to the world of esoteric knowledge and magic, further transforming him in two stages.

First, he was recast as an alchemist. This complex development stemmed from the attribution of the first of the four Byzantine alchemical poems to Heliodorus the philosopher, the title of which bears a dedication to an Emperor Theodosius, without specifying which one.[69] This was enough to connect the novelist with alchemy. Yet, Acconcia Longo, Mecella, and Friedrich Pfister before them, have argued that Heliodorus the alchemist did not originally denote the novelist, but probably the homonymous disciple of Proclus, Heliodorus of Alexandria (fifth–early sixth century), known for his expertise in astronomy and astrology. A magical divination formula preserved in the sixteenth-century *Ambrosianus* H 2 *inf.*, f. 225 recto, under the name of Heliodorus, and generally attributed to this Neoplatonic philosopher, seems to support the identification.[70] Homonymy, however, inevitably evoked the more famous Heliodorus, the novelist, linking him to alchemy and giving him the aura of a sage versed in secret knowledge. Thus, in the tenth century, the *Chronicle* of Symeon Logothetes conflates Heliodorus the author of *Aethiopica* with Heliodorus the alchemical poet:

> Ἡλιόδωρος ὁ γράψας τὰ λεγόμενα Αἰθιοπικὰ ἐπίσκοπος ἦν Τρίγκης ἐπὶ Θεοδοσίου. γράφει δὲ καὶ διὰ στίχων ἰάμβων τὴν τοῦ Χρυσοῦ ποίησιν πρὸς τὸν αὐτὸν Θεοδόσιον.

> Heliodorus, who wrote the work called *Aethiopica*, was bishop of Tricca in the time of Theodosius [the Great]. He also composed, in iambic verse, *The Making of Gold*, addressed to that same Theodosius.[71]

68 Socrates Scholasticus, *Historia ecclesiastica* 5.22.51 (ed. Hansen 1995: 302, 4–7): ἀλλὰ τοῦ μὲν ἐν Θεσσαλίᾳ ἔθους ἀρχηγὸς Ἡλιόδωρος, Τρίκκης τῆς ἐκεῖ γενόμενος <ἐπίσκοπος>, οὗ λέγεται πονήματα <εἶναι> ἐρωτικὰ βιβλία, ἃ νέος ὢν συνέταξεν καὶ Αἰθιοπικὰ προσηγόρευσεν. See further Mecella 2014: 634–635 and n3–6.

69 See Appendix. For the Greek text, see [Heliodorus], *De philosophorum arte mystica* (ed. Goldschmidt 1923: 26–34); for a German translation, see Goldschmidt 1927. All four Byzantine iambic poems on gold-making have recently been re-edited in María Belén Boned Fernández's dissertation (2025). Alexandre Roberts (2022: 175–176) – not engaging with the discussion on the transformation of Heliodorus of Emesa's image but instead focusing on how Byzantine readers might have perceived "Heliodorus the philosopher" in the *pinax* ("table of contents") of *Marcianus graecus* 299 (coll. 584) – argues that the novelist would have been the most probable association, while briefly considering and rightly dismissing the possibility that the sorcerer of the *Life of Leo of Catania* was intended by the poem's ascription.

70 Pfister 1923: col. 651; Acconcia Longo 1989: 13–14; Mecella 2014: 636n10. All cite Boll 1912 in their discussion of Heliodorus of Alexandria. Boll (1912: col. 19) explicitly rules out any identification between Heliodorus the alchemist and his Neoplatonist namesake. For a concise treatment of Heliodorus of Alexandria, see Keyser 2008. For the divination formula in *Ambrosianus* H 2 *inf.*, see Martini and Bassi 1901: 14 (no. 23) and 53, cited by Boll 1912.

71 Symeon Logothetes, *Chronicon* 95.9.39–42 (ed. Wahlgren 2006: 122); Engl. transl. mine. See Mecella 2014: 635n8. Cf. Pfister 1923: cols 651–652 and Roberts 2022: 175 and n47 (following Pfister), who cite George the Monk and not Symeon Logothetes for this passage.

This passage is important because, on the one hand, it constitutes the earliest testimony linking Heliodorus of Emesa with the reign of Theodosius I[72] (due to the dedication of the alchemical poem to an Emperor Theodosius). On the other hand, it justifies the assumption that Symeon Logothetes had access to the alchemical corpus.[73]

Later, however, according to Acconcia Longo and Mecella, Heliodorus' image took a darker turn. In the *Life of Saint Leo of Catania*, he became a demonic sorcerer, accused of blasphemous rituals, metamorphoses, counterfeit transformations of metals, and other nefarious acts, until the saint subdued and burned him. Thus, in Byzantine imagination, the author of a love romance was gradually transformed – first into an alchemist, a suspect intellectual figure associated with esoteric wisdom, and ultimately into a sorcerer, a demonized enemy of Christian order.[74]

The second stage of this suggested transformation, namely the identification of Heliodorus of Emesa with the sorcerer of the *Vita*, has received criticism from scholars who have also examined the *Life of Saint Leo of Catania*. Kazhdan, for instance, prefers to see Heliodorus the sorcerer as related instead to pre-Faustian figures, particularly Theophilus, the *oikonomos* of the Church of Adana (Asia Minor), whose *Life* (*BHG* 1319–1322) presents many similarities with the depiction of Heliodorus.[75] Alexakis has suggested that the text is anti-iconoclastic, and that Heliodorus is, in fact, a skillfully crafted allusion to John VII Grammaticus (the Grammarian), the last iconoclast patriarch of Constantinople (837–843).[76]

There is no easy resolution to this discussion. That Heliodorus the alchemist and his namesake the novelist were conflated in the tenth-century *Chronicle* of Symeon Logothetes is beyond doubt. What is especially noteworthy is the observation that

72 Mecella 2014: 635n8.

73 Lauxtermann 2019: 206n41.

74 As Acconcia Longo has observed (1989: 15–18, 31–32; 2007: 35–38), *Patria Constantinopoleos* 3.114 (ed. Preger 1901–1907: 2.253, 9–15) mentions an "impious Heliodorus", who, during the reign of Leo III (r. 717–741), set up "with much knowledge of astronomy" (Engl. transl. Berger 2013: 191) the *Anemodourin*, or *Anemodoulion*, a Constantinopolitan tetrapylon that took its name from the weather vane crowning it. See Albrecht Berger's comments on the description of the *Anemodourin* (2013: 317n121) and on the identification of this "impious Heliodorus" with the namesake sorcerer in the *Life of Leo of Catania* (p. 317n122).

75 See Kazhdan et al. 1999: 297 for a criticism of Acconcia Longo's association of Heliodorus the sorcerer with the author of the *Aethiopica*. See Kazhdan 1988: 207 and Kazhdan et al. 1999: 298 for Heliodorus' similarities with other pre-Faustian hagiographical personages. For a concise discussion of the story of Theophilus of Adana, see Alexakis 2011: 221.

76 Alexakis 2011: 63–74, 81, 83, 85, 218 (9.4), 231 (14.1). Kazhdan (1988: 207) also considers a possible allusion to John Grammaticus, at least on the grounds that the hagiographer calls Heliodorus the sorcerer "Iannes", a nickname likewise applied to John Grammaticus. Acconcia Longo (2015) finds Alexakis' identification of Heliodorus with John Grammaticus unconvincing and, as in earlier articles, rejects Kazhdan's views (see the previous footnote). On John VII Grammaticus, see Magdalino 2006a: 55–89 (*passim*); 2006b: 122–124, 128–129, 132–135.

texts which, according to our modern categorization, may appear less important or marginal than those belonging to "higher" genres could, nevertheless, permeate – and even reshape – the image of emblematic authors. The subsequent transformation of Heliodorus of Emesa into the figure of a sorcerer in the *Life of Leo of Catania* is plausible but not definitive. However attractive and uniquely inspired,[77] Acconcia Longo's rationale concerning the conflation of the novelist with the sorcerer remains difficult to assess in its entirety. Still, it remains worth considering the possibility that, even if Heliodorus is indeed a masterful allusion to John Grammaticus, this need not exclude the hagiographer's use of other stories and images in shaping the sorcerer's narrative and persona,[78] possibly including those associated with the legend of Heliodorus of Emesa.

Be that as it may, Heliodorus the sorcerer evokes alchemical undertones, with allusions to metallic transmutation. Does he, however, bear any tangible relation to alchemy? At first sight, as already noted, the answer appears negative, since there are clear discrepancies between alchemical practice and the activities described in the *Life of Leo of Catania*. If, however, we look more closely, it becomes evident that Heliodorus' gold- and silver-making would more likely fall under alchemical imitation rather than genuine transmutation. It is not without significance that in the second excerpt from the *Life of Leo* cited above, Heliodorus turns a stone into gold by his usual *manganeia*, a term employed in relation to magical arts to denote trickery.[79] Alchemy is known for its imitative and counterfeiting techniques, well attested across all traditions. Might we then speculate that alchemical imitation could be perceived in the public imagination as approximating magic, insofar as both were thought of as directed toward deception, and, ultimately, the pursuit of personal gain?

77 For example, Acconcia Longo (1989: 15) suggests that the title *Aethiopica* could evoke in the Byzantine popular imagination an immediate association with the devil, who was often designated by the name αἰθίοψ ("Ethiopian").

78 On proposed sources behind Heliodorus' story, see Alexakis 2011: 106–110; see, for example, note 55 above for one of his suggestions.

79 Liddell–Scott–Jones: s.v. μαγγανεία. Van Pelt (2021: 73n60), addressing Acconcia Longo, argues that, read in parallel with the *Acts of Peter*, the text emphasizes not Heliodorus' power to transform materials but rather "the deceptive and illusory nature of his magic". Since the text makes no mention of Heliodorus' involvement when the gold reverts to its former state, "it suggests that the nature of the stone was never truly altered".

3 Alchemical Ethics: True Alchemy versus Counterfeiting, Beneficial versus Antisocial Practice

In the *Life of Pancratius of Taormina*, Menia is praised – and, through her, the beneficial aspect of the "sacred art" – for using alchemical knowledge to produce real metals, both base and precious, that enrich her city.[80] Since Menia's alchemical activity serves the common good, there is no conflict in incorporating such a detail into a hagiographical text. A God-pleasing objective proves crucial for a favorable conception of the alchemical opus. Quite interestingly, Menia is not, however, a straightforwardly moral figure, as she had poisoned Taurus' mother out of jealousy.[81] Yet, the hagiographer never censures her,[82] and Menia is repeatedly called "wise", despite her youth – or, rather, in striking contrast to it.[83] Although the quality of "wisdom" is common in hagiographical contexts, what makes the pre-Christian Menia wise is her divinely inspired knowledge placed in the service of the common good. Wisdom also seems to be attributed to her because she is the co-founder of Taormina and the benefactor who endowed the city with immense wealth. According to the narrative, and as previously cited, during Saint Pancratius' time "of all the cities in Sicily and Calabria the city of Taurus and Menia was the richest in gold and silver and celebrated for the other types of metallurgical art".[84]

However, the *Life of Pancratius* implies that, in the end, Menia's beneficial role is a matter of perspective. Aquilinus, the rival king, whose interests suffered from her "wisdom", calls her "bewitching and filthy" (ἐπαοιδᾷ καὶ ῥυπαρᾷ) in the *Vita*.[85] This highlights the fact, already discussed, that Menia's alchemical activity – like alchemy more generally – could be perceived as "magic" by audiences unversed in, or even hostile to, the "sacred art". The author of this *Life* was likely aware of this.

Menia's beneficial role may be contrasted with that of Heliodorus. Whereas Menia produces real base and precious metals, the gold and silver created by Heliodorus are false, a diabolical deception aimed at disorder and corruption.[86] Heliodorus' case recalls that of John Isthmeus, labelled *cheimeutēs* (often, and rather conventionally, translated as "alchemist") and *epithetēs* ("imposter"). The chronicler John Malalas (sixth century) records that Isthmeus appeared in Antioch during the reign of Anastasius I

80 Cf. Acconcia Longo 1989: 57; 2001a: 10.
81 *Vita Pancratii Tauromenii* 287 (ed. Stallman-Pacitti 2018: 426, 18–19).
82 Brandes 2005: 192, 195.
83 *Vita Pancratii Tauromenii* 283; 287; 289; 290; 293; 297; 298; 299 (ed. Stallman-Pacitti 2018: 422, 9; 426, 18; 428, 27; 430, 11; 432, 28; 436, 10; 436, 22; 438, 12).
84 *Vita Pancratii Tauromenii* 151 (ed. and Engl. transl. [adapted from] Stallman-Pacitti 2018: 266, 22–24 [text], 267 [transl.]); cf. § 104, p. 198, 15–16; § 107, p. 204, 1–2; § 134, p. 240, 16–17.
85 *Vita Pancratii Tauromenii* 300 (ed. and Engl. transl. Stallman-Pacitti 2018: 440, 3 [text], 441 [transl.]).
86 Acconcia Longo 1989: 57; Brandes 2005: 195.

(r. 491–518) and deceived the *argyropratai* (goldsmiths and/or bankers, literally, "sellers of silver"), by selling them counterfeit golden products. He then fled to Constantinople, where he again tricked many *argyropratai* until he was arrested and banished to Petra by the emperor.[87] Isthmeus' characterization as a *cheimeutēs* and *epithetēs*, as well as the account of his fraudulent deeds are probably grounded on the imitative character of a range of techniques associated with alchemy. The story thus illustrates some of the reasons behind the tension between "philosopher-alchemists" and superficial practitioners who regarded the "sacred art" only as a means of making easy money through deception.[88] Moreover, it is noteworthy that in both the case of the sorcerer Heliodorus and the *cheimeutēs* Isthmeus, the everyday economic and market life of Byzantine cities was fundamentally at stake because of the circulation of counterfeit products.

Heliodorus and Isthmeus pursued their own profit to the detriment of entire cities. Menia, by contrast, sought selflessly to enrich her community, thus ensuring its prosperity. In these stories, "special" knowledge and abilities are regarded as either beneficial or harmful, depending on their purpose. The key lies in the shift from personal enrichment to communal welfare, which exalts the ethics of the practitioner. Goals deemed selfish, antisocial, and devoid of virtue rendered an alchemist no better than a sorcerer in moral terms.[89] Thus, Menia's alchemy falls into the category of "true alchemy", practiced by a "wise" person. The adjective "wise" recalls what "true alchemists" are called in both alchemical and non-specialized literature: that is, "philosophers".[90] This suggests, along with parallel views found in various alchemical texts – some already examined in an earlier study of mine – that ethics and values contributed decisively to shaping the (ideal) identity of the philosopher-alchemist in Byzantium, thereby demarcating the field.[91]

A detail in the story of Taurus and Menia could further underscore the distinction between the "sacred art" and counterfeiting techniques. As cited above, the *Life* narrates that Menia produced pure gold, as well as silver, bronze, and iron, but she did not manage to make lead and tin. In this way, she succeeded in producing the most significant metals. The absence of lead and tin in the story probably reflects an actual scarcity of these metals in the wider region.[92] Yet, when read or heard outside of Sicily, this detail

87 Ioannes Malalas, *Chronographia* 16.5 (ed. Thurn 2000: 323). I render the term *argyropratai* as "goldsmiths and/or bankers" following Dagron 2002: 435.

88 See Merianos 2024b: 153–157.

89 Cf. Greenfield 2017: 218–219.

90 Koutalis et al. 2018: 31–37; Dufault 2019: 103.

91 Merianos 2022: 284, 291–293, 302.

92 See Boulakia 1972: 139, who points out that lead was not found in Italy but was available in Attica, Macedonia, Cilicia, Rhodes, Cyprus, North Africa, Spain, Gaul, Great Britain, and Sardinia. With regard to tin – an essential component of bronze – Anastasia Drandaki (2020: 279) notes that there were few tin mines in the Roman world, and that the main sources, in Spain and England, were soon lost to Roman control; with reference to Sicily, see Penhallurick 1986: 79. The reported scarcity of tin may also function

would not necessarily have been linked to local geology. One might speculate that the author's specificity in mentioning the metals Menia could not produce may also evoke the fact that lead and tin were widely employed in metal alloys and in counterfeiting, as ancient sources attest.[93] Deliberately or unintentionally, this detail may have implied that Menia's gold (and silver) was indeed of the purest quality, ruling out any suspicion of adulteration – especially since the knowledge to produce it had been divinely granted.

This recalls once more Michael Psellus, who, in *On How to Make Gold*, writes that, like some Eurystheus, Patriarch Michael I Cerularius (1043–1058) assigned him the task of bringing the golden apples by turning lead, tin, or any other natural substance into gold.[94] Psellus' emphasis on lead and tin as base metals may therefore suggest an awareness of their central role in alloying practices, such as the adulteration of coinage. Furthermore, a hagiographical miracle tale celebrating the icon of Christ Antiphonetes ("the Guarantor"), set in the reign of Heraclius (r. 610–641) and the patriarchate of Sergius I (610–638), recounts how a shipload of tin and lead was miraculously transmuted into silver through the agency of the icon.[95] This tale may be read as recasting a well-known metallurgical reality – tin and lead employed to counterfeit silver[96] – into a miracle of authentic transmutation.

An elucidation is necessary, however, concerning the relation between the miracle account and Psellus' words, on the one hand, and Menia's story on the other. References to lead and tin in contexts such as those just mentioned often implied the possibility of

as a chronological indicator, plausibly hinting at a considerable passage of time since the Empire had lost access to the western mines. The combined reference to lead and tin, on the other hand, may echo the occasional employment of pewter in medieval times, as well as the survival of soldering practices in metallurgy (i.e., melting of tin–lead alloys for bonding metallic objects); see Mundell Mango and Fourlas 2025: 375, 387–388; also Humpston and Jacobson 2004: x, Colás 2021: 36–37.

93 See, for example, Herodotus, *Historiae* 3.56 (ed. Wilson 2015: 1.272); Engl. transl. Melville Jones 1993: 361 (no. 535): "When the Spartans had besieged Samos for forty days, and had had no success in their venture, they departed to the Peloponnese. A rather foolish story is current, that Polycrates struck a large quantity of the local coinage in lead and gilded it and gave it to them, and that they took it, and that this was the manner of their departure". See also Iulius Pollux, *Onomasticon* 9.79 (ed. Bethe 1900–1937: 2.169, 13–15); Engl. transl. Melville Jones 1993: 449 (no. 656): "Dionysius once made the Syracusans use tin instead of silver for their coinage; the small coin was worth four Attic drachmas instead of one".

94 Michael Psellus, *Epistula de chrysopoeia* 1 (ed. Bidez 1928: 26, 12–14).

95 Magdalino 1998: 220–224 (= 2007: no. VIII); 2006a: 39; 2016. Paul Magdalino (1998: 221 and n6) notes that the miraculous aspects of the story reflect a more fantastical version of an episode from the seventh-century *Life of Saint John the Almsgiver* (*BHG* 886d); see Leontius Neapolitanus, *Vita Ioannis Eleemosynarii episcopi Alexandrini* 8 (ed. Festugière and Rydén 1974: 353–354); also Mundell Mango and Fourlas 2025: 375. As for the alleged occurrence of the Antiphonetes miracle during the reign of Heraclius, this is not a trivial detail: there is evidence to suggest that the study of astrology and alchemy was actively cultivated under his rule; see Magdalino 2006a: Chapter II; Merianos 2022: 304–305.

96 See, e.g., the case of fourth- or early fifth-century vessels in the Alise-Sainte-Reine hoard, which are made of a tin–lead alloy, a combination probably employed in the imitation of silver; see Drandaki 2020: 29. On the fourth-century pewter hoard found in Oxfordshire, see Mundell Mango and Fourlas 2025: 388.

their real transmutation through chrysopoeia or argyropoeia, and not necessarily their use in counterfeit practices. The absence of lead and tin in Menia's account gains significance precisely when juxtaposed with the mention of her production of "pure gold". This detail may allude to specialized techniques of refinement familiar to both artisanal and chemical traditions, but it also implicitly excludes the possibility that the metals she produced were adulterated.

4 Concluding Remarks

The examination of Menia the Wise and Heliodorus the Sorcerer from the perspective of the history of alchemy illustrates the complex ways in which Byzantine hagiography engaged with notions of the "sacred art" and metallurgical transformation. Alchemical concepts – however loosely or inaccurately understood – proved capable of entering narratives of sanctity, contributing to the shaping of cultural imagination and moral discourse.

The *Life of Pancratius of Taormina* presents Menia as a paradigmatic figure whose divinely bestowed metallurgical skills transform her city into a place of abundance and strength. Her story exemplifies the public perception of alchemy as an art of creating metallic wealth. The context-sensitive terminology of χύμευσις and μεταλλικὴ τέχνη, the invocation of "Higher Providence", and the parallels with other traditions of divinely revealed alchemical knowledge, all highlight the permeation of alchemical ideas and images into a popular genre. Menia's "true alchemy" is depicted as a form of metallurgical wisdom divinely sanctioned, in which communal enrichment and prosperity legitimize the art.

The *Life of Leo of Catania*, by contrast, depicts Heliodorus as a destructive sorcerer and deceiver. His counterfeit transformations – stones into gold, wood into silver – possibly alluding to alchemical imitation, are acts of deception with devastating consequences for the urban economy. Here, the vocabulary of transmutation is combined with images of illusion and deception, implying broader anxieties about the boundaries between alchemy, magic, and fraud, and recalls historical anecdotes such as that of John Isthmeus. Heliodorus' story thus eloquently suggests the tensions that alchemical practice could provoke in Middle Byzantium.

In comparison, Menia and Heliodorus delineate a moral spectrum that resonates with Byzantine conceptions of alchemy. Menia's "true alchemy" aligns with the role of the philosopher-alchemist, marked by wisdom, divine inspiration, and conduct that was ethical and beneficent. Heliodorus approximates the figure of the fraudulent *cheimeutēs*, whose selfish and antisocial goals reduce the "sacred art" to charlatanism and deception. These contrasting portraits highlight the importance of ethics in shaping the Byzantine perception of alchemy: whether it was regarded as a sacred or a deceitful art depended not only on its techniques, but also on its purpose.

Collectively, the hagiographical evidence examined here – only a fragment of what probably lies within the genre, undetected or unstudied – supplements our understanding of alchemy's cultural presence in Byzantium. Saints' Lives show that alchemical notions were not confined to specialist circles but resonated with broader audiences. Their appearance in a hagiographical context is significant since such texts reached far wider readerships than technical treatises. These narratives remind us that the Byzantine imagination integrated alchemy into various contexts, offering invaluable testimony to how an allegedly "marginal" art could, in fact, appear in unexpected literary genres.

Appendix. The Four Alchemical Poems: Heliodorus' Dedication to Emperor Theodosius and the Question of Their Dating

The four Byzantine iambic poems on chrysopoeia are attributed respectively to Heliodorus, Theophrastus, Hierotheus, and Archelaus.[97] The prevailing view in current scholarship is that they were composed by a single poet, probably the first listed, Heliodorus.[98] Even so, this does not rule out the likelihood that the name was used pseudonymously.[99] The poems are usually, though not exclusively, dated to the seventh or eighth century on two main grounds, which are not mutually dependent. First, they appear to have been influenced by, or at least show parallels with, Stephanus of Alexandria (seventh century). Second, the first poem, attributed to Heliodorus the philosopher, is dedicated to an Emperor Theodosius.[100] Since three emperors bore this name – Theodosius I the Great (r. 379–395), Theodosius II (r. 408–450), and Theodosius III (r. 715–717) – the chronological reasoning has been as follows: because the poems reflect the influence of Stephanus, they cannot predate the seventh century; consequently, the dedicatee must have been Theodosius III, and "Heliodorus" would then be an early eighth-century author.[101] Other identifications, however, have also

97 Without referring to the earlier partial editions, the poems attributed to Theophrastus, Hierotheus, and Archelaus were edited by Ideler (1842: 328–352). All four poems were later edited by Goldschmidt (1923); as noted above, they were re-edited, with a modern Spanish translation, in the dissertation of Boned Fernández (2025). On the four alchemical poets, see Viano 2008 for a concise overview; see also Maria Papathanassiou's contribution to this volume.
98 Reitzenstein 1919: 28–37; Goldschmidt 1923: 11–14.
99 Pfister 1923: col. 651; Letrouit 1995: 83.
100 *Marcianus graecus* 299 (coll. 584), f. 43 verso: Ἡλιοδώρου φιλοσόφου πρὸς Θεοδόσιον τὸν μέγαν βασιλέα περὶ τῆς τῶν φιλοσόφων μυστικῆς τέχνης, διὰ στίχων ἰάμβων; see also the title of the poem in the manuscript's *pinax* ("table of contents"), f. 2 recto: Ἡλιοδώρου φιλοσόφου πρὸς Θεοδόσιον τὸν βασιλέα περὶ τῆς θείας ταύτης τέχνης, διὰ στίχων ἰάμβων.
101 Reitzenstein 1919: 28, 36; Goldschmidt 1923: 16–21.

been proposed. Berthelot, for instance, believed that Heliodorus' poem was dedicated to Theodosius I,[102] while Henri Dominique Saffrey suggested Theodosius II, "le grand organisateur de l'enseignement universitaire à Constantinople en 425".[103]

In any case, identifying the addressee to one of the three Theodosioi presupposes that the dedication is authentic. Some scholars have urged caution. Ingeborg Hammer Jensen, for example, accepts that the poem is addressed to an emperor, but suggests that one should rather think of Heraclius.[104] Friedrich Pfister draws attention to an observation by Berthelot and Ruelle:[105] in *Vaticanus graecus* 1174 (fourteenth–fifteenth century), f. 35 verso, the title of a treatise by the Anonymous (*Anepigraphos*) Philosopher – whose original title has not been preserved and is conventionally referred to as *De musica et chymica* – appears as Ἀνεπιγράφου φιλοσόφου πρὸς Θεοδόσιον τὸν μέγα βασιλέα, likewise bearing a dedication to an Emperor Theodosius. Berthelot and Ruelle concluded that the name "Theodosius" was added later, as often happened in this kind of literature.[106] Notably, this dedication is suspiciously identical to that in the poem of Heliodorus the philosopher.[107] The most probable conclusion, according to Julius Ruska, is that an unknown Byzantine of the second half of the seventh century sought to legitimize his spurious works with grand authorial names and a forged dedication to a renowned emperor of the past.[108] Charles Browne and Marc Lauxtermann, in turn, have raised the possibility that the name "Theodosius" may itself be a fabrication.[109]

Thus, if the reference to an Emperor Theodosius is spurious, then the seventh century becomes the more plausible date for the poems, based on the parallels with Stephanus of Alexandria. Maria Papathanassiou, for example, maintains that the poet-alchemists belonged to Stephanus' close environment, including his disciples, and that their poems reflect the influence of his lectures.[110]

Yet, certainty is elusive since Lauxtermann, drawing on metrical analysis and judging the parallels with Stephanus to be inconclusive, has recently proposed a significantly earlier date, placing the poems in the fifth or early sixth century.[111] This argument will need to be carefully assessed by historians of Graeco-Byzantine alchemy.

---

102 Berthelot 1885: 122, 202.
103 Saffrey 1995: 5.
104 Hammer Jensen 1921: 155n3.
105 Pfister 1923: col. 652.
106 Berthelot and Ruelle 1887–1888: 3.378.
107 Bovacqua 2025: 587
108 Ruska 1931: 265.
109 Browne 1946: 130; Lauxtermann 2019: 207.
110 Papathanassiou 2020: 492–493.
111 Lauxtermann 2019: 205–207.

Acknowledgement: I would like to thank Stephanos Efthymiadis, Zisis Melissakis, and Yannis Stoyas for their valuable remarks and suggestions.

Bibliography

Abbreviations

BHG: François Halkin, *Bibliotheca hagiographica graeca*, 3 vols (Subsidia Hagiographica 8a). Third edition. Brussels: Société des Bollandistes, 1957.

Sources

Collected Sources

Berthelot and Ruelle 1887–1888: Marcellin Berthelot and Charles-Émile Ruelle (ed. and Fr. transl.), *Collection des anciens alchimistes grecs*, 3 vols. Paris: G. Steinheil, 1887–1888 (reprint: Osnabrück: O. Zeller, 1967).

Goldschmidt 1923: Günther Goldschmidt (ed.), *Heliodori carmina quattuor ad fidem codicis Casselani* (Religionsgeschichtliche Versuche und Vorarbeiten 19.2). Giessen: A. Töpelmann, 1923.

Halleux 2021: Robert Halleux, *Traités des arts et métiers* (Les Alchimistes grecs 9.1). Paris: Belles Lettres, 2021.

Ideler 1842: Julius Ludwig Ideler (ed.), *Physici et medici Graeci minores*, vol. 2. Berlin: G. Reimer, 1842.

Melville Jones 1993: John R. Melville Jones, *Testimonia numaria: Greek and Latin Texts concerning Ancient Greek Coinage*, vol. 1: *Texts and Translations*. London: Spink, 1993.

Authors and Works

Blemmydes, Nicephorus, *De confectione auri*: ed. Berthelot and Ruelle 1887–1888: 2.452–457.

[Democritus], *Physica et mystica*: Matteo Martelli (ed. and Engl. transl.), *The* Four Books *of Pseudo-Democritus* (Sources of Alchemy and Chemistry 1; Ambix 60, Supplement 1). Leeds: Maney Publishing, 2013, pp. 78–103.

De quattuor elementis: ed. Berthelot and Ruelle 1887–1888: 2.337–342.

Galenus, *De compositione medicamentorum per genera libri VII*: Karl Gottlob Kühn (ed.), *Claudii Galeni opera omnia*, vol. 13 (Medicorum Graecorum Opera quae exstant 13). Leipzig: C. Cnobloch, 1827 (reprint: Hildesheim: Olms, 1965), pp. 362–1058.

Georgius Syncellus, *Ecloga chronographica*: Alden A. Mosshammer (ed.), *Georgii Syncelli Ecloga chronographica* (Bibliotheca Scriptorum Graecorum et Romanorum Teubneriana). Leipzig: Teubner, 1984.

[Heliodorus], *De philosophorum arte mystica*: ed. Goldschmidt 1923, pp. 26–34.
Germ. transl.: Günther Goldschmidt, "Heliodors Gedicht von der Alchemie", in Julius Ruska (ed.), *Studien zur Geschichte der Chemie. Festgabe Edmund O. v. Lippmann*. Berlin: J. Springer, 1927, pp. 21–27.

Herodotus, *Historiae*: Nigel G. Wilson (ed.), *Herodotus Historiae*, 2 vols (Oxford Classical Texts). Oxford: Oxford University Press, 2015.

[Isis Prophetissa], *Filio suo Horo*: Michèle Mertens (ed. and Fr. transl.), "Un traité gréco-égyptien d'alchimie: la *Lettre d'Isis à Horus*. Texte établi et traduit avec introduction et notes". Mémoire de licence. Université de Liège, 1983–1984.

Iulius Pollux, *Onomasticon*: Erich Bethe (ed.), *Pollucis Onomasticon*, 3 vols (Lexicographi Graeci 9.1–3). Leipzig: Teubner, 1900–1937.

Leontius Neapolitanus, ***Vita Ioannis Eleemosynarii episcopi Alexandrini*** (*BHG* 886d): André-Jean Festugière and Lennart Rydén (ed. and Fr. transl.), *Léontios de Néapolis, Vie de Syméon le Fou et Vie de Jean de Chypre* (Bibliothèque archéologique et historique 95). Paris: Librairie Orientaliste Paul Geuthner, 1974, pp. 255–632.

Lexicon chrysopoeiae: ed. Berthelot and Ruelle 1887–1888: 2.4–17.

Malalas, Ioannes, ***Chronographia***: Hans Thurn (ed.), *Ioannis Malalae Chronographia* (Corpus Fontium Historiae Byzantinae 35). Berlin and New York: De Gruyter, 2000.

Olympiodorus alchemista, ***De arte sacra***: ed. Berthelot and Ruelle 1887–1888: 2.69–104.

Patria Constantinopoleos: Theodor Preger (ed.), *Scriptores originum Constantinopolitanarum*, 2 vols. Leipzig: Teubner, 1901–1907.
Engl. transl.: Albrecht Berger, *Accounts of Medieval Constantinople: The* Patria (Dumbarton Oaks Medieval Library 24). Cambridge, MA: Harvard University Press, 2013.

Psellus, Michael, ***De Heliodoro et Achille Tatio judicium***: Andrew R. Dyck (ed. and Engl. transl.), *Michael Psellus, The Essays on Euripides and George of Pisidia and on Heliodorus and Achilles Tatius* (Byzantina Vindobonensia 16). Vienna: Verlag der Österreichischen Akademie der Wissenschaften, 1986.

Psellus, Michael, ***Epistula de chrysopoeia***: Joseph Bidez (ed.), *Catalogue des manuscrits alchimiques grecs*, vol. 6: *Michel Psellus, Épître sur la Chrysopée, Opuscules et extraits sur l'alchimie, la météorologie et la démonologie*. Brussels: M. Lamertin, 1928, pp. 1–43.
It. transl.: Francesca Albini, *Michele Psello, La crisopoea ovvero come fabbricare l'oro. Traduzione, introduzione e commento* (Nuova Atlantide, I classici del pensiero magico, esoterico e simbolico). Genova: ECIG, 1988.

Serpens Ouroboros: ed. Berthelot and Ruelle 1887–1888: 2.21–22.

Socrates Scholasticus, ***Historia ecclesiastica***: Günther Christian Hansen (ed.), with contributions by Manja Širinjan, *Sokrates, Kirchengeschichte* (Die griechischen christlichen Schriftsteller der ersten Jahrhunderte, Neue Folge 1). Berlin: Akademie Verlag, 1995.

Strabo, ***Geographica***: August Meineke (ed.), *Strabonis Geographica*, 3 vols. Leipzig: Teubner, 1877.
Engl. transl.: Horace Leonard Jones, *The Geography of Strabo*, 8 vols (Loeb Classical Library). Cambridge, MA: Harvard University Press; London: W. Heinemann, 1917–1932.

Suda lexicon: Ada Adler (ed.), *Suidae Lexicon*, 5 vols (Lexicographi Graeci 1.1–5). Leipzig: Teubner, 1928–1938.

Symeon Logothetes, ***Chronicon***: Staffan Wahlgren (ed.), *Symeonis Magistri et Logothetae Chronicon* (Corpus Fontium Historiae Byzantinae 44.1). Berlin and New York: De Gruyter, 2006.

Vita et miracula Theocleti Lacedaemonii (*BHG* 2420): Anthony Kaldellis and Ioannis Polemis (ed. and Engl. transl.), *Saints of Ninth- and Tenth-Century Greece* (Dumbarton Oaks Medieval Library 54). Cambridge, MA: Harvard University Press, 2019, pp. 163–255.

Vita Leonis Cataniae (longior) (*BHG* 981b): Alexander G. Alexakis (ed.) and Susan Wessel (Engl. transl.), *The Greek Life of St. Leo Bishop of Catania (*BHG *981b)* (Subsidia Hagiographica 91). Brussels: Société des Bollandistes, 2011.

Vita Pancratii Tauromenii (*BHG* 1410): Cynthia J. Stallman-Pacitti (ed. and Engl. transl.), *The Life of Saint Pankratios of Taormina. Greek Text, English Translation and Commentary*, ed. by John B. Burke (Byzantina Australiensia 22). Leiden and Boston, MA: Brill, 2018.
It. transl.: Francesca Angiò, "La *Vita di Tauro*: la storia di Tauro e Menia, fondatori eponimi di Taormina, dall'anonima *Vita di San Pancrazio di Taormina* (*BHG* 1410)", *Sileno* 50 (2024), pp. 247–267.

Literature

Acconcia Longo 1989: Augusta Acconcia Longo, "La Vita di S. Leone vescovo di Catania e gli incantesimi del mago Eliodoro", *Rivista di studi bizantini e neoellenici* NS 26 (1989), pp. 3–98.

Acconcia Longo 2001a: Augusta Acconcia Longo, "L'antichità pagana nell'agiografia italogreca di età iconoclasta", in Nikos Oikonomides (ed.), *L'Ellenismo Italiota dal VII al XII secolo. Alla memoria di Nikos Panagiotakis* (Convegno internazionale 8). Athens: Institute for Byzantine Research, National Hellenic Research Foundation, 2001, pp. 1–17.

Acconcia Longo 2001b: Augusta Acconcia Longo, "La data della Vita di S. Pancrazio di Taormina (*BHG* 1410)", *Bollettino della Badia Greca di Grottaferrata* NS 55 (2001), pp. 37–42.

Acconcia Longo 2007: Augusta Acconcia Longo, "Note sul *dossier* agiografico di Leone di Catania: la trasmissione della leggenda e la figura del mago Eliodoro", *Rivista di studi bizantini e neoellenici* NS 44 (2007), pp. 3–38.

Acconcia Longo 2015: Augusta Acconcia Longo, "I percorsi di una leggenda: Eliodoro Virgilio Faust", *Rivista di studi bizantini e neoellenici* NS 52 (2015), pp. 165–240.

Alexakis 2011: Alexander G. Alexakis, *The Greek Life of St. Leo Bishop of Catania (*BHG *981b)* (Subsidia Hagiographica 91). Brussels: Société des Bollandistes, 2011.

Alexakis 2019: Alexander Alexakis, "Review of *The Life of Saint Pankratios of Taormina*, ed. Cynthia Stallman-Pacitti", *Bryn Mawr Classical Review* 2019.12.30, https://bmcr.brynmawr.edu/2019/2019.12.30.

Angiò 1994: Francesca Angiò, "La *Vita di Tauro* dall'anonima *Vita di San Pancrazio di Taormina*", *Sileno* 20 (1994), pp. 117–143.

Auzépy 1992: Marie-France Auzépy, "L'analyse littéraire et l'historien: l'exemple des vies des saints iconoclastes", *Byzantinoslavica* 53 (1992), pp. 57–67.

Berthelot 1885: Marcellin Berthelot, *Les origines de l'alchimie*. Paris: G. Steinheil, 1885.

Bevacqua 2025: Flavio Bevacqua, *Les œuvres du Philosophe Anépigraphe entre alchimie, néoplatonisme et philosophie naturelle. Édition critique, traduction et commentaire*. PhD Dissertation. Sorbonne Université; Università degli Studi di Padova, 2025.

Blanco Cesteros 2022: Miriam Blanco Cesteros, "(De)Constructing an Authoritative Narrative. The Case of *The Letter of Isis*", *ARYS* 20 (2022), Special Issue: *Chemical Arts and Religion in Antiquity*, pp. 227–269.

Boll 1912: Franz Boll, "Heliodoros (13)", in August Friedrich von Pauly and Georg Wissowa (eds), *Paulys Realencyclopädie der classischen Altertumswissenschaft*, vol. 8.1. Stuttgart: J. B. Metzler, 1912, cols 18–19.

Boned Fernández 2025: María Belén Boned Fernández, *Los poemas alquímicos atribuidos a Heliodoro, Teofrasto, Hieróteo y Arquelao: edición, traducción y comentario*. PhD Dissertation. Universidad Complutense de Madrid, 2025.

Boulakia 1972: Jean David C. Boulakia, "Lead in the Roman World", *American Journal of Archaeology* 76.2 (1972), pp. 139–144.

Brandes 2005: Wolfram Brandes, "Das Gold der Menia. Ein Beispiel transkulturellen Wissenstransfers", *Millennium* 2 (2005), pp. 175–227.

Browne 1946: Charles A. Browne, "Rhetorical and Religious Aspects of Greek Alchemy. Including a Commentary and Translation of the Poem of the Philosopher Archelaos upon the Sacred Art, Part I", *Ambix* 2.3–4 (1946), pp. 129–137.

Bull 2018: Christian H. Bull, "Wicked Angels and the Good Demon: The Origins of Alchemy According to the Physica of Hermes", *Gnosis: Journal of Gnostic Studies* 3 (2018), pp. 3–33.

Cardon 2007: Dominique Cardon, *Natural Dyes: Sources, Tradition, Technology and Science*, transl. Caroline Higgitt. London: Archetype, 2007.

Casoli and Wahl 2019: Andrea Casoli and Marc Philipp Wahl, "Money and Its Technologies: Production, Distribution, and Impact", in Stefan Krmnicek (ed.), *A Cultural History of Money*, vol. 1: *A Cultural History of Money in Antiquity*. London: Bloomsbury Academic, 2019, pp. 21–42.

Colás 2021: Rafael Colás, “Ancient Lead, Mercury, and Tin”, in Patricia Silvana Carrizo (ed.), *Reverse Engineering of Ancient Metals*. Cham: Springer, 2021, pp. 29–40.

Colinet 2000: Andrée Colinet, “Le *Travail des quatre éléments* ou lorsqu’un alchimiste byzantin s’inspire de Jabir”, in Isabelle Draelants, Anne Tihon, and Baudouin van den Abeele (eds), *Occident et Proche-Orient: contacts scientifiques au temps des Croisades* (Réminisciences 5). Turnhout: Brepols, 2000, pp. 165–190.

Colinet 2010: Andrée Colinet, *Recettes alchimiques (Par. gr. 2419; Holkhamicus 109); Cosmas le Hiéromoine, Chrysopée* (Les Alchimistes grecs 11). Paris: Belles Lettres, 2010.

Dagron 2002: Gilbert Dagron, “The Urban Economy, Seventh–Twelfth Centuries”, in Angeliki E. Laiou (ed.-in-chief), *The Economic History of Byzantium: From the Seventh through the Fifteenth Century*, 3 vols (Dumbarton Oaks Studies 39). Washington, DC: Dumbarton Oaks Research Library and Collection, 2002, vol. 2, pp. 393–461.

D’Aiuto 2025: Francesco D’Aiuto, “Qualche osservazione filologica sulla *Vita metrica s. Leonis episcopi Catanensis BHG* 981c”, in Antonella Conte and Gioacchino Strano (eds), *Corolla Byzantina. Studi in onore di Carmelo Crimi*. Spoleto: Fondazione Centro italiano di studi sull’alto medioevo, 2025, pp. 75–105.

Daniels et al. 2014: Vincent Daniels, Thibaut Devièse, Marei Hacke, and Catherine Higgitt, “Technological Insights into Madder Pigment Production in Antiquity”, *The British Museum Technical Research Bulletin* 8 (2014), pp. 13–28.

Djurslev 2024: Christian Thrue Djurslev, “The Alexander Romance”, in Daniel Ogden (ed.), *The Cambridge Companion to Alexander the Great* (Cambridge Companions to the Ancient World). Cambridge: Cambridge University Press, 2024, pp. 452–470.

Drandaki 2020: Anastasia Drandaki, *Late Antique Metalware. The Production of Copper-Alloy Vessels between the 4th and 8th Centuries: The Benaki Museum Collection and Related Material* (Bibliothèque de l’Antiquité Tardive 37). Turnhout: Brepols, 2020.

Dufault 2017: Olivier Dufault, “Transmutation Theory and the Dating of the Alchemical Recipe *On the Same Divine Water*”, in Andrea Le Moli and Lela Alexidze (eds), *Prote Hyle: Notions of Matter in the Platonic and Aristotelian Traditions* (Filosofie). Palermo: Palermo University Press, 2017, pp. 67–84.

Dufault 2019: Olivier Dufault, *Early Greek Alchemy, Patronage and Innovation in Late Antiquity* (California Classical Studies 7). Berkeley, CA: Department of Classics, University of California, Berkeley, 2019.

Efthymiadis 2017: Stephanos Efthymiadis, “L’hagiographie grecque de l’Italie (VIIe–XIVe siècle)”, in Monique Goullet (ed.), *Hagiographies. Histoire internationale de la littérature hagiographique latine et vernaculaire en Occident des origines à 1550*, vol. 7 (Corpus Christianorum, Hagiographies 7). Turnhout: Brepols, 2017, pp. 345–421.

Efthymiadis 2023: Stephanos Efthymiadis, “Saints and Secondary Heroes in Byzantine Hagiography”, in Koen De Temmerman, Julie Van Pelt, and Klazina Staat (eds), *Constructing Saints in Greek and Latin Hagiography: Heroes and Heroines in Late Antique and Medieval Narrative* (Fabulae 2). Turnhout: Brepols, 2023, pp. 33–56.

Efthymiadis 2024: Stephanos Efthymiadis, “The Literary Background of the Greek Monastic Hagiography of Southern Italy (Tenth–Eleventh Century)”, in Paolo Cesaretti and Mario Re (eds), *Byzantino-Sicula 11: La scrittura agiografica: spazio di incontro e di confronto tra lingue e culture* (Quaderni 25). Palermo: Istituto siciliano di studi bizantini e neoellenici, 2024, pp. 11–25.

Escolano-Poveda 2020: Marina Escolano-Poveda, “The Representation of Kalasiris as a Priest, and Egyptian Magic in the *Aithiopika* of Heliodoros”, in Emilio Suárez de la Torre, Isabel Canzobre Martínez, and Carmen Sánchez-Mañas (eds), Ablanathanalba. *Magia, cultura y sociedad en el Mundo Antiguo* (Clásicos Dykinson, Monografías). Madrid: Dykinson, 2020, pp. 257–273.

Festugière 1944: André-Jean Festugière, *La révélation d’Hermès Trismégiste*, vol. 1: *L’astrologie et les sciences occultes* (Études bibliques). Paris: J. Gabalda, 1944.

Fraser 2004: Kyle A. Fraser, “Zosimos of Panopolis and the Book of Enoch: Alchemy as Forbidden Knowledge”, *Aries* 4.2 (2004), pp. 125–147.

Geller and Geus 2014: Mark Geller and Klaus Geus, "Esoteric Knowledge in Antiquity – Some Thoughts", in Klaus Geus and Mark Geller (eds), *Esoteric Knowledge in Antiquity* (MPIWG Preprints 454). Berlin: Max-Planck-Institut für Wissenschaftsgeschichte, 2014, pp. 3–6.

Greenfield 2017: Richard Greenfield, "Magic and the Occult Sciences", in Anthony Kaldellis and Niketas Siniossoglou (eds), *The Cambridge Intellectual History of Byzantium*. Cambridge: Cambridge University Press, 2017, pp. 215–233.

Halleux 1981: Robert Halleux, *Papyrus de Leyde. Papyrus de Stockholm. Fragments de recettes* (Les Alchimistes grecs 1). Paris: Belles Lettres, 1981.

Hammer Jensen 1921: Ingeborg Hammer Jensen, *Die älteste Alchymie* (Det Kgl. Danske Videnskabernes Selskab, Historisk-filologiske Meddelelser 4.2). Copenhagen: A. F. Høst & Søn, 1921.

Hanegraaff 2012: Wouter J. Hanegraaff, *Esotericism and the Academy: Rejected Knowledge in Western Culture*. Cambridge: Cambridge University Press, 2012.

Humpston and Jacobson 2004: Giles Humpston and David M. Jacobson, *Principles of Soldering*. Materials Park, OH: ASM International, 2004.

Idel 1986: Moshe Idel, "The Origin of Alchemy According to Zosimos and a Hebrew Parallel", *Revue des études juives* 145.1–2 (1986), pp. 117–124.

Jeffreys and Jeffreys 1991: Elizabeth M. Jeffreys and Michael J. Jeffreys, "Heliodoros", in Alexander P. Kazhdan (ed.-in-chief), *The Oxford Dictionary of Byzantium*, 3 vols. New York and Oxford: Oxford University Press, 1991, vol. 2, p. 909.

Jouanno 2002: Corinne Jouanno, *Naissance et métamorphoses du Roman d'Alexandre. Domaine grec*. Paris: CNRS Éditions, 2002.

Jouanno 2018: Corinne Jouanno, "*Alexander Romance* and Byzantine World Chronicles: History Cross-Fertilized by Fiction and the Reverse", in Richard Stoneman, Krzysztof Nawotka, and Agnieszka Wojciechowska (eds), *The* Alexander Romance*: History and Literature* (Ancient Narrative, Supplementum 25). Groningen: Barkhuis and Groningen University Library, 2018, pp. 225–243.

Kaldellis 2022: Anthony Kaldellis, "Alexander the Great in Byzantine Tradition, AD 330–1453", in Richard Stoneman (ed.), *A History of Alexander the Great in World Culture*. Cambridge: Cambridge University Press, 2022, pp. 216–241.

Kazhdan 1988: Alexander Kazhdan, "Hagiographical Notes (17–20)", *Erytheia* 9.2 (1988), pp. 197–209.

Kazhdan et al. 1999: Alexander Kazhdan, in collaboration with Lee F. Sherry and Christine Angelidi, *A History of Byzantine Literature (650–850)* (Research Series 2). Athens: Institute for Byzantine Research, National Hellenic Research Foundation, 1999.

Keyser 2008: Paul T. Keyser, "Heliodoros of Alexandria (Astron.) (475–510 CE)", in Paul T. Keyser and Georgia L. Irby-Massie (eds), *The Encyclopedia of Ancient Natural Scientists: The Greek Tradition and Its Many Heirs*. London: Routledge, 2008, p. 363.

Kieckhefer 2019: Richard Kieckhefer, "Rethinking How to Define Magic," in Sophie Page and Catherine Rider (eds), *The Routledge History of Medieval Magic* (The Routledge Histories). London: Routledge, 2019, pp. 15–25.

Koutalis et al. 2018: Vangelis Koutalis, Matteo Martelli, and Gerasimos Merianos, "Graeco-Egyptian, Byzantine and Post-Byzantine Alchemy: Introductory Remarks", in Efthymios Nicolaidis (ed.), *Greek Alchemy from Late Antiquity to Early Modernity* (De Diversis Artibus 104). Turnhout: Brepols, 2018, pp. 11–43.

Lampadaridi 2020: Anna Lampadaridi, "À la recherche de l'auteur perdu dans la *Vie de Pancrace de Taormine* (*BHG* 1410)", *Apocrypha* 31 (2020), pp. 33–59.

Lampadaridi 2022: Anna Lampadaridi, "La *Vie de Pancrace de Taormine* (*BHG* 1410) et l'histoire des images à Byzance", in Anna Lampadaridi, Vincent Déroche, and Christian Høgel (eds), *L'histoire comme elle se présentait dans l'hagiographie byzantine et médiévale* (Acta Universitatis Upsaliensis, Studia Byzantina Upsaliensia 21). Uppsala: Uppsala Universitet, 2022, pp. 75–102.

Lauxtermann 2019: Marc D. Lauxtermann, *Byzantine Poetry from Pisides to Geometres. Texts and Contexts*, vol. 2 (Wiener Byzantinistische Studien 24.2). Vienna: Austrian Academy of Sciences Press, 2019.

Letrouit 1995: Jean Letrouit, "Chronologie des alchimistes grecs", in Didier Kahn and Sylvain Matton (eds), *Alchimie: art, histoire et mythes* (Textes et travaux de *Chrysopoeia* 1). Paris: S. É. H. A.; Milan: Archè, 1995, pp. 11–93.

Lexikon zur byzantinischen Gräzität: Erich Trapp et al., *Lexikon zur byzantinischen Gräzität, besonders des 9.–12. Jahrhunderts* (Österreichische Akademie der Wissenschaften, Philosophisch-Historische Klasse, Denkschriften 238, 250, 276, 293, 326, 352, 417, 494; Veröffentlichungen der Kommission für Byzantinistik 6.1–8). Vienna: Verlag der Österreichischen Akademie der Wissenschaften, 1994–2017.

Liddell–Scott–Jones: Henry George Liddell, Robert Scott, and Henry Stuart Jones, *A Greek-English Lexicon. Ninth Edition with a Revised Supplement*. Oxford: Clarendon Press, 1996.

Lopes da Silveira 2022: Fabiana Lopes da Silveira, "In the Melting Pot: Cultural Mixture and the Presentation of Alchemical Knowledge in the *Letter from Isis to Horus*", *Ambix* 69.1 (2022), pp. 49–64.

Lourié 2021: Basil Lourié, "A Monothelete Syriac Compilation of Pseudo-Apostolic Acts Preserved only in Slavonic and the Entrance of Constans II into Rome in 663", in Maria Cioată, Anissava Miltenova, and Emanuela Timotin (eds), *Biblical Apocrypha in South-Eastern Europe and Related Areas* (Bibliothèque de l'Institut d'études sud-est européennes 16). Brăila: Editura Istros a Muzeului Brăilei "Carol I", 2021, pp. 125–217.

Magdalino 1998: Paul Magdalino, "Constantinopolitana", in Ihor Ševčenko and Irmgard Hutter (eds), *Aetos: Studies in Honour of Cyril Mango*. Stuttgart: Teubner, 1998, pp. 220–232.

Magdalino 2006a: Paul Magdalino, *L'Orthodoxie des astrologues. La science entre le dogme et la divination à Byzance (VIIe–XIVe siècle)* (Réalités byzantines 12). Paris: Lethielleux, 2006.

Magdalino 2006b: Paul Magdalino, "Occult Science and Imperial Power in Byzantine History and Historiography (9th–12th Centuries)", in Paul Magdalino and Maria Mavroudi (eds), *The Occult Sciences in Byzantium*. Geneva: La Pomme d'or, 2006, pp. 119–162.

Magdalino 2007: Paul Magdalino, *Studies on the History and Topography of Byzantine Constantinople* (Variorum Collected Studies Series 855). Aldershot: Ashgate Variorum, 2007.

Magdalino 2016: Paul Magdalino, "The Merchant of Constantinople", in Paul Magdalino and Nevra Necipoğlu (eds), with the assistance of Ivana Jevtić, *Trade in Byzantium. Papers from the Third International Sevgi Gönül Byzantine Studies Symposium*. Istanbul: Koç University Research Center for Anatolian Civilizations, 2016, pp. 181–191.

Martelli 2009: Matteo Martelli, "'Divine Water' in the Alchemical Writings of Pseudo-Democritus", *Ambix* 56.1 (2009), pp. 5–22.

Martelli 2011: Matteo Martelli, *Pseudo-Democrito, Scritti alchemici, con il commentario di Sinesio* (Textes et travaux de *Chrysopoeia* 12). Paris: S. É. H. A.; Milan: Archè, 2011.

Martelli 2013: Matteo Martelli, *The* Four Books *of Pseudo-Democritus* (Sources of Alchemy and Chemistry 1; Ambix 60, Supplement 1). Leeds: Maney Publishing, 2013.

Martelli 2014: Matteo Martelli, "The Alchemical Art of Dyeing: The Fourfold Division of Alchemy and the Enochian Tradition", in Sven Dupré (ed.), *Laboratories of Art: Alchemy and Art Technology from Antiquity to the 18th Century* (Archimedes 37). Cham: Springer, 2014, pp. 1–22.

Martelli 2018: Matteo Martelli, "Byzantine Alchemy in Two Recently Discovered Manuscripts in Saint Stephen's (Meteora) and Olympiotissa (Elassona) Monasteries", in Efthymios Nicolaidis (ed.), *Greek Alchemy from Late Antiquity to Early Modernity* (De Diversis Artibus 104). Turnhout: Brepols, 2018, pp. 99–118.

Martelli 2022: Matteo Martelli, "Transmuting Tinctures. Water of Sulphur, Quicklime, and 'Washes' in Graeco-Egyptian Alchemy", *Technai* 13 (2022), Special Issue: *Ancient Science and Technology of Colour. Pigments, Dyes, Drugs and Their Experience in Antiquity* (Part 1), pp. 115–142.

Martelli and Rumor 2014: Matteo Martelli and Maddalena Rumor, "Near Eastern Origins of Graeco-Egyptian Alchemy", in Klaus Geus and Mark Geller (eds), *Esoteric Knowledge in Antiquity* (MPIWG Preprints 454). Berlin: Max-Planck-Institut für Wissenschaftsgeschichte, 2014, pp. 37–62.

Martini and Bassi 1901: Emidio Martini and Domenico Bassi, *Catalogus codicum astrologorum graecorum*, vol. 3: *Codices Mediolanenses*. Brussels: H. Lamertin, 1901.

Mecella 2014: Laura Mecella, "L'enigmatica figura di Eliodoro e la datazione delle *Etiopiche*", *Mediterraneo antico* 17.2 (2014), pp. 633–658.

Merianos 2017: Gerasimos Merianos, "Alchemy", in Anthony Kaldellis and Niketas Siniossoglou (eds), *The Cambridge Intellectual History of Byzantium*. Cambridge: Cambridge University Press, 2017, pp. 234–251.

Merianos 2022: Gerasimos Merianos, "The Christianity of the Philosopher Christianos. Ethics and Mathematics in Alchemical Methodology", *ARYS* 20 (2022), Special Issue: *Chemical Arts and Religion in Antiquity*, pp. 271–322.

Merianos 2024a: Gerasimos Merianos, "Aspects of the Relationship between Alchemy and Christianity in Byzantium", in Kostas Tampakis and Ronald Numbers (eds), *Orthodox Christianity and the Study of Nature: Histories of Interaction* (Science and Orthodox Christianity 4). Turnhout: Brepols, 2024, pp. 21–54.

Merianos 2024b: Gerasimos Merianos, "Insights on Alchemy, Deception, and Artisanal Knowledge in Byzantium", in Antje Bosselmann-Ruickbie, Zachary Chitwood, Johannes Pahlitzsch, and Martin Marko Vučetić (eds), *Byzanz am Rhein. Festschrift für Günter Prinzing anlässlich seines 80. Geburtstags* (Mainzer Veröffentlichungen zur Byzantinistik 18). Wiesbaden: Harrassowitz, 2024, pp. 145–164.

Merianos and Sakorrafou 2013: Gerasimos Merianos and Sandy Sakorrafou, "Μαρτυρίες περί αλχημείας στο Βυζάντιο σε μη αλχημικά κείμενα", in Eirini Mergoupi-Savaidou, Gerasimos Merianos, Faidra Papanelopoulou, and Christiana Christopoulou (eds), *Επιστήμη και Τεχνολογία. Ιστορικές και Ιστοριογραφικές Μελέτες* (Εταιρεία Μελέτης και Διάδοσης της Ιστορίας των Επιστημών και της Τεχνολογίας 1). Athens: Ekdotike Athenon, 2013, pp. 45–65.

Mertens 1988: Michèle Mertens, "Une scène d'initiation alchimique: la 'Lettre d'Isis à Horus'", *Revue de l'histoire des religions* 205.1 (1988), pp. 3–23.

Mertens 1989a: Michèle Mertens, "Pourquoi Isis est-elle appelée προφῆτις?", *Chronique d'Égypte* 64.127–128 (1989), pp. 260–266.

Mertens 1989b: Michèle Mertens, "Sur la trace des anges rebelles dans les traditions ésotériques du début de notre ère jusqu'au XVIIe siècle", in Julien Ries and Henri Limet (eds), *Anges et démons* (Homo Religiosus 14). Louvain-la-Neuve: Centre d'histoire des religions, 1989, pp. 383–398.

Moureau 2020: Sébastien Moureau, "*Min al-Kīmiyā' ad Alchimiam*. The Transmission of Alchemy from the Arab-Muslim World to the Latin West in the Middle Ages", *Micrologus* 28 (2020), pp. 87–141.

Mundell Mango and Fourlas 2025: Marlia Mundell Mango and Benjamin Fourlas, "Metalwork", in Leonard V. Rutgers, Neil Christie, Robin M. Jensen, and Jodi Magness (eds), *The Cambridge Encyclopaedia of Late Antique Art and Archaeology*, vol. 1: *Architecture, Artefacts and Evidence*. Cambridge: Cambridge University Press, 2025, pp. 374–397.

Noyé 2014: Ghislaine Noyé, "L'économie de la Calabre de la fin du VIe au VIIIe siècle", *Cahiers de recherches médiévales et humanistes* 28 (2014), pp. 323–388.

Noyé 2021: Ghislaine Noyé, "Byzantine Calabria", in Salvatore Cosentino (ed.), *A Companion to Byzantine Italy* (Brill's Companions to the Byzantine World 8). Leiden and Boston, MA: Brill, 2021, pp. 434–452.

Papathanassiou 2020: Maria K. Papathanassiou, "The Occult Sciences in Byzantium", in Stavros Lazaris (ed.), *A Companion to Byzantine Science* (Brill's Companions to the Byzantine World 6). Leiden and Boston, MA: Brill, 2020, pp. 464–495.

Penhallurick 1986: Roger David Penhallurick, *Tin in Antiquity: Its Mining and Trade throughout the Ancient World with Particular Reference to Cornwall*. London: Institute of Metals, 1986.

Pfister 1923: Friedrich Pfister, "Review of *Heliodori carmina quattuor ad fidem codicis Casselani* edidit Günther Goldschmidt", *Philologische Wochenschrift* 43.28 (1923), cols 651–654.

Rapp 2014: Stephen H. Rapp, Jr., "The Georgian Nimrod", in Kevork B. Bardakjian and Sergio La Porta (eds), *The Armenian Apocalyptic Tradition: A Comparative Perspective. Essays Presented in Honor of Professor Robert W. Thomson on the Occasion of His Eightieth Birthday* (Studia in Veteris Testamenti Pseudepigrapha 25). Leiden and Boston, MA: Brill, 2014, pp. 188–216.

Re 2011: Mario Re, "Italo-Greek Hagiography", in Stephanos Efthymiadis (ed.), *The Ashgate Research Companion to Byzantine Hagiography*, vol. 1: *Periods and Places* (Ashgate Research Companions). Farnham: Ashgate, 2011, pp. 227–258.

Reed 2014: Annette Yoshiko Reed, "Gendering Heavenly Secrets? Women, Angels, and the Problem of Misogyny and 'Magic'", in Kimberly B. Stratton and Dayna S. Kalleres (eds), *Daughters of Hecate: Women and Magic in the Ancient World*. Oxford: Oxford University Press, 2014, pp. 108–151.

Reitzenstein 1919: Richard Reitzenstein, "Zur Geschichte der Alchemie und des Mystizismus", *Nachrichten von der Königlichen Gesellschaft der Wissenschaften zu Göttingen, Philologisch-Historische Klasse* (1919), pp. 1–37.

Repath and Whitmarsh 2022: Ian Repath and Tim Whitmarsh (eds), *Reading Heliodorus'* Aethiopica. Oxford: Oxford University Press, 2022.

Rives 2009: James Rives, "Apion *Περὶ μάγου* and the Meaning of the Word μάγος", *MHNH* 9 (2009), pp. 121–134.

Roberts 2022: Alexandre M. Roberts, "Hierotechnicians by Name and Their Middle Byzantine Fame", *Journal of Late Antique, Islamic and Byzantine Studies* 1.1–2 (2022), pp. 167–199.

Roberts 2023: Alexandre M. Roberts, "Thinking about Chemistry in Byzantium and the Islamic World", *Journal of the History of Ideas* 84.4 (2023), pp. 595–619.

Ruska 1931: Julius Ruska, *Turba philosophorum. Ein Beitrag zur Geschichte der Alchemie* (Quellen und Studien zur Geschichte der Naturwissenschaften und der Medizin 1). Berlin: J. Springer, 1931.

Saffrey 1995: Henri Dominique Saffrey, "Historique et description du manuscrit alchimique de Venise *Marcianus Graecus* 299", in Didier Kahn and Sylvain Matton (eds), *Alchimie: art, histoire et mythes* (Textes et travaux de *Chrysopoeia* 1). Paris: S. É. H. A.; Milan: Archè, 1995, pp. 1–10.

Skouvaras 1967: Evangelos Skouvaras, *Ὀλυμπιώτισσα. Περιγραφὴ καὶ ἱστορία τῆς Μονῆς. Κατάλογος τῶν χειρογράφων – Χρονικὰ σημειώματα. Ἀκολουθία Παναγίας τῆς Ὀλυμπιωτίσσης. Ἔγγραφα ἐκ τοῦ ἀρχείου τῆς Μονῆς (1336–1900)*. Athens: Academy of Athens, Research Centre for Medieval and Modern Hellenism, 1967.

Sofianos 1986: Demetrios Z. Sofianos, *Τὰ χειρόγραφα τῶν Μετεώρων*, vol. 3: *Τὰ χειρόγραφα τῆς Μονῆς Ἁγίου Στεφάνου*. Athens: Academy of Athens, Research Centre for Medieval and Modern Hellenism, 1986.

Sophocles 1914: Evangelinos Apostolides Sophocles, *Greek Lexicon of the Roman and Byzantine Periods (From B. C. 146 to A. D. 1100)*. Memorial Edition. Cambridge, MA: Harvard University Press, 1914.

Stallman-Pacitti 2018: Cynthia J. Stallman-Pacitti, *The Life of Saint Pankratios of Taormina. Greek Text, English Translation and Commentary*, ed. by John B. Burke (Byzantina Australiensia 22). Leiden and Boston, MA: Brill, 2018.

Steiner 2020: Shannon Steiner, *Byzantine Enamel and the Aesthetics of Technological Power, Ninth to Twelfth Centuries*. PhD Dissertation. Bryn Mawr College, 2020.

Stephanides 1909: Michael K. Stephanides, *Ψαμμουργικὴ καὶ χυμεία. Συμβολαὶ εἰς τὴν ἱστορίαν τῆς χυμείας*. Mytilene: M. Nikolaidis, 1909.

Theodoropoulos 2020: Panagiotis Theodoropoulos, "The Migration of Syrian and Palestinian Populations in the 7th Century: Movement of Individuals and Groups in the Mediterranean", in Johannes Preiser-Kapeller, Lucian Reinfandt, and Yannis Stouraitis (eds), *Migration Histories of the Medieval Afroeurasian Transition Zone: Aspects of Mobility between Africa, Asia and Europe, 300–1500 C. E.* (Studies in Global Social History 39; Studies in Global Migration History 13). Leiden and Boston, MA: Brill, 2020, pp. 261–287.

Tuval 2016: Michael Tuval, "'Συναγωγὴ γιγάντων' (Prov 21:16): The Giants in the Jewish Literature in Greek", in Matthew Goff, Loren T. Stuckenbruck, and Enrico Morano (eds), *Ancient Tales of Giants from Qumran and Turfan: Contexts, Traditions, and Influences* (Wissenschaftliche Untersuchungen zum Neuen Testament 360). Tübingen: Mohr Siebeck, 2016, pp. 41–57.

Van Esbroeck and Zanetti 1988: Michel van Esbroeck and Ugo Zanetti, "Le dossier hagiographique de S. Pancrace de Taormine", in Salvatore Pricoco (ed.), *Storia della Sicilia e tradizione agiografica nella tarda antichità*. Soveria Mannelli: Rubbettino, 1988, pp. 155–171.

Van Pelt 2021: Julie Van Pelt, “From the *Acts of Peter* to the *Life of Leo of Catania*: Distinguishing Magic and Miracle”, in Tobias Nicklas, Janet E. Spittler, and Jan N. Bremmer (eds), *The Apostles Peter, Paul, John, Thomas and Philip with Their Companions in Late Antiquity* (Studies on Early Christian Apocrypha 17). Leuven: Peeters, 2021, pp. 55–80.

Veselovskiy 1886: Aleksander N. Veselovskiy, “Iz istorii romana i povesti, II. Epizod o Tavre i Menii v apokrificheskom zhitii sv. Pankratiya”, *Sbornik otdeleniya russkogo yazyka i slovesnosti Imperatorskoy Akademii Nauk* 40.2 (1886), pp. 65–128.

Viano 2008: Cristina Viano, “Hēliodōros (pseudo?) et alii (700–800 CE)”, in Paul T. Keyser and Georgia L. Irby-Massie (eds), *The Encyclopedia of Ancient Natural Scientists: The Greek Tradition and Its Many Heirs*. London: Routledge, 2008, p. 364.

Viano 2022: Cristina Viano, “Noir alchimique; questions d’étymologie et de transmutation”, *Technai* 13 (2022), Special Issue: *Ancient Science and Technology of Colour. Pigments, Dyes, Drugs and Their Experience in Antiquity* (Part 1), pp. 143–155.

Wessels and Klooster 2022: Antje B. Wessels and Jacqueline J. H. Klooster (eds), *Inventing Origins? Aetiological Thinking in Greek and Roman Antiquity* (Euhormos: Greco-Roman Studies in Anchoring Innovation 2). Leiden and Boston, MA: Brill, 2022.

Zuretti 1932: Carlo Oreste Zuretti, *Catalogue des manuscrits alchimiques grecs*, vol. 8: *Alchemistica signa*. Brussels: M. Lamertin, 1932.

Alexandre M. Roberts

Alchemy, History, and Myth in the *Suda* Lexicon

This chapter examines several interconnected entries in the *Suda* lexicon as evidence for how a learned Middle Byzantine reader might have understood the term *chēmeia* (or *chymeia*), the legitimacy of the techniques to which it apparently referred, and its place on the Byzantine cultural landscape. Building on recent work, I argue that such a reader, while closely associating *chēmeia* with the production of gold and silver, would also have been comfortable thinking of *chēmeia* as a legitimate *technē* that could be located both in history and in myth: as the target (alongside Christianity) of Emperor Diocletian's censorship and general crackdown in Egypt, and as the true meaning behind the myth of the Golden Fleece. In both cases, the effect is to reinforce the image of *chēmeia* as a *technē* that works and is explicable in rational terms.

The entries I will discuss below are referenced relatively frequently in historical work on Greek alchemy (including, more recently, Byzantine alchemy), but they have rarely been studied synoptically.[1] Recent research into how different Byzantine readers perceived alchemical literature, chemistry, and metallurgy has highlighted the importance of the *Suda* as a source, citing its relevant entries to address aspects of the topic.[2] In the present essay, I bring the *Suda* itself to the foreground, as part of the project to understand the wider cultural context in which texts typically labeled "alchemical" today were studied and perceived in the past[3] and, at the same time, to trace the cultural ramifications of theories, practices, and textual traditions that discussed or sought to effect the transformation of matter.

---

1 E.g., Viano 2006: 469n5, 471n12, 472; 2018: 945, 957; Mertens 2006: 215, 219, 229; Papathanassiou 2006: 165n4. See also the introduction in Martelli 2013, where he makes extensive use of a wide range of *Suda* entries not addressed in the present chapter.

2 Especially important is Merianos and Sakorrafou 2013: 49–50, providing a brief but insightful analysis of the *Suda*'s entries on χημεία and the Golden Fleece, and concluding that *chēmeia* is presented primarily as a practical technique. See also Merianos 2014: 250–251; Merianos 2017: 238, 239, 240, citing (as part of a lucid exposition of alchemy and how it was viewed in Byzantium) the *Suda*'s definition of *chēmeia*, account of Diocletian's book-burning, and description of the Golden Fleece, and observing that the *Suda*'s entries "point to three features of 'the sacred art' that probably would have been known by the educated Byzantine of the middle period", namely "its technical orientation [...] antique authoritative tradition [...] [and] textual character".

3 For examples of such work, see Merianos and Sakorrafou 2013 (a survey of how alchemy is depicted in historiographical, hagiographical, and other Byzantine sources); Sakorrafou and Merianos 2014; Dufault 2019. Merianos and Sakorrafou (2013: 49) articulated the key question: "But what was 'chēmeia' for the Byzantines?" (Τι ήταν όμως η «χημεία» για τους Βυζαντινούς;). All translations are mine unless otherwise noted.

https://doi.org/10.1515/9783111589619-006

In what follows, I will use the modern term "alchemy" sparingly. As Gerasimos Merianos has observed,[4] there is no single Greek term that corresponds to "alchemy", and the closest, *chēmeia*, had a semantic range that varied considerably over time and according to author and context. Part of the problem is the modern term "alchemy" itself, which also has a wide semantic range, being used narrowly to describe attempts to manufacture or imitate gold and other precious metals but also more widely to designate any techniques or theories related to the artificial transformation of matter, and much else besides.[5] Even more problematically, the term in its modern sense, distinguished from the term "chemistry", was born out of polemics beginning in the late seventeenth century by professional chemists in Europe against what even then continued to be an important part of their field, the aim of artificially producing gold from less expensive ingredients; previously, there had been no substantive distinction between the terms "alchemy" and "chemistry" in European languages.[6]

Rather than take for granted even a rough equivalence between the term *chēmeia* as used in the Middle Byzantine period and the capacious and problematic term "alchemy" (or *Alchemie, alchimie*, αλχημεία, etc.) as used in modern languages of the last two centuries, I will first examine the evidence and only then inquire how it might contribute to our knowledge of this volume's theme, "alchemy in Byzantium".

1 Preparation of Gold and Silver

The *Suda* lexicon is an anonymous compilation of Greek literary and historical knowledge, arranged in a Middle Byzantine version of Greek alphabetical order (in which homophonous letters and digraphs, as pronounced in the medieval period, are adjacent to each other).[7] Probably compiled during the reign of Basil II (r. 976–1025), perhaps ca. 1000, its entries combine lexicographical and topical information compiled from earlier lexica, the *Historical Excerpts* of the tenth-century emperor Constantine VII (r. 913–959), and the rich library of classical literature and scholia that was clearly at the compiler's

4 E.g., in his presentation, "Terms, Concepts, and Perceptions of Alchemy in Byzantium", at the panel "Byzantine Alchemy between Art and Science" at the International Medieval Congress, University of Leeds, 5 July 2021.

5 See also Roberts 2022a: 559–560.

6 Newman and Principe 1998: 32, 63. I proposed an alternative conceptual vocabulary in my lecture "Disentangling Alchemy" for the Mary Jaharis Center for Byzantine Art and Culture Lecture Series, co-sponsored by the Committee on Medieval Studies at Harvard University, 7 April 2022; now published in revised form as Roberts 2025. For preliminary statements see also Roberts 2022b: 169–170; Roberts 2023: 598–599.

7 *Suda lexicon* (ed. Adler 1928–1938). Adler's *apparatus fontium* is my starting point for the observations below concerning the *Suda*'s sources.

(or compilers') disposal.[8] It proved popular among educated Byzantines and was copied frequently.[9] These copies often preserve additional glosses and cross-references interpolated after the work's original production.[10]

The *Suda*'s entry for *chēmeia* defines it as follows:

> Χημεία: ἡ τοῦ ἀργύρου καὶ χρυσοῦ κατασκευή, ἧς τὰ βιβλία διερευνησάμενος ὁ Διοκλητιανὸς ἔκαυσεν. ὅτι διὰ τὰ νεωτερισθέντα Αἰγυπτίοις Διοκλητιανῷ τούτοις ἀνημέρως καὶ φονικῶς ἐχρήσατο. ὅτε δὴ καὶ τὰ περὶ χημείας χρυσοῦ καὶ ἀργύρου τοῖς παλαιοῖς αὐτῶν γεγραμμένα βιβλία διερευνησάμενος ἔκαυσε πρὸς τὸ μηκέτι πλοῦτον Αἰγυπτίοις ἐκ τῆς τοιαύτης προσγίνεσθαι τέχνης μηδὲ χρημάτων αὐτοὺς θαρροῦντας περιουσίᾳ τοῦ λοιποῦ Ῥωμαίοις ἀνταίρειν.
>
> *Chēmeia*: the preparation of silver and gold. Diocletian tracked down the books about it and burned them. [It is said] THAT[11] because of revolutionary measures taken by the Egyptians, Diocletian treated them savagely and murderously. At which time he also tracked down the books on *chēmeia* of gold and silver written by the ancient [Egyptians] and burned them so that no more wealth would accrue to the Egyptians from such an art [*technē*], and so that henceforth they would not rebel against the Romans, emboldened by their surplus of funds.[12]

---

8 Wilson (1983: 145–147) provides a concise analysis of these features of the *Suda* and its date, authorship, and sources. Wilson's primary interest in the *Suda* is as a witness to ancient literature: "it can be reckoned a valuable source for the literary history of the Roman empire" (Wilson 1983: 147). See also the literature cited in Kazhdan 1991. In particular, for the debate over the etymology of the *Suda*'s name (a debate which Wilson declared "unprofitable" [1983: 145]), see Lavagnini 1962: 441–444, who cites the theories of Paul Maas (not the author's name – Souidas – as Eustathius of Thessaloniki suggests, but rather the work's title, derived from *suda*, the Latin imperative "sweat!"), Franz Dölger (a title derived from a Latin word for "palisade"), Henri Grégoire (an acronym which also referred to a Latin word, not for "palisade" but for "moat"), and finally, in greater detail, that of Silvio Giuseppe Mercati (that the lexicon was originally nameless, but then an "Italo-Greek" gave it the name *guida*, an Italian word for "guide" of Germanic origin, but wrote it in Greek characters ΓΟΥΙΔΑ, later misread as ϹΟΥΙΔΑ).

9 Wilson 1983: 145: the *Suda* was "[f]ar less ambitious and much more popular than the emperor Constantine [VII]'s [encyclopedic] enterprises". See also Wilson 1983: 146.

10 Adler explains that the glosses do not all appear in all manuscripts: "The old hand of codex Parisinus 2626 omits many types of glosses, and F always does, while T V frequently do; the same glosses are often relegated to the margin in the codices Paris. 2625, I M; sometimes they appear in a different order in different codices" (*Multa glossarum genera codicis Parisini 2626 vetus manus et F semper, T V plerumque omittunt, eadem saepe in codicibus Paris. 2625, I M in marginem reiecta sunt; interdum alio in aliis codicibus ordine feruntur*); Adler 1928–1938: 1.xv. Glosses appear in small print in the text in Adler's edition (1928–1938: 1.xv).

11 This expression typically introduces the beginning of a new excerpt from a source.

12 *Suda lexicon*, s.v. Χημεία (X 280) (ed. Adler 1928–1938: 4.804). Papathanassiou 2002: 123, in her discussion of this *Suda* passage, points to the real threat that technical ability with precious metals and their imitation might pose to a central government in the form of counterfeiting, and notes that this particular narrative "should probably be related to Diocletian's fiscal reform, [...] which [...] included the monetary system of Egypt" and strictly enforced the emperor's monopoly on minting coins bearing his likeness. See also Halleux 1981: 24n4. Thorndike (1923–1958: 1.194–195) considered the account of Diocletian burning alchemical books to be a fiction; cited by Merianos 2017: 238n42.

We may begin by noting a seemingly trivial fact: the *Suda*'s chosen spelling for the word itself is *chēmeia* (as opposed to one of the various homophonous forms also attested in Greek texts, such as *chymeia, cheimeia*, and *chimeia*). Scholars of the past two millennia have discussed the possible etymological origins of this term, understood as closely related to the origins of the discipline or practice to which it refers, and orthography has played an important role.[13] The *Suda*'s choice may align it with etymologies tracing the term back to the title of a book (Χημεῦ) in which the fallen angels of the Book of Enoch had, according to Zosimus of Panopolis, expounded *chēmeia*, or perhaps to an epithet for Egypt, Χημία, mentioned by Plutarch as evoking the moist, fertile blackness of its soil.[14] The manuscript tradition of the *Suda* indicates at least later interest in this orthographical question. While in most *Suda* manuscripts the lemma on the subject is spelled *chēmeia*, two witnesses have *cheimeia* and also preserve a marginal note reading: "Check if perhaps *chimeia* is spelled with an *iota* [in the other manuscript: 'with an *ēta*']".[15] Some versions of the *Suda* acknowledge a different alternative spelling of *chēmeia*, namely *cheimeia*, for which there is a brief cross-reference entry interpolated in most manuscripts: "*Cheimeia*: fusion [*cheimeusis*]". Some manuscripts add: "See under *chēmeia*".[16] The definition of *cheimeia* as *cheimeusis*, which is clearly meant to represent the word *chymeusis*, is the closest the *Suda* tradition comes to suggesting the

13 See Merianos and Sakorrafou 2013: 48; Merianos 2017: 238–239.

14 Zosimus Panopolitanus *apud* Georgius Syncellus, *Ecloga chronographica* (ed. Mosshammer 1984: 14, 2–14); Plutarchus, *De Iside et Osiride* 33 (= *Moralia* 364C) (ed. Froidefond 1988: 207). Both cited, with references, by Merianos 2017: 239. For the Egyptian word to which Χημία might correspond, see Froidefond 1988: 285 (207n2); Crum 1939: 109b–110a.

15 *Suda lexicon*, s.v. Χημεία (X 280) (ed. Adler 1928–1938: 4.804): ζήτει μὴ πῶς [read μή πως] τὸ χιμεία διὰ τοῦ ι [**S** : η **M**] γράφεται. **S** = *Vaticanus graecus* 1296 (copied in 1205 CE; Diktyon 67927); **M** = Venice, *Marcianus graecus* 448 ([coll. 1047] thirteenth century; Diktyon 69919), not to be confused with the ca.-late-tenth-century alchemical codex usually referenced with the siglum **M**, Venice, *Marcianus graecus* 299 ([coll. 584] Diktyon 69770). **S** reads χειμεία for the entry lemma, while *Suda* codex **M**'s lemma (originally χημεία) was corrected to χειμεία. Similarly in the entry on the emperor Diocletian (quoted in notes 23–24, 27–28, 32–34 below), the reading χημείας given by most manuscripts becomes χειμείας in *Suda*-**M** (although a corrector then changed it to χημείας) and *Suda*-**A** (= *Parisinus graecus* 2626 early hand, twelfth century; Diktyon 52261). "Diktyon" numbers refer to the respective entries in the *Pinakes* database of Greek manuscripts (https://pinakes.irht.cnrs.fr/); the URL for a given entry is https://pinakes.irht.cnrs.fr/notices/cote/69770/ (where the final number should be replaced by the "Diktyon" number in question). It is also possible to enter the Diktyon number on the manuscript-search page, https://pinakes.irht.cnrs.fr/recherche-manuscrit.html, in the field labeled "Numéro diktyon".

16 *Suda lexicon*, s.v. Χειμεία (X 227) (ed. Adler 1928–1938: 4.800), all printed in the small type which Adler reserves for interpolated glosses (see note 10 above): Χειμεία: ἡ χείμευσις. ζήτει ἐν τῷ χημεία. Manuscripts **G** and **S** omit the portion from ζήτει to the end, while **A** omits the entire entry. For the word χύμευσις in the sense of "fusion" or "casting", see Sophocles 1914: 1175.

etymology favored by most modern scholars from χέω (to pour) and its derivatives such as χύμα (a fluid).[17]

The definition of *chēmeia* with which the *Suda* entry begins is simple and narrow: it is the making of gold and silver. The narrative excerpt that follows, however, implies a broader sense for *chēmeia*: the phrase "*chēmeia* of gold and silver" suggests that *chēmeia* was a process or technique that could be used to produce substances other than gold and silver as well.

This phrase, and the narrative passage where it appears, is an excerpt from a fragment ascribed to John of Antioch's chronicle (a work written in the sixth or seventh century, of which only fragments survive) in the volume *On Virtues and Vices* (*De virtutibus et vitiis*) of Constantine VII Porphyrogenitus' *Historical Excerpts*,[18] presumably the *Suda*'s source for the passage.[19] There is a related report in the prefatory portion of the *Life of Saint Procopius*, a Greek hagiographical narrative preserved in an eighth-century manuscript but thought to have been written earlier (as early as the fourth century according to some modern scholars).[20]

---

17 For which see Merianos 2017: 239, with references, esp. Mertens 2006: 227n69 and n70. The *chy*- spelling (and associated etymology) is also favored by at least some Late Byzantine writers: Merianos 2017: 240.

18 Fragments of John of Antioch were gathered by Karl Müller (1841–1873: 4.538–622). There are two recent editions of the fragments of Ioannes Antiochenus' *Historia chronica*: Roberto 2005 and Mariev 2008. Both must be used with caution, as explained by Van Nuffelen 2012, because the selection of the excerpts that are presented as fragments was based not only on explicit ascriptions in the tradition (almost all of which appear in Constantine's *Historical Excerpts*) but also the editors' own (contrasting) reconstructions of what the original text must have contained and when it must have been written. Thus for example, Mariev's view that the "the composition of the main part of the historical narrative" from which genuine fragments were excerpted dates "to the first half of the sixth century" (Mariev 2008: 8*), based on stylistic and other significant but ultimately inconclusive criteria, leads him to omit excerpts explicitly attributed to John of Antioch but narrating events after that date or overlapping with the chronicle of John Malalas, completed (soon) after when John of Antioch is supposed to have finished his. In the present case, however, we are dealing with an unproblematic fragment that is explicitly attributed to John of Antioch and recounts events from long before the sixth century.

19 For the fragment in question, see Ioannes Antiochenus, *Fragmenta* (Io. Antioch., frag. 165, ed. Müller 1841–1873: 4.601–602 = ed. Roberto 2005: frag. 248, 1–7, p. 428 = ed. Mariev 2008: frag. 191, p. 348). Cited by Adler's edition; and by Merianos and Sakorrafou 2013: 50.

20 *Acta Sanctorum Julii 2* (8 July): §1.4, 557C (referenced by Halleux 1981: 23–24; and by Merianos and Sakorrafou 2013: 50n21): Τάτε γὰρ τείχη αὐτῶν ἠρήμωσε, καὶ τῶν συνασπισάντων ἐκείνῳ συχνοὺς ἀπέκτεινε. Τὰ ἐπίσημά τε τῶν ἀναθημάτων ἀφείλετο, προσῆλθε δὲ καὶ μέχρι Βουσίρεως, τὴν ἀντίπαλον χεῖρα χολάζων ἅπασαν. Οὐ μὲν, ἀλλὰ καὶ τὰς βίβλους, ὅσαι περὶ χειμείας ἀργύρου τε καὶ χρυσοῦ τοῖς παλαιοτέροις τῶν Αἰγυπτίων κατὰ σπουδὴν ἐγράφησαν, ἀνάλωμα πυρὸς αὐτὰς ἔθηκεν, εἴργων [read εἴργων] Αἰγυπτίους πορισμοῦ χρημάτων· ὥστε μὴ ἐκ τῆσδε τῆς τέχνης εὐκοπώτατα χρηματιζομένους, ῥᾳδίως πρὸς νεωτερισμοὺς ὑπάγεσθαι, εἰδὼς ὅτι πλούτου γένημα ὕβρις, ὡς ἵππου τρυφῶντος ἀτεχνῶς σκίρτημα, καὶ λογισμοί, σῶφρον οὐδὲν ἔχοντες, οὐδὲν μέτριον, ἀλλὰ κενῶν καὶ ἀλλοκότων ἐπιθυμίας πραγμάτων ("He laid waste to their walls, he killed many of those who had fought side-by-side with him, and he tore down the most distinguished monuments; he went all the way to Bousiris and punished every hand [of

The fact that the *Suda* entry is excerpting from another source should not lead us to disassociate the *Suda* entry's redactor from the excerpted passage, but rather to ask in what way his or her work constituted an interpretation of that passage. For one thing, the redactor seems to have inserted a sentence that is not present in the *Historical Excerpts*, perhaps excerpted from elsewhere, that serves to make Diocletian seem more ferocious: the explanation that "Diocletian treated" the Egyptians "savagely and murderously" on account of "revolutionary measures" they had taken is missing from the Constantinian excerpt. The definition with which the *Suda*'s entry begins is a reasonable contextual interpretation of the word *chēmeia* as it appears in the excerpted passage – an interpretation that we can probably ascribe to the redactor. Perhaps when pressed, the redactor would have admitted that *chēmeia* could "prepare" things other than gold and silver, but these two precious metals were what he or she considered, based at least in part on the excerpted passage, to be the typical products of *chēmeia*.[21]

Entirely missing in this definition is any hint of the connotations of folly or occultism often associated with the present-day English term "alchemy". In particular, there is no suggestion that *chēmeia* fails to produce gold and silver, nor that it requires any kind of hidden revealed knowledge.[22] On the contrary, the narrative's logic requires *chēmeia*'s success in the hands of the Egyptians: the abundant wealth they manage to produce is cited as the reason for Diocletian's burning of books on making gold and silver, and, in general, the entry would seem to imply, for his violent treatment of the Egyptians. Diocletian's book-burning as a remedy for counterfeiting makes most sense if the books are understood to communicate technical know-how, not secret revelations about attaining human perfection couched in metallurgical allegories.

Chēmeia is presented here as an effective process used to produce gold and silver. It is also censored by none other than Diocletian, the late-third-century Roman emperor remembered in the Middle Ages as a cruel persecutor of Christians. This image of Diocletian was still very much alive in Byzantium when the *Suda* was compiled, as the *Suda* itself attests.

that city?] that opposed him. Not only that, but also the books that the ancient Egyptians had laboriously written about the *cheimeia* of silver and gold he consigned to the fire, preventing the Egyptians from providing themselves with money, such that they could not effortlessly make money from that craft and so easily be induced towards revolutions. For he knew that the fruit of wealth is insolence, like the leap of the spoiled horse, and thoughts that have no temperance or moderation but only yearnings for vain and extraordinary things"). For Bousiris, see Montanari 2015: 402 (s.v.).

21 By comparison, the seventeenth-century scholar Andreas Libavius divides *alchemia* into *encheria* (i.e., ἐγχειρία, "manipulation") and *chemia* (i.e., χημία): cited by Newman and Principe 1998: 44. As Newman and Principe describe, *encheria* "comprises manual operations such as the use of apparatus and the regulation of the fire"; *chemia* is "the preparation of chemical substances". This definition of *chemia* seems quite similar to the sense implied by this passage quoted by the *Suda*.

22 See also Merianos and Sakorrafou 2013: 49–50.

2 Diocletian

The *Suda*'s entry on Diocletian (drawing on an excerpt from George the Monk's chronicle found in the *Historical Excerpts*, in the volume *On Virtues and Vices*) begins: "Diocletian, Emperor of the Romans. In his and his brother-in-law Maximian's reign, a most horrible persecution against Christians was set in motion".[23] The agency of these emperors in persecuting Christians is stressed: "For they ordered the churches of Christ in country and city to be overturned, their sacred scriptures burned, and those discovered as Christians to be forced to sacrifice to the *daimones*".[24] But an overwhelming number of Christians are killed,[25] or arrested,[26] as a result, so the emperors decree that those found to be Christians are to have their right eye gouged out (a Byzantine punishment for treason), "not only for the pain", but also to dishonor them, make an example of them, and make them alien to "the polity of the Romans".[27] It is a standard medieval Christian hagiographical motif that persecutors are punished for their sins, and here too "divine justice" eventually catches up with Diocletian and Maximian, for "one of them was slain by the Senate, and the other was hanged".[28] Here ends the excerpt from George the Monk in Constantine's *On Virtues and Vices* that was the *Suda*'s source, neatly pairing

23 *Suda lexicon*, s.v. Διοκλητιανός (Δ 1156) (ed. Adler 1928–1938: 2.104, 8–10): Διοκλητιανός, βασιλεὺς Ῥωμαίων. ἐπὶ τούτου καὶ Μαξιμιανοῦ γαμβροῦ αὐτοῦ διωγμὸς κατὰ Χριστιανῶν ἐκινήθη φρικωδέστατος. Cf. Georgius Monachus, *Chronicon* (ed. de Boor and Wirth 1978: 2.477, 9–15 and 481, 12–19) = Constantinus VII Porphyrogenitus, *Excerpta de virtutibus et vitiis 1*, ex Georg. Mon. nos. 13–14 (ed. Büttner-Wobst 1906: 146, 14–25); both cited by Adler's apparatus as sources for up to line 18. Bleckmann 2006: 509: "Diocletian received him [Maximian] as a brother in his family". For the entry's various factual errors, see the notes to Ross Scaife's translation of this *Suda* entry: *Suda On Line*, s.v. "Dioklētianos". 7 November 2002, https://www.cs.uky.edu/~raphael/sol/sol-entries/delta/1156. In what follows, these errors are of no concern, since my aim is to use the *Suda* as a source not for the reign of Diocletian but for how he and his reign were portrayed.

24 *Suda lexicon*, s.v. Διοκλητιανός (Δ 1156) (ed. Adler 1928–1938: 2.104, 10–12): προσέταξαν γὰρ κατὰ χώραν καὶ πόλιν τὰς Χριστοῦ ἐκκλησίας καταστρέφεσθαι καὶ τὰς θείας αὐτῶν γραφὰς κατακαίεσθαι, τοὺς δὲ Χριστιανοὺς εὑρισκομένους ἀναγκάζεσθαι θύειν τοῖς δαίμοσιν.

25 The expected sense of ἀναιρουμένων (see Liddell–Scott–Jones: s.v. ἀναιρέω II), supported by the full context in George the Monk, in which a mention of martyrdoms and then a series of detailed martyrdom narratives intervene between the forced sacrifice and the report that the emperors were "overwhelmed": Georgius Monachus, *Chronicon* (ed. de Boor and Wirth 1978: 2.477, 15–481, 11).

26 As a reader of the excerpted version in the *Suda* or the Constantinian *Historical Excerpts*, without the series of martyrdom narratives that intervene, might theoretically have surmised. This is the interpretation of Scaife's translation (see note 23 above).

27 *Suda lexicon*, s.v. Διοκλητιανός (Δ 1156) (ed. Adler 1928–1938: 2.104, 13–16): ἡττηθέντες δὲ τῷ πλήθει τῶν ἀναιρουμένων Χριστιανῶν ἐξέθεντο δόγμα ὥστε τοὺς εὑρισκομένους Χριστιανοὺς ἐξορύττεσθαι τὸν δεξιὸν ὀφθαλμόν, οὐ μόνον διὰ τὸ ὀδυνηρόν, ἀλλὰ διὰ τὸ ἄτιμόν τε καὶ πρόδηλον καὶ τῆς τῶν Ῥωμαίων πολιτείας ἀλλότριον.

28 *Suda lexicon*, s.v. Διοκλητιανός (Δ 1156) (ed. Adler 1928–1938: 2.104, 16–18): οὓς ἡ θεία δίκη ἐνδίκως μετελθοῦσα δικαίως ἐξέκοψε· καὶ ὁ μὲν ἐσφάγη ὑπὸ τῆς συγκλήτου, ὁ δὲ ἀπήγξατο.

Diocletian's and Maximian's violence against Christians with the violence of their divine punishment.

But the *Suda*'s text continues, now drawing on another passage from *On Virtues and Vices* – the very same passage from John of Antioch that is the basis for the *chēmeia* entry.[29] Whereas in the *chēmeia* entry, the *Suda* redactor modified the beginning of the passage, the entry on Diocletian reproduces the text of *On Virtues and Vices* almost verbatim; the only exception is that the redactor has replaced the first words "[It is reported] THAT Diocletian" (Ὅτι Διοκλητιανός) with the decidedly more tendentious "This irrational enemy of Christ" (οὗτος ὁ ἄνους καὶ μισόχριστος).[30] This beginning – corresponding in its content to the *chēmeia* entry from "Diocletian tracked down [...]" to "[...] savagely and murderously"[31] – reads (in the entry on Diocletian):

> οὗτος ὁ ἄνους καὶ μισόχριστος μνήμῃ καὶ ὀργῇ τῶν περὶ τὴν ἀρχὴν νεωτερισθέντων περὶ τὴν Αἴγυπτον οὐ μετρίως οὐδὲ ἡμέρως τῷ κρατεῖν ἀπεχρήσατο, ἀλλὰ προγραφαῖς τε καὶ φόνοις τῶν ἐπισήμων μιαίνων ἐπῆλθε τὴν Αἴγυπτον. ὅτε δὴ καὶ [...].
>
> This irrational enemy of Christ, in memory and wrath at the revolution attempted against his rule, in the case of Egypt did not moderately or in a civilized fashion avail himself of his power, but defiling the place with proscriptions and murders of the notables, descended upon Egypt. At which time he also [...].[32]

At this point follows the rest of the *chēmeia* entry printed above, on how Diocletian destroyed books on *chēmeia* to prevent the Egyptians from becoming wealthy enough to revolt. In the context of the entry on Diocletian, in which Diocletian also burns Christian scripture, the emperor's destruction of chemical writings is no longer a neutral or at least politically justified act. The emperor is clearly the worst sort of tyrant, such that the *Suda* redactor felt the need to add that he was an "irrational enemy of Christ", that is, an impious madman. His attacks on books about *chēmeia* are framed as part of this madness, parallel to his assassinations of Egyptian notables on the one hand, and to the burning of the Christian scriptures on the other. The effect is to leave *chēmeia* as an entirely legitimate craft, which only a tyrant would seek to stifle.

Although the *chēmeia* entry stops here, the entry on Diocletian continues with the rest of the same excerpt from John of Antioch – where the emphasis on Diocletian's

---

29 Ioannes Antiochenus, *Fragmenta* (Io. Antioch., frag. 165, ed. Müller 1841–1873: 4.601–602 = ed. Roberto 2005: frag. 248, 1–7, p. 428 = ed. Mariev 2008: frag. 191, p. 348); see note 19 above. Adler signals this overlap in her apparatus to both entries.

30 For μισόχριστος as "enemy of Christ", see Lampe 1961: s.v.

31 From ἧς τὰ βιβλία to ἐχρήσατο. See above at note 12.

32 *Suda lexicon*, s.v. Διοκλητιανός (Δ 1156) (ed. Adler 1928–1938: 2.104, 18–21). For the rest (p. 104, 21–25), beginning with "ὅτε δὴ καὶ", see note 12 above.

duplicity and transgression of Roman custom does nothing to improve his image.[33] Only the end of the *Suda*'s entry on Diocletian softens the emperor's portrait. This final passage is another excerpt, based on Eutropius' Latin *Breviarium ab urbe condita* (fourth century), with two phrases added from Zosimus' *New History* (fifth–sixth century); it has been argued that this passage too, as it appears (in Greek) in the *Suda*, might derive from John of Antioch. In it, Diocletian and Maximian exchange the reins of empire for the life of a private citizen. Maximian soon regrets this choice, but Diocletian "grew old in peace for three years, displaying his excellent merit", although "he did not by any means renounce Hellenic worship".[34] This epilogue sets Diocletian's tyranny in relief. He was only a bad ruler, the entry seems to imply, not an entirely evil man. Once he stopped destroying churches, burning holy books, murdering the leading men of Egypt, and destroying books on *chēmeia*, he made quite a decent private citizen. The final line qualifying this conclusion then implicitly reframes the entire entry in terms of religious adherence: he never became fully virtuous, it suggests, because he continued to be a pagan. In this implied schema, *chēmeia* is grouped with Christianity and the customs of the Roman polity (two cherished components of Middle Byzantine political ideology)[35] in opposition to Hellenic (pagan) worship.[36]

---

33 *Suda lexicon*, s.v. Διοκλητιανός (Δ 1156) (ed. Adler 1928–1938: 2.104, 25–30): ἦν δὲ τὸ ἦθος ποικίλος τις καὶ πανοῦργος, τῷ δὲ λίαν συνετῷ καὶ ὀξεῖ τῆς γνώμης ἐπεκάλυπτε πολλάκις τὰ τῆς οἰκείας φύσεως ἐλαττώματα, πᾶσαν σκληρὰν πρᾶξιν ἑτέροις ἀνατιθείς. ἐπιμελὴς δὲ ὅμως καὶ ταχὺς ἐν ταῖς τῶν πρακτέων ἐπιβολαῖς καὶ πολλὰ τῶν τῆς βασιλικῆς θεραπείας ἐπὶ τὸ αὐθαδέστερον παρὰ τὰ καθεστηκότα Ῥωμαίοις πάτρια μετεσκεύασεν ("In disposition, he was a changeful and wicked man, but by his intelligence and keenness of wit, he would frequently hide the defects of his own nature, attributing every cruel act to others. All the same, attentive and quick in tackling what needed to be done, he also changed many of the [customs] of imperial service to suit his great stubbornness, in violation of ancestral [customs] prevailing among the Romans").

34 *Suda lexicon*, s.v. Διοκλητιανός (Δ 1156) (ed. Adler 1928–1938: 2.104, 31–105, 2): ὅτι Διοκλητιανὸς καὶ Μαξιμιανὸς τὴν βασιλείαν ἀφέντες τὸν ἰδιώτην μετῆλθον βίον. καὶ ὁ μὲν ἐς Σάλωνας, πόλιν Ἰλλυρικήν, ὁ δὲ ἐς τὴν Λευκανῶν ἀφίκετο. καὶ ὁ μὲν Μαξιμιανὸς πόθῳ τῆς ἀρχῆς ἐς μεταμέλειαν ἦλθε, Διοκλητιανὸς δὲ ἐν ἡσυχίᾳ κατεγήρα ἐν ἔτεσι τρισίν, ὑπερβάλλουσαν ἀρετὴν ἐνδειξάμενος, τῆς δὲ Ἑλληνικῆς θρησκείας οὐδ' ὅλως ἀποστάς. Adler's apparatus: "cf. Eutrop. 167, fort. Io. Antioch". Included as a fragment of John of Antioch by Roberto 2005: frag. 251.1, p. 432 (with details of the passage's sources in the *apparatus fontium*); and by Mariev 2008: frag. 193, p. 350. The basis of these editions is the *Suda*'s entry, which includes no ascription to John of Antioch.

35 Ahrweiler 1975: 16–17.

36 This surprisingly positive view may ultimately derive from an appreciation, even on the part of the Christian historiographical tradition, of the foundational administrative changes that took place under Diocletian, such as the indiction cycle. I owe this suggestion to Maria Mavroudi.

3 The Golden Fleece

If we return to the *Suda*'s entry on *chēmeia*, we find that it ends with a cross-reference to the entry on "the hide",[37] that is, the mythical Golden Fleece – or, to be more precise, the golden "hide" (δέρας). In that entry, the *Suda* explains that "the hide" refers to the "golden-fleeced hide" which Jason and the Argonauts "took once they had come through the Pontus Sea to Colchis", but that this is just the fanciful tale of poets. The real Golden Fleece

> […] βιβλίον ἦν ἐν δέρμασι γεγραμμένον, περιέχον ὅπως δεῖ γίνεσθαι διὰ χημείας χρυσόν. εἰκότως οὖν οἱ τότε χρυσοῦν ὠνόμαζον αὐτὸ δέρας, διὰ τὴν ἐνέργειαν τὴν ἐξ αὐτοῦ.
>
> […] was a book written on hides [i.e., parchment], containing (instructions for) how gold must come to be through *chēmeia*. Now, it is fitting that people at that time called it a golden hide, on account of the result that it produced.[38]

This entry seems to be based on a historical excerpt preserved in a fourteenth-century manuscript in Paris,[39] with one significant difference: missing from the Paris manuscript is the entire last sentence of the *Suda*'s entry ("Now, it is fitting […]"). Umberto Roberto's edition of the fragments of John of Antioch's chronicle treats this last sentence as the original reading of the text from which it is excerpted (which his edition considers to be John of Antioch's chronicle), but the sentence could also be the *Suda*'s own, or its source's, addition. Whatever the source, this final sentence attempts to justify this interpretation of the myth of the Golden Fleece and thus suggests a commitment to that interpretation.

37 *Suda lexicon*, s.v. Χημεία (X 280) (ed. Adler 1928–1938: 4.804, 8–9): ζήτει ἐν τῷ δέρας.

38 *Suda lexicon*, s.v. Δέρας (Δ 250) (ed. Adler 1928–1938: 2.24). The whole entry reads: Δέρας· τὸ χρυσόμαλλον δέρας, ὅπερ ὁ Ἰάσων διὰ τῆς Ποντικῆς θαλάσσης σὺν τοῖς Ἀργοναύταις εἰς τὴν Κολχίδα παραγενόμενοι ἔλαβον, καὶ τὴν Μήδειαν τὴν Αἰήτου τοῦ βασιλέως θυγατέρα. τοῦτο δὲ ἦν οὐχ ὡς ποιητικῶς φέρεται, ἀλλὰ βιβλίον ἦν ἐν δέρμασι γεγραμμένον, περιέχον ὅπως δεῖ γίνεσθαι διὰ χημείας χρυσόν. εἰκότως οὖν οἱ τότε χρυσοῦν ὠνόμαζον αὐτὸ δέρας, διὰ τὴν ἐνέργειαν τὴν ἐξ αὐτοῦ. On this *Suda* entry, see also Matteo Martelli's chapter in this volume.

39 This is one of the controversial passages appearing in *Parisinus graecus* 1630 (fourteenth century; Diktyon 51252) and ascribed by some scholars to John of Antioch (Io. Antioch., frag. 15.3, ed. Müller 1841–1873: 4.548 = Roberto 2005: frag. 26.3, 11–16, pp. 64–66) but not by others. It is not included by Mariev in his edition of the fragments. On this problem, still unresolved, see Van Nuffelen 2012: 440–441. A more concise version of this excerpt (apparently based on the text as it appears in the *Suda*) appears in *Vaticanus graecus* 305 (thirteenth century; Diktyon 66936), as part of an anonymous collection of incredible stories known as the *Excerpta Vaticana*: see Anonymus paradoxographus, *De incredibilibus* (*Excerpta Vaticana*) 3 (ed. Festa 1902: 89); cited by Lordkipanidze 2001: 26. This collection seems to draw on an older collection of "rationalizing myths" that had circulated for centuries before; in its present form, it was compiled no earlier than the fifth century: Hawes 2014: Chapter 3, esp. p. 119, and, for a reference to the collection's interpretation of the Golden Fleece, 125n19.

The *Suda*'s entry as it stands speaks of *chēmeia* of gold in matter-of-fact terms. It presents this as a technical procedure, in contrast to the fanciful myth of Jason. In the account's de-mythologizing construction, effected by an allegorical interpretation that parallels not only ancient but also modern attempts to work out the real story behind anecdotes deemed outlandish,[40] *chēmeia* is a craft of deep antiquity, since it is the origin for that old story about the Argonauts. It is presented not as irrational, fallacious, or impious charlatanry – quite the opposite: production of gold is the hard-headed reality that lies behind the myths of pagan Hellenism.[41]

This approach to mythology should be placed in its proper context, which is to say, not only the long history of Platonism, but also the development of Byzantine intellectual culture in the Early Middle Ages. Allegorical interpretation of ancient texts was a prominent and influential mode of Byzantine intellectual production, not only in biblical exegesis but also in texts about philosophy, including alchemical texts, as Maria Mavroudi's chapter in the present volume argues.[42] Modern interpreters of the myth of the Golden Fleece seeking to understand the true origin of the story (not unlike the *Suda* and its sources) have tended to emphasize more prosaic possibilities.[43] Sharing the cultural rather than material emphasis of the *Suda*'s allegorical interpretation is the alternative approach that seeks to find the story's origins not in a specific historical sheep's hide, but rather in symbolism.[44]

40 Such as Diocletian's burning of books on *chēmeia* in Egypt; see note 12 above. For the approach to myth of Plato and other ancient philosophers, see Morgan 2000; Collobert et al. 2012. For a study of this approach to myth from Palaephatus (fourth century BCE) to Pausanias (second century CE), see Hawes 2014. In particular, for Socrates' comments (in the *Phaedrus*) on attempts to explain away myths, see Morgan 2000: 214–215; Hawes 2014: 15–16.

41 See also Merianos and Sakorrafou 2013: 50: "The attempt to rationalize a myth by distilling it to alchemical techniques that, for the Byzantines, were plausible is indisputably interesting" (Είναι αναμφισβήτητα ενδιαφέρουσα η απόπειρα εξορθολογισμού ενός μύθου μέσω της αναγωγής του σε αλχημικές τεχνικές που για τους Βυζαντινούς ήταν αληθοφανείς).

42 See also Mavroudi 2020: 455–456; and Martelli 2017: 231–233, including a reference to the *Suda* entry on the Golden Fleece (in note 11).

43 Ryder (1991: 57–60) weighed several proposed theories: that the fineness of the wool on the hide in question made it precious, so that the description "golden" was applied to it metaphorically, or that the fine wool made it useful for filtering gold from silt or water (Ryder argues that these two are not mutually exclusive); that a diet high in olive leaves caused liver damage in a ram, whose wool was discolored yellow as a result (a hypothesis so implausible, Ryder asserts, that it can be dismissed); or (as Ryder suggests) that the hide was tan-colored and so came to be called "golden". Smith and Smith 1992: 119–120, responding to Ryder, argue in favor of the "liver-damaged sheep hypothesis", on the grounds that the vocabulary for colors in ancient Greek was limited, such that Greeks might plausibly have called a yellow fleece "golden", a term also applied to yellow wool "in modern times".

44 In this vein, Lordkipanidze 2001, who refers briefly to the *Suda*'s entry (p. 26), argues that the myth of the Golden Fleece had two stages: first, Archaic, when the fleece symbolized royal power; and second, Hellenistic, when it came to symbolize the wealth (especially wealth in gold and other metals) of Colchis on the Black Sea, with which it had come to be associated. His approach is to contextualize both stages in

All these approaches, including the one appreciated and preserved by the *Suda*'s redactor, share the conviction that the story of the Argonauts and the Golden Fleece is fanciful and cannot be literally true. From this starting point, they each construct what they consider likely to be the true referent of the story's subject. This process depends on the interpreter's intuition for what is plausible and what is not, conditioning the kinds of questions he or she asks. Modern interpreters were puzzled by what made a sheep's hide valuable: fine wool? utility for filtering out gold dust from water? legitimation of claims to royal authority? For the author of the interpretation in the *Suda*, however, a sheep's hide quite naturally evoked the writing support typically used in early medieval Byzantium for important, highly prized books: parchment.[45] And what might make such a hide golden? Liver damage? A vaguer association with wealth? In the logic of the *Suda*'s account, it was the chrysopoetic contents of the book itself.[46]

4 Books by Zosimus

Finally, the *Suda* lexicon also includes a few lines about the famous "alchemical" author Zosimus of Panopolis (ca. third–fourth century CE):

> Ζώσιμος, Ἀλεξανδρεύς, φιλόσοφος. Χημευτικὰ πρὸς Θεοσεβίαν τὴν ἀδελφήν· ἔστι δὲ κατὰ στοιχεῖον ἐν βιβλίοις κη′, ἐπιγράφεται δὲ ὑπό τινων Χειρόκμητα· καὶ ὁ Πλάτωνος βίος.
>
> Zosimus, an Alexandrian, a philosopher. [He wrote] *chēmeutika* [addressed] to his sister Theosebia – they are arranged by letter [or: alphabetically] in 28 books and entitled by some the *Handiworks* – and the life of Plato.[47]

This brief entry describes Zosimus as a "philosopher" and ascribes to him a biography of Plato, which serves to emphasize that his philosophy is in the same tradition as that of Plato. At the same time, his "writings about *chēmeia*" (*chēmeutika*) are described in neutral terms and are clearly considered compatible with, or even an illustration of,

terms of material culture and surrounding cultures. Thus, for the royal symbolism, he adduces evidence for the association of the ram with royal power in Hittite and ancient Greek culture. For Colchis, he points to material remains that suggest early Greek contact with Colchis, and to evidence of prospecting for gold by filtering gold dust from the local sediment, continuing into the modern period.

45 The word δέρμα seems to have been used as a term for parchment, at least in the fourth/fifth century. See Bülow-Jacobsen 2009: 11n20; and Lampe 1961: s.v. δέρμα 3, citing Palladius (d. before 431), *Dialogus de vita Ioannis Chrysostomi* 20.3–4 (ed. Malingrey 1988: 1.392): "[...] τῶν λεχθέντων, ὧν [...] γραφῇ παραδώσω ἐν δέρματι ἀκμαίῳ [...]".

46 The association of the Golden Fleece with texts on gold-making would have a lively afterlife in early modern Europe. See Faivre 1993, cited by Merianos 2017: 239n51; and Matton 1995.

47 *Suda lexicon*, s.v. Ζώσιμος (Z 168) (ed. Adler 1928–1938: 2.515). See also Mertens 1995: XVI, XCVII–CI; Mertens 2006: 219; Merianos 2017: 241.

his work as a philosopher. The title that the *Suda* mentions for these *chēmeutika*, the *Handiworks*, implies a "hands-on" or practical focus for that text.[48]

5 Conclusion

Taken together, these entries in the *Suda* treat *chēmeia* as a *technē*, the craft of artificially producing (or imitating, or even simply working) precious substances – a legitimate craft *as craft* (that is, a craft that successfully did what it was supposed to do) but one which Diocletian – an emperor remembered not only for persecuting Christians but also for his administrative and monetary reforms[49] – wished to restrict or stamp out entirely. This craft was envisioned as having a very long history, such that it would be plausible for Jason and the Argonauts to have traveled in search of books describing it. The contexts in which the *Suda* discusses it suggest a narrow scope for this craft, which seems to involve a technical procedure that is supposed to result in gold, silver, and the like. Chemical theory, cosmologies incorporating chemical theory, the redemption of the human soul, and other topics understood by some modern scholars to be part of something called "alchemy", are absent.

The closest we come to a broader view is the association of Zosimus with *chēmeia*, in the form of the *chēmeutika* that he wrote, and the fact that he is called a philosopher, suggesting not only craft but also the science of explaining what the craft accomplishes and how. Here we have a slight hint that *chēmeia*, while a craft, was of interest to philosophers. Yet even in this case, the entry gives the impression (regardless of whether that impression is correct) that the *chēmeutika* he wrote were quite practical, since they were also known as the *Cheirokmēta*, "things wrought by hand". So here too in the *Suda*'s presentation *chēmeia* remains narrowly a craft, a technical discipline concerned with effecting, more than interpreting, material transformation.[50]

48 That the twenty-eight books of *chēmeutika* are said to have been arranged "by letter" could simply mean that they were arranged alphabetically (perhaps by subject heading), but it might also suggest that a letter was assigned to each of the books. Since the classical Greek alphabet only has twenty-four letters, Reitzenstein hypothesized that the four extra letters were derived from the Coptic alphabet. Riess proposed that only the first twenty-four had letters (just like the Homeric epics, as Mertens points out), and then four further texts were added to that. See the discussion of these hypotheses in Mertens 1995: xcviii–c, who prefers Riess' hypothesis. As Maria Mavroudi pointed out to me, if the Greek alphabet were being used, it could have included the three additional letters employed in Greek numerals, *stigma*, *koppa*, and *sampi*, bringing the total to just one short of twenty-eight.

49 I thank Maria Mavroudi for stressing this point.

50 This is not to say that the *Suda*'s portrait of *chēmeia* is an accurate or fair characterization of the interrelation between technology and science, which are usually deeply implicated in one another: new technologies are developed based at least on a tacit working interpretation of existing techniques, and new interpretations are often inspired by existing techniques, which often also make it possible to test

What can this tell us about "alchemy in Byzantium"? First, it is a key piece of evidence that *chēmeia* referred, at least in the Middle Byzantine period, to only part of what modern scholars tend to mean when they speak of "alchemy". This should make us hesitate to translate *chēmeia* as "alchemy" (whether explicitly or tacitly); a better approximation, at least according to the *Suda*'s usage, might be something like "practical chemistry". But this doesn't mean that *chēmeia* was unrelated to what we call "alchemy". For the practices of *chēmeia* were understood as effecting phenomena that were appropriate objects of theorizing for a "philosopher" like Zosimus. After all, it was the task of the natural philosopher "to discuss bodies and their kinds and capacities", as the *Suda*'s entry on philosophy assumed its readers would agree.[51] A craft that seemed to make gold – or in any case to effect remarkable and unexpected transformations resulting in substances that shared many properties with gold – would surely interest anyone seeking to understand the underlying properties and potencies of material bodies.

Bibliography

Sources

Collected Sources

Acta Sanctorum Julii 2: *Acta Sanctorum, Julii*, vol. 2. Antwerp: J. du Moulin, 1721.

Müller 1841–1873: Karl Müller (ed.), *Fragmenta Historicorum Graecorum*, 5 vols. Paris: A. Firmin-Didot, 1841–1873.

new interpretations. For a well-explained example of the interplay between science and technology, see Barnes' (1982. 1–7) discussion of Kuhn's account of how and why Carnot was able to develop his thermodynamic theory known as the "Carnot cycle".

51 *Suda lexicon*, s.v. Φιλοσοφία (Φ 416) (ed. Adler 1928–1938: 4.733, 17–18): εἴγε φυσικοῦ μέν ἐστιν ἔργον τὸ περὶ τῶν σωμάτων διαλεχθῆναι καὶ τῶν εἰδῶν αὐτῶν καὶ δυνάμεων.

Acknowledgement: I am grateful to Maria Mavroudi for her comments on an earlier version of this essay drafted in 2014. But for several minor changes, the essay printed here was completed in August 2021.

Authors and Works

Anonymus paradoxographus, *De incredibilibus* (*Excerpta Vaticana*): Nicola Festa (ed.), *Mythographi Graeci*, vol. 3.2: *Palaephati Περὶ ἀπίστων, Heracliti qui fertur libellus Περὶ ἀπίστων, Excerpta Vaticana (vulgo anonymus De incredibilibus)* (Mythographi Graeci 3.2). Leipzig: Teubner, 1902, pp. 88–99.

Constantinus VII Porphyrogenitus, *Excerpta de virtutibus et vitiis 1*: Theodor Büttner-Wobst (ed.), *Excerpta de virtutibus et vitiis, pars 1* (Excerpta historica iussu imp. Constantini Porphyrogeniti confecta 2.1). Berlin: Weidmann, 1906.

Georgius Monachus, *Chronicon*: Carl de Boor and Peter Wirth (eds), *Georgii Monachi Chronicon*, 2 vols (Bibliotheca Scriptorum Graecorum et Romanorum Teubneriana). Second edition. Stuttgart: Teubner, 1978.

Georgius Syncellus, *Ecloga chronographica*: Alden A. Mosshammer (ed.), *Georgii Syncelli Ecloga chronographica* (Bibliotheca Scriptorum Graecorum et Romanorum Teubneriana). Leipzig: Teubner, 1984.

Ioannes Antiochenus, *Fragmenta*: Umberto Roberto (ed. and It. transl.), *Ioannis Antiocheni fragmenta ex Historia chronica* (Texte und Untersuchungen zur Geschichte der altchristlichen Literatur 154). Berlin and New York: De Gruyter, 2005.
Sergei Mariev (ed. and Engl. transl.), *Ioannis Antiocheni fragmenta quae supersunt omnia* (Corpus Fontium Historiae Byzantinae 47). Berlin and New York: De Gruyter, 2008.

Palladius, *Dialogus de vita Ioannis Chrysostomi*: Anne-Marie Malingrey (ed.), *Palladios, Dialogue sur la vie de Jean Chrysostome*, 2 vols (Sources chrétiennes 341–342). Paris: Les Éditions du Cerf, 1988.

Plutarchus, *De Iside et Osiride*: Christian Froidefond (ed. and Fr. transl.), *Plutarque, Œuvres morales*, vol. 5.2: *Isis et Osiris*. Paris: Belles Lettres, 1988.

Suda lexicon: Ada Adler (ed.), *Suidae Lexicon*, 5 vols (Lexicographi Graeci 1.1–5). Leipzig: Teubner, 1928–1938. Engl. transl.: *Suda On Line: Byzantine Lexicography*, ed. David Whitehead, https://www.cs.uky.edu/~raphael/sol/sol-html/.

Literature

Adler 1928–1938: Ada Adler, *Suidae Lexicon*, 5 vols (Lexicographi Graeci 1.1–5). Leipzig: Teubner, 1928–1938.

Ahrweiler 1975: Hélène Ahrweiler, *L'idéologie politique de l'Empire byzantin* (Collection SUP, L'historien 20). Paris: Presses universitaires de France, 1975.

Barnes 1982: Barry Barnes, *T. S. Kuhn and Social Science*. London: Macmillan, 1982.

Bleckmann 2006: Bruno Bleckmann, "Maximianus [1]", in Hubert Cancik and Helmuth Schneider (eds), *Brill's New Pauly: Encyclopaedia of the Ancient World*, vol. 8. Leiden and Boston, MA: Brill, 2006, cols 509–511.

Bülow-Jacobsen 2009: Adam Bülow-Jacobsen, "Writing Materials in the Ancient World", in Roger S. Bagnall (ed.), *The Oxford Handbook of Papyrology*. Oxford: Oxford University Press, 2009, pp. 3–29.

Collobert et al. 2012: Catherine Collobert, Pierre Destrée, and Francisco J. Gonzalez (eds), *Plato and Myth: Studies on the Use and Status of Platonic Myths* (Mnemosyne, Supplements 337). Leiden: Brill, 2012.

Crum 1939: Walter E. Crum, *A Coptic Dictionary*. Oxford: Clarendon Press, 1939.

Dufault 2019: Olivier Dufault, *Early Greek Alchemy, Patronage and Innovation in Late Antiquity* (California Classical Studies 7). Berkeley, CA: Department of Classics, University of California, Berkeley, 2019.

Faivre 1993: Antoine Faivre, *The Golden Fleece and Alchemy* (Western Esoteric Traditions). Albany, NY: SUNY Press, 1993.

Halleux 1981: Robert Halleux, *Papyrus de Leyde. Papyrus de Stockholm. Fragments de recettes* (Les Alchimistes grecs 1). Paris: Belles Lettres, 1981.

Hawes 2014: Greta Hawes, *Rationalizing Myth in Antiquity*. Oxford: Oxford University Press, 2014.

Kazhdan 1991: Alexander Kazhdan, "Souda", in Alexander P. Kazhdan (ed.-in-chief), *The Oxford Dictionary of Byzantium*, 3 vols. New York and Oxford: Oxford University Press, 1991, vol. 3, pp. 1930–1931.
Lampe 1961: G. W. H. Lampe, *A Patristic Greek Lexicon*. Oxford: Clarendon Press, 1961.
Lavagnini 1962: Bruno Lavagnini, "Suida, Suda o Guida", *Rivista di filologia e di istruzione classica* NS 40 (1962), pp. 441–444.
Liddell–Scott–Jones: Henry George Liddell, Robert Scott, and Henry Stuart Jones, *A Greek-English Lexicon. Ninth Edition with a Revised Supplement*. Oxford: Clarendon Press, 1996.
Lordkipanidze 2001: Otar Lordkipanidze, "The Golden Fleece: Myth, Euhemeristic Explanation and Archaeology", *Oxford Journal of Archaeology* 20.1 (2001), pp. 1–38.
Magdalino and Mavroudi 2006: Paul Magdalino and Maria Mavroudi (eds), *The Occult Sciences in Byzantium*. Geneva: La Pomme d'or, 2006.
Martelli 2013: Matteo Martelli, *The* Four Books *of Pseudo-Democritus* (Sources of Alchemy and Chemistry 1; Ambix 60, Supplement 1). Leeds: Maney Publishing, 2013.
Martelli 2017: Matteo Martelli, "Hippocrates in Two Syriac Alchemical Collections", *Aramaic Studies* 15.2 (2017), pp. 230–251.
Matton 1995: Sylvain Matton, "L'interprétation alchimique de la mythologie", *Dix-huitième siècle* 27.1 (1995), pp. 73–87.
Mavroudi 2020: Maria Mavroudi, "Homer in Greece from the End of Antiquity 1: The Byzantine Reception of Homer and His Export to Other Cultures", in Corinne Ondine Pache (ed.), *The Cambridge Guide to Homer*. Cambridge: Cambridge University Press, 2020, pp. 444–472.
Merianos 2014: Gerasimos Merianos, "Oil and Wine in Byzantine Alchemical Recipes", in Antonella Pellettieri (ed.), *Identità euromediterranea e paesaggi culturali del vino e dell'olio* (Collana MenSALe, Documenta et Monumenta 2). Foggia: Centro Grafico, 2014, pp. 249–259.
Merianos 2017: Gerasimos Merianos, "Alchemy", in Anthony Kaldellis and Niketas Siniossoglou (eds), *The Cambridge Intellectual History of Byzantium*. Cambridge: Cambridge University Press, 2017, pp. 234–251.
Merianos and Sakorrafou 2013: Gerasimos Merianos and Sandy Sakorrafou, "Μαρτυρίες περί αλχημείας στο Βυζάντιο σε μη αλχημικά κείμενα", in Eirini Mergoupi-Savaidou, Gerasimos Merianos, Faidra Papanelopoulou, and Christiana Christopoulou (eds), *Επιστήμη και Τεχνολογία. Ιστορικές και Ιστοριογραφικές Μελέτες* (Εταιρεία Μελέτης και Διάδοσης της Ιστορίας των Επιστημών και της Τεχνολογίας 1). Athens: Ekdotike Athenon, 2013, pp. 45–65.
Mertens 1995: Michèle Mertens, *Zosime de Panopolis, Mémoires authentiques* (Les Alchimistes grecs 4.1). Paris: Belles Lettres, 1995.
Mertens 2006: Michèle Mertens, "Graeco-Egyptian Alchemy in Byzantium", in Magdalino and Mavroudi 2006, pp. 205–230.
Montanari 2015: Franco Montanari, *The Brill Dictionary of Ancient Greek*. Leiden and Boston, MA: Brill, 2015.
Morgan 2000: Kathryn A. Morgan, *Myth and Philosophy from the Presocratics to Plato*. Cambridge: Cambridge University Press, 2000.
Newman and Principe 1998: William R. Newman and Lawrence M. Principe, "Alchemy vs. Chemistry: The Etymological Origins of a Historiographic Mistake", *Early Science and Medicine* 3.1 (1998), pp. 32–65.
Papathanassiou 2002: Maria K. Papathanassiou, "Metallurgy and Metalworking Techniques", in Angeliki E. Laiou (ed.-in-chief), *The Economic History of Byzantium: From the Seventh through the Fifteenth Century*, 3 vols (Dumbarton Oaks Studies 39). Washington, DC: Dumbarton Oaks Research Library and Collection, 2002, vol. 1, pp. 121–127.
Papathanassiou 2006: Maria Papathanassiou, "Stephanos of Alexandria: A Famous Byzantine Scholar, Alchemist and Astrologer", in Magdalino and Mavroudi 2006, pp. 163–203.
Roberts 2022a: Alexandre M. Roberts, "Byzantine Engagement with Islamicate Alchemy", *Isis* 113.3 (2022), pp. 559–580.
Roberts 2022b: Alexandre M. Roberts, "Hierotechnicians by Name and Their Middle Byzantine Fame", *Journal of Late Antique, Islamic and Byzantine Studies* 1.1–2 (2022), pp. 167–199.

Roberts 2023: Alexandre M. Roberts, “Thinking about Chemistry in Byzantium and the Islamic World”, *Journal of the History of Ideas* 84.4 (2023), pp. 595–619.

Roberts 2025: Alexandre M. Roberts, “Disentangling Alchemy”, *Isis* 116.4 (2025), pp. 623–644.

Ryder 1991: Michael L. Ryder, “The Last Word on the Golden Fleece Legend?”, *Oxford Journal of Archaeology* 10.1 (1991), pp. 57–60.

Sakorrafou and Merianos 2014: Sandy Sakorrafou and Gerasimos Merianos, “John Kanaboutzes' Commentary on Dionysios of Halikarnassos: A Perception of Alchemy in a Fifteenth-Century Greek Text”, in Gianna Katsiampoura (ed.), *Scientific Cosmopolitanism and Local Cultures: Religions, Ideologies, Societies*. Athens: Institute of Historical Research, National Hellenic Research Foundation, 2014, pp. 86–94.

Smith and Smith 1992: G. J. Smith and A. J. Smith, “Jason's Golden Fleece”, *Oxford Journal of Archaeology* 11.1 (1992), pp. 119–120.

Sophocles 1914: Evangelinos Apostolides Sophocles, *Greek Lexicon of the Roman and Byzantine Periods (From B. C. 146 to A. D. 1100). Memorial Edition*. Cambridge, MA: Harvard University Press, 1914.

Thorndike 1923–1958: Lynn Thorndike, *A History of Magic and Experimental Science*, 8 vols. New York: Macmillan, 1923–1958.

Van Nuffelen 2012: Peter Van Nuffelen, “John of Antioch, Inflated and Deflated. Or: How (Not) to Collect Fragments of Early Byzantine Historians”, *Byzantion* 82 (2012), pp. 437–450.

Viano 2006: Cristina Viano, “Greco-Egyptian Alchemy”, in Katharine Park and Lorraine Daston (eds), *Early Modern Science* (The Cambridge History of Science 3). Cambridge: Cambridge University Press, 2006, pp. 468–482.

Viano 2018: Cristina Viano, “Byzantine Alchemy, or the Era of Systematization”, in Paul Keyser and John Scarborough (eds), *The Oxford Handbook of Science and Medicine in the Classical World*. Oxford: Oxford University Press, 2018, pp. 943–964.

Wilson 1983: Nigel G. Wilson, *Scholars of Byzantium*. London: Duckworth, 1983.

Günter Prinzing

On the Relationship between the Two Versions of the Late Byzantine Alchemical Recipe Collection on the Art of Goldsmithing

Research on the transmission of expert knowledge in written form concerning the technical aspects of the Middle or Late Byzantine art of goldsmithing was, up to around 2015, largely centered on a special, anonymous collection of alchemical recipes. This collection is completely contained under the heading *Περὶ τῆς τιμιωτάτης καὶ πολυφήμου χρυσοχοϊκῆς* (On the Most Praiseworthy and Illustrious Art of Goldsmithing) in the manuscript *Parisinus graecus* 2327 [Diktyon 51958],[1] on ff. 280 recto – 291 recto. The codex, consisting of 299 paper folia, was, according to its colophon on f. 291 recto, copied in 1478 on Crete, at Chandax (Herakleion), by the Greek Theodore Pelecanus from Kerkyra (Corfu). Later, around the year 1537, it was acquired by the Corfiot humanist Antonius Eparchus (1491–1571), who sold it in 1540 to Bishop Guillaume Pélicier of Montpellier, the ambassador of King Francis I (r. 1515–1547) in Venice. Soon afterwards, it entered the Royal Library at Fontainebleau, which under King Charles IX (r. 1560–1574) was transferred to Paris. Since then, the codex has been preserved in the Bibliothèque nationale de France, where it was catalogued as part of the collection of ancient Greek alchemists and designated by the *siglum* **A**.[2]

1 The State of Research on the Recipe Collection

When this recipe collection – seemingly of 58, but in fact 56 recipes, as we shall see – was first edited and translated into French by Marcellin Berthelot and Charles-Émile Ruelle in 1887–1888,[3] the editors described it as a *traité technique*, that is, a *treatise* or

1 See https://diktyon.irht.cnrs.fr/en.
2 See Lebègue 1924: 54–61 (no. 2) for the recipe collection (280 recto – 290 verso). This description is partly incomplete, as becomes clear from the overview below, 125–133. Cf. also Mertens 1995: xxxi–xxxviii, especially xxxiv–xxxvii for a list of **A**'s content; note, in particular, xxxvii (no. 64), where the recipe collection is briefly mentioned. See also Colinet 2010: lviii and cxxx–cxxxiii; Bosselmann-Ruickbie et al. 2018: 231. On the history and cultural background of the manuscript, see most recently Merianos (Forthcoming).
3 See Berthelot and Ruelle 1887–1888: 2.321–337 for the Greek text of recipes nos. 1–56 from **A**. Concerning no. 57, however, the editors merely printed its heading and incipit, accompanied by a note indicating that the full text is provided according to manuscript **M** (= *Marcianus graecus* 299 [coll. 584]; second half of the tenth century) in Berthelot and Ruelle 1887–1888: 2.141–143. Only at the end of the French translation of recipes nos. 1–56 (see Berthelot and Ruelle 1887–1888: 3.322) it is noted that no. 57 forms part of a

https://doi.org/10.1515/9783111589619-007

discourse on the art of goldsmithing. Since then, and especially after the publication of Jochem Wolters' German translation (and commentary) "Der Traktat 'Über die edle und hochberühmte Goldschmiedekunst' (11. Jahrhundert)" in 2004, this unique collection has attracted increasing attention in the history of alchemy and the study of medieval and/or Byzantine metalworking.[4]

At this point, before examining more closely the question of how this special collection was handed down, I would like to briefly address whether it is appropriate to regard the collection of recipes in question as a *traité technique* – a "technical treatise" – or merely a "treatise" or "discourse". Berthelot and Ruelle not only assigned this recipe collection to the general category of "technical treatises" in Part Five of their edition, but also included 31 additional alchemical texts of partly similar content and structure under the same label. It is therefore no surprise that their classification was widely adopted in later scholarship. In my view, however, this label is not justified in the case of the present recipe collection, for the text as a whole lacks the well-known characteristics of a treatise.[5] On the contrary, even a quick glance at the recipes clearly shows that we are dealing with a straightforward sequence of recipes on various techniques of goldsmithing, arranged (or compiled) by an anonymous author (or perhaps a few *anonymi*) for practical purposes in connection with the growing interest in goldsmithing techniques or even alchemical practices. It would therefore be appropriate to abandon the terms *traité*, "treatise", "discourse", or *Traktat*, which have frequently been used in the secondary literature, in favour of "recipe collection" or, as Matteo Martelli put it, "recipe book".[6]

Turning now to the transmission of this special collection, it is useful to keep the following in mind. For a long time, **A** was believed to be the only manuscript that contained this exceptional collection of recipes in its entirety, though it was already known to Berthelot and Ruelle that a few (six) recipes from this collection are also found in a slightly later alchemical manuscript, preserved in the Biblioteca Medicea Laurenziana

text, written by Zosimus of Panopolis (ca. late third–fourth century), and that its translation appears in Berthelot and Ruelle 1887–1888: 3.143–145. See also the next note and note 7 below.

4 The first German translation (with a commentary), based on the French translation by Berthelot and Ruelle, was published by Wolters 2004: 163–179. Wolters provides only a brief summary of the content of no. 56 (p. 179), while for no. 57 he merely repeats Berthelot and Ruelle's reference to Zosimus as the author. He also cites "Mertens 1995, p. 30–33", quoted below (see note 7), but overlooks p. 201; cf. also Wolters 2006 and Wolters 2008. Most recently, see Martelli (Forthcoming). On the importance of the recipe-collection, see Bosselmann-Ruickbie 2014: 337, 339, 342, 345, 354, 356, 358, 360; Bosselmann-Ruickbie 2024: 254–256; and on its cultural background, see Merianos (Forthcoming).

5 It may suffice to cite the concise definition offered by the Merriam Webster online dictionary: "a systematic exposition or argument in writing including a methodical discussion of the facts and principles involved and conclusions reached" (https://www.merriam-webster.com/dictionary/treatise).

6 Martelli 2018: 109–112.

at Florence, *Laurentianus Pluteus* 86.16 [Diktyon 16802], known under the *siglum* **L**. It was copied by the scribe Antonius Dranganas in the year 1492, at an unknown location.[7]

Yet, apart from the existence of **L**, what is far more significant in the context of the special collection's transmission is that a recent search for alchemical manuscripts in Greece, undertaken by Martelli (Bologna), led to the discovery of two previously unconsidered alchemical manuscripts. One of these, codex *Meteora, Mone Hagiou Stephanou* 97, happens to contain another version of the special collection of recipes transmitted in **A**. This manuscript, counting 203 folia, is preserved in the Monastery of St. Stephen, which belongs to the Greek monastic federation of the Meteora region near Kalambaka in Thessaly. According to its colophon on f. 203 recto, the manuscript was written by a certain Georgius in *anno mundi* 7012, that is, 1503/1504.[8] It remains an open question whether this Georgius should be identified with the priest (*pa*[*pas*?]) Georgius, who, according to another colophon on f. 179 verso, had donated the manuscript to the Monastery of St. Stephen. Likewise, it is unclear whether the manuscript (**Met**) was copied in the Monastery of St. Stephen or elsewhere, outside the monastery.

These recipes, compiled, like those of the special collection in **A**, by one or more anonymous authors, are found on folia 182 verso – 201 verso (see overview below), under the general title ἑρμηνεία σὺν Θεῷ τῆς τιμιωτάτης καὶ πολυφήμου χρυσοχοϊκῆς / Explanation with God of the Most Praiseworthy and Illustrious Art of Goldsmithing.[9] The fact that both headings (in **A** and **Met**) are identical, apart from the often-used phrase *syn Theō* in **Met**, points to the assumption that it was perhaps the scribe Georgius (or his unknown dictator), who inserted this phrase, probably influenced by the clerical or monastic environment in which **Met** originated.

Now, looking more closely at the question of the relationship between the two versions of the special recipe collections in **A** and **Met**, it goes without saying that **A** could

7 See the most recent overview of the content of **L** in relation to this particular collection, provided by Martelli 2018: 99, 101–102, and especially 109. Martelli states that **L** contains a total of fifteen technical recipes, nine of which remain unedited. Six of these are also found in **A** as nos. 40–44, 49, while two further recipes are likewise included in **A** as nos. 56 and 57. The latter is, as noted above (note 3), a work by Zosimus of Panopolis (*Hypomnemata authentica* 9 [*De eadem aqua divina*] [ed. Mertens 1995: 30–33]; cf. **A**, ff. 289 verso–290 verso) while no. 56 is a work by the author Christianus (late sixth–seventh century), as demonstrated by Merianos 2018; on Christianus, see Merianos 2022. Although the relationship between **A** and **L** appears to be close, it has not yet been possible to determine with certainty whether **L** was copied from **A**; see Mertens 1995: XLII; Colinet 2010: CXXXIII; and Merianos 2017: 235n6.

8 See Sofianos 1986: 305–309; Martelli 2018: 99–100, on the origin of **Met**, particularly with regard to its scribe and the two colophons, mentioning the scribe – perhaps the priest – Georgius (f. 203 recto); cf. also Merianos (Forthcoming).

9 See Martelli 2018: 101–109 on the content of **Met**, compared not only to another alchemical manuscript – preserved at Elassona (Greece) in the library of the Olympiotissa monastery as MS 197 (paper, 105 folia, copied by three different copyists, partly in 1507 and partly by the priest Margarites in 1741) – but also to **L**. See also the next note. On the title's wording, see Martelli 2018: 109 and especially 112 (read by autopsy); for the detailed list of recipes under the heading in question, pp. 112–117.

not have been copied from the younger manuscript **Met**. Moreover, a rough comparison of the content and form of the two versions **A** + **Met** shows that, despite certain differences, one can clearly recognize a close relationship between them.[10] But what about the degree of similarity or difference between corresponding parts of the recipes? In answer to this question, it seemed appropriate to create an *Overview* of the two versions based on their general and particular headings (see below, "Overview by Headings of the Recipe Collection in Versions **A** and **Met**").

Regarding the number of recipes in **A** and **Met** listed in this overview, it is reasonable, firstly, to exclude the two recipes missing in **Met**, namely nos. 56 and 57, from the special collection. As shown above (in notes 3, 4, and 7), these were written by Early- or Middle-Byzantine authors. This fact suggests that their inclusion in the anonymously compiled special collection under examination should be sought "outside" its original framework. Secondly, it seems reasonable to regard recipe no. 39m* on *smardōma* (in **Met** only) as a later addition to, or broadening of, no. 39, which refers at its end to *smardos* (in **A/Met**).[11] Nevertheless, one must recognize that this recipe (39m) is undoubtedly part of the special collection (version **Met**).

Hence, we can conclude that version **A** consists of 56 recipes (nos. 1–55 + 14b), while version **Met** comprises 55 (nos. 1–5 and 8–55 + 14b + 39m*), as it lacks recipes nos. 6 and 7.[12] Their headings are, in a few cases, identical; otherwise, they are more or less similar. Not only for this reason, but also in view of the form and content of both versions **A** and **Met**, it is reasonable to assume that these versions can, to a defined extent, be traced back to a common textual template.

Incidentally, Martelli observes that, in contrast to **A**, version **Met** also contains five additional technical recipes – each with a particular heading (on ff. 180 verso – 182 verso, l. 13), thus *before* the beginning of the special collection (on 182 verso, l. 13) – as well as four additional recipes *after* recipe no. 55 of our list (on ff. 201 verso, l. 8–202 verso, l. 17), two with headings and two without any heading.[13] This fact indirectly

---

10 On the content of **A**, see note 2 above; on the content of **Met**, see Martelli 2018: 101–117, especially 102–104, 106–107. More important, regarding the recipes of the particular collection, is his comparison of the corresponding parts in **A** and **Met** on 112–117. Also noteworthy is that Martelli not only edits (on 112–113 and 115) the Greek texts of the recipes no. 3 and no. 30 (29) in the version of **Met** (with English translation), but also correctly observes – on 113 (12) and 114 (13) – that recipe no. 14 in both **A** and **Met** in fact consists of two distinct parts, which he labels nos. 14/1 and 14/2. Hence, they are presented below, in the section "Overview by Headings of the Recipe Collection in Versions **A** and **Met**", as nos. 14a and 14b.

11 Cf. Merianos 2018 and Martelli (Forthcoming).

12 Cf. Martelli 2018: 112–117, who rightly notes on page 112 that "this recipe book [= **Met**'s version of the collection] substantially overlaps with the version edited by Berthelot–Ruelle, except for the omission of four recipes" [nos. **A** 6, 7, 56, 57] "and the inclusion of a few texts that are not in the *Parisinus*". On the other hand, Martelli also counted on page 116, under no. 39, the recipe no. 39m* in the section "Overview by Headings of the Recipe Collection in Versions **A** and **Met**", below.

13 Cf. Martelli 2018: 110–112 (nos. 1–5 [with edition and translation of the texts]) and 117 (nos. 56–57 [headings only] and 58–59).

reinforces the correctness of the conclusion that the recipes grouped under the general heading in both **A** and **Met** form a nearly congruent special collection. Thus, Martelli's statement that "the Meteora manuscript stands out as the testimony that includes the richest selection of recipes, which is opened by five recipes (not grouped under any general heading) that are not handed down in the *Parisinus* [...]",[14] could rightly be supplemented by the observation that it likewise ends with the five additional recipes mentioned above.

2 Overview by Headings of the Recipe Collection in Versions A and Met

What follows is an overview of the content and variants of the special recipe collection in versions **A** and **Met** according to the (main or specific) headings. The wording is given first in the scribe's spelling and, after the slash, in normalized orthography (according to the rules of the forthcoming edition, omitting, e.g., the *iota subscriptum*, since it appears in neither **A** nor **Met**).

Main Heading

A περὶ τῆς τιμιοτάτης καὶ πολυφήμου χρυσοχωεικῆς / περὶ τῆς τιμιωτάτης καὶ πολυφήμου χρυσοχοϊκῆς
Met ἑρμηνεία σὺν Θεῷ τῆς τημιωτάτης καὶ πολυφήμου χρυσωχωικῆς / ἑρμηνεία σὺν Θεῷ τῆς τιμιωτάτης καὶ πολυφήμου χρυσοχοϊκῆς[15]

Particular Headings

Recipe 1
A περὶ τοῦ λαγαρῆσαι τὸ χρυσίον / περὶ τοῦ λαγαρῖσαι τὸ χρυσίον
Met ἀρχ<ὴ> εἰς λαγάρισμ(α) χρυσαφήου / ἀρχὴ εἰς λαγάρισμα χρυσαφίου

2
A εἰς τὸ λαγαρῆσαι ἄργυρρον / εἰς τὸ λαγαρῖσαι ἄργυρον
Met εἰς τὸ λαγάρισμα τοῦ ασημείου / εἰς τὸ λαγάρισμα τοῦ ἀσημίου

---

14 Martelli 2018: 109.

15 In **A** and **Met**, the main heading is immediately followed by the heading of recipe 1, separated only by a colon in **A** or a comma in **Met**. Both headings are rubricated in **A** and **Met**.

3

A ἑρμηνεία τοῦ χρυσόματος / ἑρμηνεία τοῦ χρυσώματος
Met ἑρμηνεία τοῦ χρυσώματος[16]

4

A χρύσομαν ἄλλον κλαποτόν / χρύσωμαν ἄλλον κλαπωτόν
Met χρύσομα ἄλλον κλαπόντον / χρύσωμα ἄλλον κλαπωτόν

5

A ἑρμηνεία εἰς τὴν ἔγκαψην / ἑρμηνεία εἰς τὴν ἔγκαψιν
Met ἑρμη(νεία) εἰς τὴν ἔγκαψιν

6[17]

A ἑρμηνεία τοῦ σμαράγδου [σμαρδου *ante correctionem*]
Met *missing*

7

A ἑρμηνεία τοῦ σαπουνίου
Met *missing*[18]

8

A ἑρμηνεία ἑτέρου σαπουνίου
Met ἑρμη(νεία) ἑτέρ(α) τοῦ σαπου(νίου)[19]

9

A ἑρμηνεία τῆς βασιλικῆς κολλήσεως
Met † ἑρμη(νεία) · τῆς βασιληκ(ῆς) · κώλυσης · τῆς χρυσῖς / † ἑρμηνεία τῆς βασιλικῆς κόλλησης[20] τῆς χρυσῆς

---

16 Cf. Martelli 2018: 112–113 (no. 3), where the recipe from **Met** is edited and translated.
17 Since **Met** lacks recipes nos. 6 and 7, these are also absent from the corresponding list provided by Martelli 2018: 112–113; see also Martelli (Forthcoming). Hence the discrepancy in his subsequent numbering of the recipes, from his no. 6 (corresponding to no. 8 here) to his no. 38 (corresponding to no. 39).
18 See previous note.
19 The adjective ἑτέρα indicates that recipe no. 7 must also have existed in the model of **Met**.
20 Martelli 2018: 113 (no. 7) proposes to read κολλήσεως.

10 (a+b)
A περὶ τῆς βασιλικῆς κολλίσεως τῆς ἀργυρής / περὶ τῆς κολλήσεως τῆς ἀργυρῆς
Met περi τῆς βασιλικῆς κώλησις τῆς ἀργυρεῖς / περi τῆς βασιλικῆς κόλλησης[21] τῆς ἀργυρῆς

11
A ἄλλη ἑρμηνεία τῆς ἀργυροκολλίσεως / ἄλλη ἑρμηνεία τῆς ἀργυροκολλήσεως
Met ἄλλη · ἑρμη(νεία) · εἰς τὴν ἀργυροκώλυσιν / ἄλλη ἑρμηνεία εἰς τὴν ἀργυροκόλλησιν

12
A ἑτέρα κόλλισ(ι)ς ταχυτάτη ἡ ἀλαμάρσα / ἑτέρα κόλλησις ταχυτάτη ἡ ἀλαμάρσα
Met ἄλλη κώλυσις ἐγλήγορη · ἡ ἀλαμάρσα / ἄλλη κόλλησις ἐγλήγορη ἡ ἀλαμάρσα

13
A ἑρμηνεία εἰς τὸ ποιῆσαι χρώαν χρυσαφίου / ἑρμηνεία εἰς τὸ ποιῆσαι χρόαν χρυσαφίου
Met ἑρμηνεία · να ποιήσεις, χρῶαν χρυσαφίου / ἑρμηνεία να ποιήσεις χρόαν χρυσαφίου

14a
A εἰς τὸ ποιῆσαι χρόαν ἐν ἀργυρω σκεύη χρύσομαν / εἰς τὸ ποιῆσαι χρόαν ἐν ἀργυρῶ σκεύει χρύσωμαν
Met εἰς τὸ ποίσαι χρώαν · ἐν ἀργυρῶ σκεύει[22] χρυσομ(ένην) / εἰς τὸ ποιῆσαι χρόαν ἐν ἀργυρῶ σκεύει χρυσωμένην

14b[23]
A εἰς τὴν ἔγκαψην / εἰς τὴν ἔγκαψιν
Met εἰς τὴν ἔγκαψην / εἰς τὴν ἔγκαψιν

15
A εἰς τὸ γανόσαι ἄργυρον / εἰς τὸ γανῶσαι ἄργυρον
Met εἰς τὼ ληανεῖσε [*ante correctionem* λανεῖσε] ἄργυρον / εἰς τὸ ληανεῖσαι ἄργυρον

16 (a–c)
A σημεῖωσις / σημείωσις
Met σημείωσεις / σημείωσις

21 Martelli 2018: 113 (no. 8) proposes again to read κολλήσεως.
22 According to the photograph of the manuscript page I consulted, the heading's reading is entirely clear, with the exception of the four letters following σκ[....]. See also Martelli 2018: 113, whose reading is partially inaccurate.
23 Recipes 14a+b are counted as nos. 12 and 13 in Martelli 2018: 113–114; see note 17 above.

17
A μυστικόν
Met εἰς στά(λαγ)μ(α)[24] / εἰς στάλαγμα (?)

18
A περὶ τοῦ ποιῆσαι φούρμας[25]
Met να ποιήσεις φούρμας

19[26]
A περὶ χρυσογραμμίας ἕτερον
Met περὶ χρυσογραμείας / περὶ χρυσογραμμίας

20 (a+b)[27]
A διήγησις [*supra scriptum in margine addidit*] περὶ τοῦ ποιῆσαι χρυσὰ κεφάλαια ἐν βιβλίοις
Met διήγησις να ποιήσις χρυσα καίφάλ(αι)α εἰς βιβλίον / διήγησις να ποιήσεις χρυσὰ κεφάλαια εἰς βιβλίον

21
A διήγησις [*addidit in margine*] περὶ τοῦ χρυσόσαι ζώα εἰς κόπαν ἢ κλαδὴν ἢ ἄλλον ἕτερον καὶ τὸ ἄλλον ἀχρύσοτον / διήγησις περὶ τοῦ χρυσῶσαι ζῶα εἰς κόπαν ἢ κλαδὶν ἢ ἄλλον ἕτερον καὶ τὸ ἄλλον ἀχρύσωτον
Met δή ἥγησις · ὅταν θέλ(η) να χρυσόση[28] ζῶα · ἢ κούπ(αν) · ἢ κλαδὴν · ἤ ἄλλον ἕτερ(ον) · καὶ τὸ ἄλον αχρύσωτον / διήγησις ὅταν θέλη να χρυσώση ζῶα ἢ κοῦπαν ἢ κλαδὶν ἢ ἄλλον ἕτερον καὶ τὸ ἄλλον ἀχρύσωτον

22
A περὶ τῆς ἐγκαύσεως
Met περί της ἐγκαύσεως / περὶ τῆς ἐγκαύσεως

23
A εἰς τὸ χρυσόσαι ζῶα εἰς κούπαν, καὶ ὁ κάμπος να ἔναι ἄσπρος / εἰς τὸ χρυσῶσαι ζῶα εἰς κοῦπαν, καὶ ὁ κάμπος να ἔναι ἄσπρος
Met εἰς τ(ὸ) χρυσώσε ζώα εἰς κούπαν · καὶ ὁ κάμ[πος να ἔναι ἄσπ(ρος) [*partim aqua corruptus et difficile lectu*, *sed* cf. Martelli 2018: 114, no. 22] / εἰς τὸ χρυσῶσαι ζῶα εἰς κοῦπαν · καὶ ὁ κάμπος να ἔναι ἄσπρος

24 On the difficult reading of this passage, see Martelli 2018: 114n63.
25 On *phourmai* (i.e., molds for the imitation of coinage), see Merianos 2024: 147, 152.
26 Cf. also the edition by Schreiner and Oltrogge 2011: 67 (no. 58).
27 Cf. also the edition by Schreiner and Oltrogge 2011: 58–59 (no. 44); the word διήγησις has been overlooked by Schreiner.
28 Martelli 2018: 114 (no. 20) proposes a different reading of this passage.

24
A εἰς τὴν χρυσοκόλλησιν
Met εἰς τὴν χρυσὴν τὴν κώλυσιν / εἰς τὴν χρυσὴν τὴν κόλλησιν [*vel* κόλλησην]

25
A περὶ τοῦ χρυσώσαι χαλκὸν με τὸν ἄσημον / περὶ τοῦ χρυσῶσαι χαλκὸν με τὸν ἄσημον
Met περι του χρυσώσαι χαλκὸν · μετα ασήμ(ην) / περὶ τοῦ χρυσῶσαι χαλκὸν μετὰ ἀσήμιν

26
A περὶ τοῦ χρυσόματος τοῦ ἀέρος τοῦ μαλάγματος / περὶ τοῦ χρυσώματος τοῦ ἀέρος τοῦ μαλάγματος
Met εἰς τ(ὸ) χρύσωμαν τοῦ ἀἕρος τοῦ μαλάγματος / εἰς τὸ χρύσωμαν τοῦ ἀέρος τοῦ μαλάγματος

27
A κόλλισις ἀν κοπύρινη /κόλλησις, ἂν κοπῇ, ῥινεῖ
Met κόλυσις ἐὰν κοπή ρινοί / κόλλησις, ἐὰν κοπῇ, ῥινεῖ

28
A ὅταν χρυσώνης ἄσημον καὶ οὐδὲν ποιάνη / ὅταν χρυσώνης ἄσημον καὶ οὐδὲν πιάνη
Met ὅταν χρυσώνης ἀσήμην · καὶ οὐδὲν πιάνει / ὅταν χρυσώνης ἀσήμιν καὶ οὐδὲν πιάνει [*vel* πιάνη]

29
A εἰς τὸ χρυσόσε ζῶα εἰς κάμπον κούμπας καὶ ὁ κάμπος ἕναι ἀχρύσοτος / εἰς τὸ χρυσῶσαι ζῶα εἰς κάμπον κούμπας καὶ ὁ κάμπος ἕναι ἀχρύσωτος
Met εἰς τ(ὸ) χρυσώσε ζώα εἰς κούπ(αν) · καὶ ὁ κάμπος ἀχρύσωτος / εἰς τὸ χρυσῶσαι ζῶα εἰς κοῦπαν καὶ ὁ κάμπος ἀχρύσωτος

30
A περὶ τοῦ ποιῆσαι χρόαν ὡραιότατην εἰς ἄσημον χρυσομένον / περὶ τοῦ ποιῆσαι χρόαν ὡραιότατην εἰς ἄσημον χρυσωμένον[29]
Met εἰς το δώσαι χρώαν εὔμορφην · εἰς ἀσήμην χρυσωμένον / εἰς τὸ δῶσαι χρόαν εὔμορφην εἰς ἀσήμιν χρυσωμένον[30]

31
A ὅταν σκάζει τὸ ἀσήμην / ὅταν σκάζει τὸ ἀσήμιν
Met ὅταν σκάζει τὸ ἀσήμην / ὅταν σκάζει τὸ ἀσήμιν

---

29 For the translation of the recipe as preserved in **A**, see Merianos 2025: 602.
30 For the edition and translation of the recipe as preserved in **Met**, see Martelli 2018: 115 (no. 29).

32
A εἰς κόλλισιν τοῦ σμάρδου / εἰς κόλλησιν τοῦ σμάρδου[31]
Met Εἰς τ(ὴν) κώλυσιν τοῦ σμάρδου / εἰς τὴν κόλλησιν τοῦ σμάρδου

33

A περὶ τοῦ ποιῆσαι σήρμαν κούφιον / περὶ τοῦ ποιῆσαι σύρμαν κούφιον
Met εις τ(ὸ) ποιήσαι σήρμ(αν) κούφιον / εἰς τὸ ποιῆσαι σύρμαν κούφιον

34
A περὶ τοῦ ποιῆσαι ἔγκαυσιν
Met να κάμις ἔγκαυσιν / να κάμεις (*vel* κάμης) ἔγκαυσιν

35
A περὶ τοῦ ποιῆσαι χρόαν εὔμορφην εἰς χρυσομένον ἀσήμην / περὶ τοῦ ποιῆσαι χρόαν εὔμορφην εἰς χρυσωμένον ἀσήμιν
Met εις τ(ὸ) ποιήσαι χρώαν · ἔμορφον · εἰς χρυσωμέν(ον) ἀσήμην / εἰς τὸ ποιῆσαι χρόαν ἔμορφον εἰς χρυσωμένον ἀσήμιν

36
A περὶ τοῦ ποιῆσαι κόλλησιν κακαβίων, καὶ εἰς σουληνάρια λουτρόν / περὶ τοῦ ποιῆσαι κόλλησιν κακαβίων, καὶ εἰς σουληνάρια λουτρῶν
Met εις τ(ο) ποιήσαι κώλσιν των κακαβίων · εἰς λουτρὸν · καὶ εἰς λινάρια / εἰς τὸ ποιῆσαι κόλλησιν τῶν κακαβίων εἰς λουτρὸν καὶ εἰς λινάρια[32]

37
A εἰς τὸ λαμπρήναι μαργαριτάριν / εἰς τὸ λαμπρῦναι μαργαριτάριν
Met εις τ(ο) λαμπρύναι μαργαρίτην /εἰς τὸ λαμπρῦναι μαργαρίτην

38
A ἄλλον
Met ἕτερον εἰς αὐτό

31 The heading of no. **A** 32 appears directly – without any rubrication or emphasis – after the final word of no. 31, whereas in **Met** it is properly rubricated.
32 The scribe of **Met** misunderstood the final word σουληνάρια ("pipes") and wrote λινάρια ("flax") instead.

39 (a+b)
A περὶ τῶν συρμάτων τῆς χρυσοχωηκῆς τέχνης / περὶ τῶν συρμάτων τῆς χρυσοχοϊκῆς τέχνης
Met ἐξήγησης τ(ῆς) τιμιωτάτ(ης) καὶ ψυλωτ(ά)τ(ης) · καὶ ἀχειροτ(ά)τ(ης) τέχνης τῶν συρμάτ(ων) · τ(ῆς) χρυσοχοϊκῆς τέχνης / ἐξήγησις τῆς τιμιωτάτης καὶ ὑψηλοτάτης[33] καὶ ἀχειροτάτης τέχνης τῶν συρμάτων · τῆς χρυσοχοϊκῆς τέχνης

39m*
A *missing*
Met σμάρδωμαν · ἄλλον, εἰς ἀχαμν(ὸν) ἀσήμην, ὅ λέγεται φαλτζηκῆ · κόλλησις παναλία (?: proposed by Nikolaos Vryzidis [e-mail, 10 September 2025]) · καὶ ὅ διπλῶς πᾶσα λόγος[34] / σμάρδωμαν ἄλλον εἰς ἀχαμνὸν ἀσήμιν, ὃ λέγεται φαλτζηκὴ κόλλησις παναλία (?), καὶ ὃ διπλὸς πᾶσα λόγος

40 (a+b)
A ἄλλη μέθοδος μυστικῆ / ἄλλη μέθοδος μυστική
Met ἑτέρα δουλία

41
A ἕτερον ὕδωρ θείον / ἕτερον ὕδωρ θεῖον
Met ἕτε(ρον) ὕδωρ θεῖ(ον) 

42
A ὕδωρ ἵνα ἐκβάλης χρύσομαν ἀπὸ ἀσήμην / ὕδωρ, ἵνα ἐκβάλης χρύσωμαν ἀπὸ ἀσήμιν
Met ἕτερον θεῖον (*per signum sulphuris*) ὕδωρ να ξεβάλ(ης) τὸ χρύσωμα ἀπὸ το ἀσήμ(ιν) / ἕτερον θεῖον ὕδωρ να ξεβάλης τὸ χρύσωμα ἀπὸ τὸ ἀσήμιν

43
A ἕτερον ὡς αὔτως / ἕτερον ὡσαύτως
Met ἕτερον εἰς αὐτό

44
A Εἰς λαγάρισμαν χρυσαφίου
Met λαγάρισμ(α) · χρυσαφήου / λαγάρισμα χρυσαφίου

---

33 Convincing reading proposed by Martelli 2018: 115 (no. 38), in contrast to his following proposal ἀχειροθέτου; see note 49 below.
34 Martelli 2018: 116 (no. 39) overlooked the following part of the rubricated heading: εἰς ἀχαμν(ὸν) – λόγος.

45
A ἕτερον ὅμοιον εἰς ἄσημον
Met ὅμ(οιον) · εἰς ἀσήμην χρυσῶ / ὅμοιον εἰς ἀσήμιν χρυσω<μένον>[35]

46
A ὅταν σκάζει τὸ ἀσήμην ἢ τὸ χρυσάφην / ὅταν σκάζει τὸ ἀσήμιν ἢ τὸ χρυσάφιν
Met ὅταν σκάζη τὸ ἀσήμ · εἲ τὸ χρυσάφην / ὅταν σκάζη τὸ ἀσήμ<ιν> ἢ τὸ χρυσάφιν

47
A Τό που σταμάτισις ὑδραργύρου / Τὸ ποῦ σταμάτισις ὑδραργύρου
Met τὸ πῶς να σταματήσης δυἄργυρον / τὸ πῶς να σταματίσης διάργυρον

48
A ἄλλον
Met ἄλλον

49[36]
A περὶ τοῦ ποιῆσαι γάμματα χρυσά
Met περὶ τοῦ ποιῆσαι γάμματα χρυσά

50
A περὶ τοῦ ποιῆσαι ὡραιόχαλκον ὥσπερ χρυσόν
Met εἰς τ(ὸ) ποιῆσαι · ὥραιον χαλκον · ὥσπερ χρυσόν (*per signum aurei*) / εἰς τὸ ποιῆσαι ὡραιόχαλκον ὥσπερ χρυσόν

51
A περὶ τοῦ σαπουνίου
Met ἕτερόν εἰΤς σαπούνιον / ἕτερον εἰς τὸ σαπούνιον

52
A ἕτερον
Met ἄλλον

53
A ὁ ὕελος
Met τὸ γεαλή

35 Cf. Martelli 2018: 116 (no. 45) for the corrected reading χρυσώ<μενον>, but read χρυσωμένον.
36 Cf. also the edition by Schreiner and Oltrogge 2011: 50–51 (no. 31).

54 (a+b)
A περὶ τοῦ λευκάναι κασίτηρον (*per symbolum*) / περὶ τοῦ λευκᾶναι κασίτηρον
Met εἰς το λευκά(ναι) χαλκ(όν) / εἰς τὸ λευκᾶναι χαλκόν

55
A περὶ τοῦ ποιῆσαι χαλκὸν (*per symbolum*) ὥσπερ χρυσόν (*per symbolum*)
Met εἰς τ(ο) ποιῆσαι · χαλκὸν (*per symbolum, partim deletum*) ὥσπερ χρυσόν (*per symbolum*) / εἰς τὸ ποιῆσαι χαλκὸν ὥσπερ χρυσόν

56
A ὕδωρ πιστῆς οἰκονομίας (text, in part written by Christianus, see note 7).
Met *missing*

57
A καὶ ἄλλος φησίν (= Zosimus of Panopolis, *Hypomnemata authentica* 9 (*De eadem aqua divina*) [ed. Mertens 1995]); see notes 3, 4, and 7 above.
Met *missing*

3 General Observations

In what follows, the first subsection examines the external layout and overall structure of the collection, while the second turns to a different level of analysis, presenting formal and philological observations that clarify its linguistic characteristics and textual features.

On the External Layout and Structure of the Collection

1) The main heading of the special collection, as well as the subsequent individual headings for each recipe, were almost always rubricated in **A**, and without exception in **Met**. In the case of **A**, however, there are some exceptions: in the headings of nos. 14b, 16, 28 and 31, only the first letter was rubricated. Additionally, the heading of no. 14b ends with a comma, while that of no. 16 is not distinguished from the rest of the text at all. In the case of no. 32, however, the heading – written in normal ink – is not immediately recognizable as such, because it directly follows the end of the text of no. 31; it is identifiable as a heading only by the colon at its end.

2) In both versions **A** and **Met**, the recipes are strung together without any commentary but separated from each other by the punctuation sign (: -), a "colon with dash".[37] The scribe(s) placed this mark after the end of each recipe, after which the heading of the next recipe follows, ending with a colon.

3) The reason for the absence of recipes nos. 6 and 7 in **Met** is partly unclear. However, their omission could be seen in the case of no. 6 in that its content deals with another aspect of *enkapsis*, which already formed the topic of no. 5. In the case of nos. 7 and 8, however, their headings are almost identical, a fact that might have suggested from the outset that no. 7 need not be included.

4) Regarding the additional recipe 39m*, preserved only in **Met**,[38] it was clearly inserted later as an *additamentum* after no. 39, because it is, in terms of content (*smardōma*), at least partly connected to the latter.

5) The special marking of recipe no. 14 as 14 (a+b) requires a brief explanation, as it indicates that the recipe visibly consists of two parts, each with its own heading in both manuscripts, **A** and **Met**. It was Martelli, who rightly observed this,[39] whereas Berthelot and Ruelle overlooked it in their edition.[40] Lebègue, in his description of **A**,[41] likewise failed to notice the punctuation [:-] of no. 14a following the word χρυσόματος (*sic*) on f. 282 verso, l. 10, as well as the letter ε of the following εἰς (*ibid.*, l. 11), which is written in red and thus clearly marks the beginning of no. 14b's heading: εἰς τὴν ἔγκαψην (*sic*). All the other letters of this heading, however, were mistakenly written in black. That this occurred in error is confirmed by the fact that, in **Met**, the entire heading of 14b – like all other headings in that manuscript – is rubricated, as already mentioned above. If, in the case of recipes 14a–14b, this inherently illogical numbering is nevertheless retained in the forthcoming Mainz edition, despite the fact that we are dealing with two distinct recipes, it is solely in order to preserve consistency with the numbering in Berthelot and Ruelle.

6) In the case of recipes nos. 10, 16, 20, 39, 40, and 54, the subdivisions (a+b or a–c), indicated in brackets, have no basis in the manuscripts. Rather, they were introduced by Wolters, based solely on the content of the recipes.[42] Accordingly, his additional subdivisions are not taken into account here or in the forthcoming Mainz edition.

---

37 Tocci 2011: 63.
38 See Martelli 2018: 116 (no. 39).
39 See note 10 above.
40 Berthelot and Ruelle 1887–1888: 2.325, 23–25.
41 Lebègue 1924: 56.
42 Wolters 2004: 163 (explanation of his subdivision), 173–174, and 177–178.

Formal and Philological Observations

The text of the recipes (including their headings) contained in **Met** (see nos. 12, 20, 21, 34, 42, and 47) is written in somewhat more vernacular Greek compared to those in **A**.

At first glance, the affinity between the two versions of the special collection seems to be relatively close. Apart from the nearly identical main heading, among the remaining 55 individual recipe headings, found in versions **A** and **Met**, 11 are identical (= **Id**), 26 nearly identical (= **Id n**), and 17 more or less different (= **D**).[43] Two headings present in version **A**[44] are missing from version **Met**, while one heading in the latter (39m) is absent from **A**. However, as we shall see, it is particularly interesting to consider the subtle differences in the wording of the six headings of nos. 1, 20, 21, 39, 40, and 42.

Concerning the differences (**D**) in the headings of the recipes nos. **A/Met** 1, 20, 21, 39, 40, and 42:

1) With regard to the main heading and the specific title of recipe no. 1: although in **A** no distinction other than a colon is made between the two headings, the distinction becomes obvious in **Met**, f. 182 verso, l. 3 (from below), due to its wording, [...] χρυσωχωϊκῆς, ἀρχ(ὴ) εἰς [...]. Here, the comma, in combination with the word *archē*, marks the following text as the beginning of recipe no. 1. Interestingly, the word *archē* also appears once in version **A**, specifically in recipe 27, immediately after the heading, introducing the main text of the recipe.

2) In the case of recipe no. 20, the difference at first glance appears to be merely linguistic. A closer examination of the manuscripts, however, reveals that in **A**, f. 283 verso,[45] the word διήγησις was added, probably by the same scribe, in black ink, *supra lineam et in margine superiore*, above the initial heading written in red ink: περὶ τοῦ ποιῆσαι χρυσὰ κεφάλαια ἐν βιβλίοις. Thus, the addition apparently intended to correct and complete the heading, since **A**'s scribe, Pelecanus (or his dictator), noticed during a final proofreading that the word διήγησις had originally been overlooked. In **Met**, by contrast, the complete heading διήγησις να ποιήσις etc. was written correctly in red ink from the outset on f. 189 verso. This observation leads to the conclusion that, in this case, the model for **A** and **Met** seems to have had, at least regarding the word *diēgēsis*, the corresponding formulation.

Yet, theoretically, one could, relying on this single observation, entertain the possibility that **Met** was copied from **A**. As will be shown below, however, this must certainly

43 Group **D** consists of the headings of nos. 1, 11, 15, 17, 20, 21, 26, 29, 30, 33, 34, 36, 38, 39, 40, 51, 54, group **Id** of nos. 3, 5, 9, 14b, 16, 22, 31, 37, 41, 46, 48, and group **Id n** of nos. 2, 4, 8, 10, 12, 13, 14a, 18, 19, 23, 24, 25, 27, 28, 32, 35, 42, 43, 44, 45, 47, 49, 50, 52, 53, 55.

44 Namely the headings of nos. 6 and 7.

45 In his overview on the headings of the recipes in **Met** and **A**, Martelli (2018: 114) overlooks the fact that the word διήγησις was added by **A**'s scribe in the upper margin, an addition noted not only by Berthelot and Ruelle (1887–1888: 2.327, *apparatus criticus* to line 7), but also, and more clearly, by Lebègue (1924: 57), referring to f. 283 verso.

be excluded on the grounds of further philological peculiarities in the transmission of the recipes of **A** and **Met**.

3) The same seems to have occurred with the heading of no. 21 as in the case of no. 20: διήγησις περὶ τοῦ χρυσῶσαι ζῷα εἰς κόπαν ἢ κλαδὶν ἢ ἄλλον ἕτερον καὶ τὸ ἄλλον ἀχρύσωτον. Here, again, in **A**, f. 283 verso, the word διήγησις was added in black ink on the left side *in margine.*[46] Unlike the marginal note in no. 20,[47] however, it is not accompanied by any additional signs indicating its function as a correction. Thus, it remains uncertain whether the addition διήγησις should be understood as a hint that the heading, as in no. 20, should be corrected to begin with this word. Otherwise, one might interpret it as nothing more than a mere pointer, intended to draw attention to noteworthy content. Viewed in this way, the marginal note διήγησις would serve the same function as the note χρυσόκολλα (or: χρυσόκολλαν), written *in margine* of the heading of no. 9 on f. 281 verso, and in the same red ink as the heading – thus serving as a hint to the content of the recipe.[48]

In **Met**, however, the heading of no. 21 (διήγησις περὶ τοῦ χρυσῶσαι ζῷα εἰς κόπαν ἢ κλαδὶν ἢ ἄλλον ἕτερον καὶ τὸ ἄλλον) was written from the outset in this form, in red ink, on f. 190 verso. Hence, also **Met**'s model clearly already included the word διήγησις in its heading. One might therefore interpret the marginal addition of διήγησις in **A**, no. 21, as another correction comparable to that in no. 20, although it is not clearly marked as such.

4) As for recipe no. 39, it is remarkable that both headings, the one in **A** (περὶ τῶν συρμάτων τῆς χρυσοχοϊκῆς τέχνης) and the one in **Met** (ἐξήγησις τῆς τιμιωτάτης καὶ ὑψηλοτάτης καὶ ἀχειροτάτης τέχνης τῶν συρμάτων τῆς χρυσοχοϊκῆς τέχνης),[49] not only refer to the content of the recipe but also clearly allude to the main heading of "our" special collection. However, both headings also differ considerably in this, since **Met**'s

---

46 In his overview on the headings of the recipes in **Met** and **A**, Martelli (2018: 114) also overlooks the fact that the word διήγησις was added by **A**'s scribe in the left margin, though this had already been noted by Berthelot and Ruelle (1887–18882: 2.327, *apparatus criticus* to line 26), and, less clearly by Lebègue (1924: 57), referring to f. 283 verso: "[…] (*E regione tituli* διήγησις *nigro atrameno*)".

47 In **A**, further marginal notes indicating corrections are also found, specifically in recipes nos. 5 (f. 281 recto) and 41 (f. 286 verso); in each case, the notes are clearly marked by additional signs denoting corrections. On corrections – whether carried out or not – in **A** (or **Met**) by Pelecanus, see Merianos 2018 (concerning *smardos/smaragdos* or *smardōma*). See also Merianos (Forthcoming) on the meaning of the verb γανόω in recipe no. 15 (and the different wording of the recipe in **A** and **Met**).

48 See Berthelot and Ruelle 1887–1888: 2.324, *apparatus criticus* to line 15, where it reads χρυσόκολλαν; and Lebègue 1924: 55, note to f. 281 verso, where, in connection with the heading of no. 9, he wrote: "(*In margine sinistra, litteris rubris*: Χρυσόκομαν)", clearly an erroneous reading.

49 Martelli 2018: 115 (no. 38) partly erroneously wrote: "ἐξήγησης καὶ τῆς τιμιωτάτης καὶ (ὑ)ψηλωτάτης [ψυ-MS] καὶ ἀχειροθέτου (? αχειρο$^{\tau\tau}$ MS; perhaps ἀχειροτμήτου or -ποιήτου) τέχνης τῶν συρμάτων τῆς χρυσοχοϊκῆς τέχνης, "Explanation of the most precious and sublime and divine (i.e., not belonging to human beings) art of (making metallic) wires for the art of goldsmiths". Yet, the reading ἀχειροθέτου, or even ἀχειροτμήτου or ἀχειροποιήτου, instead of ἀχειροτ[ά]τ[ης] must be excluded.

heading, through the choice of words in the passage mentioned above, emphasises the special value of producing wires in the context of the art of goldsmithing. This interpretation is based on the very helpful advice I have received from Professor Erich Trapp[50] on the meaning of the adjective ἀχείρωτος. The meanings offered in major dictionaries – "untamed", "unconquered",[51] "invincible", "impregnable", also "not to be overcome/mastered", or "unaffected",[52] and *unbearbeitet, unbehandelt*[53] – all proved inapplicable in the context mentioned above. Professor Trapp replied kindly to my request,[54] and drawing on research in the *Thesaurus graecae linguae* (*Tgl*) and the *Thesaurus Linguae Graecae* (*TLG*),[55] proposed an additional meaning: ἀχείρωτον = ἄμωμον (faultless, excellent, splendid, perfect). Accordingly, the heading in **Met**'s recipe no. 39 may be translated: "Explanation of the most praiseworthy and sublime and splendid art of [making metallic] wires [in the context] of the art of goldsmithing". Be this as it may, the remarkable, though not contradictory, difference between the two headings of no. 39 reinforces the impression that, in relation to our special collection, **A** and **Met** derive from only seemingly different models, as the variation is not substantial. This impression is further supported by the headings of recipes nos. 40 and 42 – no. 40: **A** ἄλλη μέθοδος μυστική / **Met** ἑτέρα δουλία; no. 42: **A** ὕδωρ, ἵνα ἐκβάλῃς [...] / **Met** ἕτερον θεῖον ὕδωρ [...]. These differences, too, appear superficial. In the case of no. 42, the omission of the sign for *theion* in **A** becomes apparent from the following context.

4 A Preliminary *recensio*

The analysis that follows is an attempt at a preliminary *recensio*, aimed at clarifying the relationship between **A** and **Met** in the transmission of "our" special collection.

Seemingly or Real Common Mistakes in A and Met

Both versions mention the Lord's prayer four times, associating it with a specific time period – rendered as (ὥρα) πάτερ ἡμοῦ in **A** or πατερημοῦ in **Met** – in recipes nos. 3, 14a, 30, 35 and 38. The seemingly irregular or "incorrect" genitive ending -οῦ, instead of the expected πάτερ ἡμῶν / πατερημῶν, suggests a derivation from the genitive of the noun

---

50 E-mail from 22 March 2022; see note 54.
51 Liddell–Scott–Jones: s.v. ἀχείρωτος.
52 Lampe 1961: s.v. ἀχείρωτος.
53 *Lexikon zur byzantinischen Gräzität*: s.v. ἀχείρωτος.
54 *"ein schwieriger Fall, zusätzlich fand ich nur im Tgl und danach im TLG: acheirwton: amwmon (tadellos). Könnte das helfen?"*
55 *Tgl*: Estienne et al. 1831–1865; *TLG*: https://stephanus.tlg.uci.edu/.

τὸ πατερημόν. This form is, in fact, attested in Kriaras' dictionary.[56] We are thus dealing with an admittedly less frequent, but nonetheless "correct", wording.

In recipes nos. 4, 8, 9, 10, 25, the adjective ὑψηλὸς appears instead of ψιλὸς in contexts where the (anonymous) author(s) intended to express the meaning "fine" or "light" (as opposed to "difficult" or "heavy"). Though both words are phonetically identical – especially given that, in this case, the initial vowel (as is often the case) remains unvoiced – the use of ὑψηλὸς seems less a matter of grammatical ignorance than a possible attempt at hypercorrect usage by the author(s).

Gaps and Mistakes

Gaps occur several times in **A** and **Met**, as shown by comparison between the two versions, with **Met** often supplying the missing text, the correct word or the correct numeral.

As for the gaps, mistakes, and omissions in **A**, the first error, caused by *saut du même au même*, appears already in recipe no. 2, lines 4–5. There, the nonsensical phrase (in the plural) ὥστε να ψυχρανθοῦν ἀφ' ἑαυτοῖς βράζωντες, instead of the correct ὥστε να ψυχρανθῇ βράζοντα found in **Met**, is clearly due to the scribe's eye being misled by the *homoeoteleuton* in lines 3–4. More conspicuous omissions then follow in no. 3, namely that of the number ι΄ (10) after the word λίτρας, as well as the passage καὶ σβέσον αὐτά · εἰθ'οὕτως χύσον τὸ χρυσάφιν μέσον after κάρβουνα ζωντανὰ μέσον, which are both the result of a *saut du même au même*.[57] Other gaps of this kind occur in nos. 5, 12, 18, and 39. As for errors, it may suffice to mention the form πετζόλλαν in no. 29, instead of the correct πετζόκολλαν (as in **Met**).

Similar examples of gaps and considerably more errors are found in **Met**. For instance, in no. 3, the omission of the passage ἆρον αὐτὸ καὶ τρίψον μετὰ βρούτζου χοιρείαν · καὶ πάλιν βαλὼν αὐτὸ ἐν τῇ πύρα following χρυσωτήριον · καὶ βαλὼν αὐτὸ ἐν τῇ πύρα, and in no. 4, where the passage πύρωσον αὐτὸ εἰς τὸν σύρτην εἰς πᾶσαν φορὰν β΄ ἢ γ΄ · εἰθ' οὕτως is missing after ἑπταπλασίονα· εἰθ' οὕτως, are likewise attributable to *saut du même au même*. Similar cases occur in nos. 12, 18, and 39.

A particularly curious case is found in no. 41, where **Met** not only lacks the passage τὸ κινούμενον ἐκ τῆς ῥυτῆς τοῦ ἰαμβίκου, but also continues as follows: ὅπου θέλης (read: θέλεις) να κκνήσει (*sic*, instead of κοκκινίσῃ, cf. κοκινήσει **A**) μὴ κοιτάση το καπούτζην (read: καπούτζιν) τοῦ λυμπικήου. Here, the inserted admonition μὴ κοιτάση,

56 Kriaras: vol. 15, s.v. πατερημόν.

57 Martelli 2018: 112n54 rightly notes: "[...] the MS *Parisinus* seems to omit only the words: καὶ σβέσον αὐτά· εἰθ'οὕτως χύσον τὸ χρυσάφιν", but he overlooks the final μέσον in this passage, which provides the evidence for the *saut*.

directed at the sleepy scribe who recorded it automatically, is much more noteworthy than the mistakenly written κκνήσει.

It is hardly surprising that one also encounters those kinds of mistakes most likely caused by the scribe's mishearing of dictated text.

An example of this is found in no. 9, where the reconstructed sentence of our edition – καὶ κόλλησον μετὰ σαπουνίου καμένου μέρη β' – appears in **A** as καὶ κόλλησον μετὰ πανίου καμένου μέρη β' in **A**, but in **Met** as κόλλησον με σαπούνηον καμένου μέρη β'. Apparently, **A**'s version reflects a mishearing of μετὰ σαπουνίου, as confirmed by **Met**, despite its wording με σαπούνηον, because **Met** retains the form καμένου found in the common model of **A** and **Met**. Further examples of mistakes caused by mishearing include: ἔντωπε (**Met**) instead of ἔκτοτε (**A**) in no. 2; πετάλιν (**Met**), instead of πασπάλιν (**A**) in no. 20; κόλλα τὰ (**Met**) instead of κέρατα (**A**) in no. 39; and probably also ἐπληρώθην (**A**) instead of ἐπληρώθη (**Met**) in no. 42.

One should bear in mind, however, that some additional recipes were meanwhile copied into the model of **Met**; thus, five are placed before the general heading of the special collection, and four after the final recipe, no. 55.[58]

If one attempts a preliminary *recensio*, based on the above observations concerning the structure, form, and characteristic features of the versions **A** and **Met**, the following picture emerges.

With regard to the transmission of the special collection, both versions **A** and **Met** can be traced back to an older, now lost, ancestor (**Ψ***), which (among other texts) consisted of recipes 1–55, each with its own rubricated heading, following a rubricated main heading. From **Ψ*** – or perhaps already from a no longer extant copy – manuscript **A** (= *Parisinus graecus* 2327) was copied in Crete in 1478. **Met**, by contrast, was probably not derived directly from **Ψ***, but rather from a more recent copy, in the preparation of which the headings of the special collection's recipes were partially reformulated and their number slightly changed, as described above. This manuscript was produced in the monastery of St. Stephen in the Meteora region in 1503/1504. Given the clearly recognizable proximity of **A** and **Met** to **Ψ***, it follows that, for the forthcoming edition, the text will in principle be based on **A** as the leading manuscript from a linguistic point of view. However, erroneous or corrupted passages in both **A** and **Met** will be corrected wherever possible by cross-consulting the two manuscripts, as they are at least partially equivalent.

---

58 See Martelli 2018: 110–112 (nos. 1–5) and 117 (nos. 56–59), where, as Martelli notes, the last two recipes lack headings.

Bibliography

Sources

Collected Sources

Berthelot and Ruelle 1887–1888: Marcellin Berthelot and Charles-Émile Ruelle (ed. and Fr. transl.), *Collection des anciens alchimistes grecs*, 3 vols. Paris: G. Steinheil, 1887–1888 (reprint: Osnabrück: O. Zeller, 1967).

Schreiner and Oltrogge 2011: Peter Schreiner and Doris Oltrogge, *Byzantinische Tinten-, Tuschen- und Farbrezepte* (Österreichische Akademie der Wissenschaften, Philosophisch-Historische Klasse, Denkschriften 419; Veröffentlichungen der Kommission für Schrift und Buchwesen des Mittelalters, Reihe IV, Monographien 4). Vienna: Verlag der Österreichischen Akademie der Wissenschaften, 2011.

Authors and Works

Zosimus Panopolitanus, *Hypomnemata authentica*: Michèle Mertens (ed. and Fr. transl.), *Zosime de Panopolis, Mémoires authentiques* (Les Alchimistes grecs 4.1). Paris: Belles Lettres, 1995.

Literature

Bosselmann-Ruickbie 2014: Antje Bosselmann-Ruickbie, "Das Verhältnis der 'Schedula diversarum artium' des Theophilus Presbyter zu byzantinischen Goldschmiedearbeiten: Grenzüberschreitende Wissensverbreitung im Mittelalter", in Andreas Speer (ed.), *Zwischen Kunsthandwerk und Kunst: "Die Schedula diversarum artium"* (Miscellanea Mediaevalia 37). Berlin and Boston, MA: De Gruyter, 2014, pp. 333–368.

Bosselmann-Ruickbie et al. 2018: Antje Bosselmann-Ruickbie, Susanne Greiff, and Günter Prinzing, "[Katalog-Objekt] nr. 193", in *Byzanz & der Westen. 1000 Jahre vergessene Jahre*. Schallaburg: Schallaburg Kulturbetriebsges, 2018, p. 231.

Bosselmann-Ruickbie 2024: Antje Bosselmann-Ruickbie, "Das Forschungsprojekt zum Goldschmiedetraktat 'Über die hochgeschätzte und berühmte Goldschmiedekunst'", in Antje Bosselmann-Ruickbie, Zachary Chitwood, Johannes Pahlitzsch, and Martin Marko Vučetić (eds), *Byzanz am Rhein. Festschrift für Günter Prinzing anlässlich seines 80. Geburtstags* (Mainzer Veröffentlichungen zur Byzantinistik 18). Wiesbaden: Harrassowitz, 2024, pp. 251–261.

Colinet 2010: Andrée Colinet, *Recettes alchimiques (Par. gr. 2419; Holkhamicus 109) – Cosmas le Hiéromoine, Chrysopée* (Les Alchimistes grecs 11). Paris: Belles Lettres, 2010.

Estienne et al. 1831–1865: Henri Estienne et al., *Thesaurus graecae linguae*, 9 vols. Paris: A. Firmin-Didot, 1831–1865.

Kriaras: Emmanuel Kriaras et al., *Λεξικό τῆς Μεσαιωνικῆς Ἑλληνικῆς Δημώδους Γραμματείας, 1100–1669*, 21 vols. Thessaloniki: Kentro Hellenikes Glossas, 1968–.

Lampe 1961: G. W. H. Lampe, *A Patristic Greek Lexicon*. Oxford: Clarendon Press, 1961.

Lebègue 1924: Henri Lebègue, *Catalogue des manuscrits alchimiques grecs*, vol. 1: *Les* Parisini. Brussels: M. Lamertin, 1924.

Lexikon zur byzantinischen Gräzität: Erich Trapp et al., *Lexikon zur byzantinischen Gräzität, besonders des 9.–12. Jahrhunderts* (Österreichische Akademie der Wissenschaften, Philosophisch-Historische Klasse, Denkschriften 238, 250, 276, 293, 326, 352, 417, 494; Veröffentlichungen der Kommission für Byzantinistik 6.1–8). Vienna: Verlag der Österreichischen Akademie der Wissenschaften, 1994–2017.

Liddell–Scott–Jones: Henry George Liddell, Robert Scott, and Henry Stuart Jones, *A Greek-English Lexicon. Ninth Edition with a Revised Supplement*. Oxford: Clarendon Press, 1996.

Martelli 2018: Matteo Martelli, "Byzantine Alchemy in Two Recently Discovered Manuscripts in Saint Stephen's (Meteora) and Olympiotissa (Elassona) Monasteries", in Efthymios Nicolaidis (ed.), *Greek Alchemy from Late Antiquity to Early Modernity* (De Diversis Artibus 104). Turnhout: Brepols, 2018, pp. 99–118.

Martelli (Forthcoming): Matteo Martelli, "Technical Recipes in Byzantine Alchemical Compendia: An Overview on the Manuscript Tradition", in Stefan Albrecht, Antje Bosselmann-Ruickbie, Susanne Greiff, Vassiliki Papadopoulou, Stephan Patscher, Günter Prinzing, and Roland Schwab (eds), *"Über die höchst verehrungswürdige und berühmte Goldschmiedekunst" – Eine byzantinische Sammlung von Werkstattrezepten für Goldschmiede: Edition, Übersetzung, Kommentar* (Byzanz zwischen Orient und Okzident 30). Mainz: LEIZA (Forthcoming).

Merianos 2017: Gerasimos Merianos, "Alchemy", in Anthony Kaldellis and Niketas Siniossoglou (eds), *The Cambridge Intellectual History of Byzantium*. Cambridge: Cambridge University Press, 2017, pp. 234–251.

Merianos 2018: Gerasimos Merianos, "Looking at the Recipe Collection on Goldsmithing in the Cultural Context of *Par. gr.* 2327" (unpublished paper presented at the workshop *Testing and Translating Textual Evidence: A Greek Medieval Treatise on Goldsmithing*, Mainz, 12–15 March 2018).

Merianos 2022: Gerasimos Merianos, "The Christianity of the Philosopher Christianos. Ethics and Mathematics in Alchemical Methodology", *ARYS* 20 (2022), Special Issue: *Chemical Arts and Religion in Antiquity*, pp. 271–322.

Merianos 2024: Gerasimos Merianos, "Insights on Alchemy, Deception, and Artisanal Knowledge in Byzantium", in Antje Bosselmann-Ruickbie, Zachary Chitwood, Johannes Pahlitzsch, and Martin Marko Vučetić (eds), *Byzanz am Rhein. Festschrift für Günter Prinzing anlässlich seines 80. Geburtstags* (Mainzer Veröffentlichungen zur Byzantinistik 18). Wiesbaden: Harrassowitz, 2024, pp. 145–164.

Merianos 2025: Gerasimos Merianos, "Byproducts of Winemaking in Greek Alchemy: *Pheklē* and *Trygia*", in Steven M. Oberhelman (ed.), *Manuscripts, Plants, Remedies in the Mediterranean Traditions: Studies across Disciplines for Alain Touwaide*, vol. 3: *Remedies: Pharmacy, Drugs, Archaeology, Tradition* (Medical Traditions 6.3). Berlin and Boston, MA: De Gruyter, 2025, pp. 587–610.

Merianos (Forthcoming): Gerasimos Merianos, "The Recipe Collection on Goldsmithing in the Cultural Context of *Parisinus graecus* 2327", in Stefan Albrecht, Antje Bosselmann-Ruickbie, Susanne Greiff, Vassiliki Papadopoulou, Stephan Patscher, Günter Prinzing, and Roland Schwab (eds), *"Über die höchst verehrungswürdige und berühmte Goldschmiedekunst" – Eine byzantinische Sammlung von Werkstattrezepten für Goldschmiede: Edition, Übersetzung, Kommentar* (Byzanz zwischen Orient und Okzident 30). Mainz: LEIZA (Forthcoming).

Mertens 1995: Michèle Mertens, *Zosime de Panopolis, Mémoires authentiques* (Les Alchimistes grecs 4.1). Paris: Belles Lettres, 1995.

Sofianos 1986: Demetrios Z. Sofianos, *Τὰ χειρόγραφα τῶν Μετεώρων*, vol. 3: *Τὰ χειρόγραφα τῆς Μονῆς Ἁγίου Στεφάνου*. Athens: Academy of Athens, Research Centre for Medieval and Modern Hellenism, 1986.

Tocci 2011: Raimondo Tocci, "Kopist vs. Verfasser: zur Interpunktion im Codex *Vatic. gr.* 1889 (13. Jh.)", *Parekbolai* 1 (2011), pp. 61–86.

Wolters 2004: Jochem Wolters, "Der Traktat 'Über die edle und hochberühmte Goldschmiedekunst' (11. Jahrhundert)", *Das Münster. Zeitschrift für christliche Kunst und Kunstwissenschaft* 57 (2004), pp. 162–181.

Wolters 2006: Jochem Wolters, "Der byzantinische Traktat 'Über die edle und hochberühmte Goldschmiedekunst' aus dem 11. Jahrhundert", in Christoph Stiegemann and Hiltrud Westermann-Angerhausen (eds), *Schatzkunst am Anfang der Romanik. Der Paderborner Tragaltar und sein Umkreis*. Munich: Hirmer, 2006, pp. 222–243.

Wolters 2008: Jochem Wolters, "Written Sources on the History of Goldsmithing Techniques from the Beginnings to the End of the 12th Century", *Jewellery Studies* 11 (2008), pp. 46–66.

Part II: **Cultural Crossroads: Byzantine Alchemy between East and West**

Maria Mavroudi

Alchemy, Philosophy, and Related Disciplines between the Greek and the Islamicate World, Seventh–Ninth Centuries: Towards a New Narrative on the Medieval Transmission of Philosophy and Science

Between the sixteenth and the twentieth centuries, the transmission of philosophy and science from antiquity into modernity has been synthesized into a grand narrative, as follows:[1] the sciences were born in the ancient Near East and Egypt, were endowed with a theoretical background through the development of ancient Greek philosophy, and were cultivated in the Graeco-Roman world until, at the very latest, the late sixth or early seventh century. After the Muslim conquests of the seventh century, Latin and Greek culture is presumed to have entered a period of intellectual (and other) decline that lasted into the ninth century. This period is generally labeled as the "dark centuries". Philosophy and the sciences would have disappeared, had they not been picked up by the Arabs, especially through the Greek-to-Arabic translation movement that took place in ninth- and tenth-century Baghdad. In the Muslim world, the Graeco-Roman heritage was enriched and developed further. The result is thought to have greatly contributed to what came to be called "classical Islamic civilization". When the Muslim world started declining towards the end of the eleventh and the twelfth centuries, philosophy and the sciences were presumably saved by the Latin-speaking world through an Arabic-to-Latin translation movement, which transferred the knowledge of the ancient Greeks into Latin via Arabic intermediaries. The next stride for the rise of Western civilization came in the fourteenth and fifteenth centuries, when Byzantine intellectuals fleeing the advancing Ottomans brought with them manuscripts of the ancient authors to the West and taught Westerners how to read and translate Greek into Latin.

1 The next few paragraphs summarize the more detailed discussion of the creation and dissemination of this narrative in Mavroudi 2006: 47–56; 2015. For a more detailed tracing of this narrative to the sixteenth century, especially as it involves the discipline of astronomy, see Mavroudi 2023: 233–238.

https://doi.org/10.1515/9783111589619-008

1 The Problems of the Old Narrative on the Medieval Transmission of Philosophy and Science

From the second half of the twentieth century onwards, scholarship has identified several problems with this narrative. For example, scholars of Islamic philosophy and science have insisted on their continued vitality past the twelfth century, while Western medievalists have reconsidered the intellectual outlook of the twelfth century.[2] As for literary production in Greek after Late Antiquity, the old narrative deems it as largely irrelevant for any developments outside of Byzantine territory. Philosophy and the sciences are supposed to have been transmitted in linear fashion through language, religion, and geography by way of three distinct translation movements: Greek-to-Arabic, Arabic-to-Latin, Greek-to-Latin. In both cases where the prototypes for the translations were Greek, Byzantium is presumed to have passively supplied the manuscripts but not the active selection of texts, let alone any interpretation of their meaning. Agency belongs exclusively to the receiving civilization. The result is a marginalization of Byzantium that Byzantinists have increasingly discussed during the last twenty years.[3]

Key to Byzantium's absence is the intellectual void presumed for the "dark centuries": Byzantium cannot play a role because with the advent of the Middle Ages philosophy and the sciences died out in Greek. In some versions of the narrative (influenced by the anti-religious and anti-clerical streak in Enlightenment thought), Christianity facilitated their demise.[4] Philosophy and the sciences were reborn in Arabic in the ninth century. Ya'qūb b. Isḥāq al-Kindī (ca. 800–870), the first self-identified philosopher in the Arabic tradition, is frequently portrayed as at the forefront of this development – a lonely genius pioneer accessing the ancient classics in translation and creatively spinning them into pure philosophical gold.[5] Instrumental for the presumed rebirth of

2 For more details, see Mavroudi 2023: 227–230.

3 Averil Cameron has repeatedly focused on the marginalization of Byzantium in modern historiography; for a recent contribution with reference to her earlier publications, see Cameron 2019. Other problems with this old narrative already pointed out by recent scholarship include the continued vitality of Islamic philosophy and science past the twelfth century. These cannot be analyzed here. For more details, see Mavroudi 2015b: 54; 2023: 228–229.

4 This claim is made in medieval Arabic sources that were taken at face value by modern Western scholars. For details, see Mavroudi 2015b: 38–39.

5 Franz Rosenthal (1956: 456) pointed out that Western European enthusiasm for al-Kindī's contribution to philosophy goes at least as far back as the sixteenth century and was re-circulated in the middle of the nineteenth through the work of the influential orientalist Gustav Flügel (1802–1870). More recently, the death of philosophy for two centuries and its resurrection in Baghdad at the hands of al-Kindī was reiterated in several essays by Dimitri Gutas; see, for example, Gutas 2004: 195–197. The work of Adamson and Pormann on al-Kindī, which is very important for making him accessible to a broader readership, also helps recycle some of the old historiographical tropes: a quotation from Gutas' essay on the death of philosophy around 610 is placed at the beginning of a section titled "The Birth of Philosophy" in Adamson and Pormann 2012: XXIII and sets the background for substantiating the claim of the title. The ensuing

philosophy and the sciences in the Arab world was the Greek-to-Arabic translation movement. Modern scholarship has presumed that this movement expressed an Arabic interest in *ancient* Greek learning. Although several translators were Christians of the same variety as Byzantium, the translation movement is supposed to have been uninterested in its contemporary or near contemporary Byzantine learning because there was none. The translators of the Greek texts, when they were working out of Greek and not Syriac originals, are presumed to have found the manuscripts that contained them mostly within Muslim territory. An important source were the libraries of the Christian monasteries in which Greek works had been collected before the Muslim conquests of the seventh century, and where they had remained unread since then. A further assumption is that the texts chosen for translation reflect the Greek school curricula of Late Antiquity that were no longer cultivated in Byzantium but were preserved in the Syriac schools that continued to flourish in Muslim lands.[6]

Objections to these formulations are already expressed in scholarship.[7] As for the image of Byzantium that this understanding of the Greek-to-Arabic translation movement presupposes, it is directly indebted to Edward Gibbon's influential masterpiece, *The Decline and Fall of the Roman Empire*. Gibbon also suggested that the Arabic interest in philosophy and the sciences jump-started a rediscovery of the ancient classics in Byzantium shortly before and especially after the restoration of the icons in 843 – the intellectual phenomenon generally referred to as "the Macedonian Renaissance".[8]

Gibbon's ideas have held up remarkably well into the twenty-first century. This survival was decisively helped by their re-iteration in a widely used work, Dimitri Gutas' *Greek Thought, Arabic Culture* (1998).[9] Easily accessible and reasonably comprehensive,

---

narrative attempts to straddle the nineteenth and the twenty-first century by presenting al-Kindī as both a pioneer and a product of his time. At the same time, an attraction to broader nineteenth-century formulations is in evidence. For example, the narrative paraphrases Ernest Renan's well-known quotation (pronounced in 1851) that "Islam was born in the full light of history" as "Arabic philosophy was born in the full light of history"; see Adamson and Pormann 2012: XXIII. The same narrative is further disseminated through the popular podcasts of Peter Adamson, titled "History of Philosophy Without Any Gaps", available at https://historyofphilosophy.net/.

6 For a discussion of the problems with these assumptions, see Mavroudi (Forthcoming b).

7 See, for example, Cottrell 2012: 11–12. Cottrell argues that the "translation movement" is not a sudden phenomenon but belongs to a continuum that encompasses a geography larger than Baghdad.

8 See Gibbon 1909–1913: 6.103: "In the ninth century we trace the first dawnings of the restoration of science. After the fanaticism of the Arabs had subsided, the caliphs aspired to conquer the arts, rather than the provinces, of the empire: their liberal curiosity rekindled the emulation of the Greeks, brushed away the dust from their libraries, and taught them to know and reward the philosophers, whose labours had been hitherto repaid by the pleasure of study and the pursuit of truth".

9 Gibbon's idea was refuted in Lemerle 1971: 22–42. Lemerle outlined what he was opposing but did not name the publications in which he had encountered it, with the single exception of Hemmerdinger 1962. Gutas (1998: 175–186) appears to have encountered Gibbon's argument in Lemerle because he repeats it with reference only to Hemmerdinger. This means that, regarding the impetus for the so-called "Macedonian Renaissance", Gutas chose to side with the opinion of Hemmerdinger (ultimately traceable

Gutas' book is the only such account in a Western European language. It has become the "go-to" reference on the Greek-to-Arabic translation movement, both for outsiders and insiders to the field. Gutas' reiteration of Gibbon's argument on the Arabic impetus for the "Macedonian Renaissance" was enthusiastically received by Graeco-Arabists, Syriacists, and even some Byzantinists.[10] Given Gibbon's critical examination during the last two decades, one cannot but wonder whether this recent resonance would have been the same, had the ultimate ancestry of its fundamental arguments on Greek literary culture been known.[11]

The denial of a Byzantine role in the medieval transmission of philosophy and science is part of a broader invention of the "Middle Ages" (whether "East" or "West") as a dark period overwhelmed by religion that began in the sixteenth century – a response to the battle between Protestantism and Catholicism and part of the sixteenth century's self-definition as a new and modern era.[12] In the eighteenth and early nineteenth centuries, the perpetuation of this image was facilitated by the anti-religious and anti-clerical streak in Enlightenment thought.[13] In the second half of the nineteenth century, when the study of Byzantium became a modern academic discipline and radical revisions to fundamental notions of science took place, it became commonplace to write of an inherent conflict between religion and science. Although the trend did not explicitly involve the study of Byzantine science, it certainly influenced it implicitly.[14] A more pos-

to Edward Gibbon) and not of Lemerle. Hemmerdinger recanted the idea that the Byzantine intellectual developments and the ninth and tenth centuries were inspired by the Islamic world in an article published two years after the one referenced by Gutas. On the recanting, see Hemmerdinger 1964. In this article, Hemmerdinger briefly outlines the historiography on the so-called Byzantine "Dark Ages" beginning with Gibbon. It is therefore clear that Hemmerdinger had been reading Gibbon and likely published the 1962 article under his influence, which he rejected in 1964.

10 The route of Gibbon's recent re-circulation is revealed when the only publication cited to support it is Hemmerdinger's retracted article. This is the case with Gutas 1998: 175–186. All subsequent followers of Gibbon's thesis cite Gutas, whether they also refer to Hemmerdinger or not. Gutas reiterated Gibbon's thesis in several essays, most recently in the joint article Gutas et al. 2017. Further re-iterations of Gibbon that cite Gutas (and sometimes also Hemmerdinger, but nothing else) can be found in several recent contributions. Examples include: Debié 2017; Jokisch 2007: 321–515 (proposing, inter alia, that major Byzantine iconoclast intellectuals like Leo the Mathematician and John the Grammarian grew up and were trained in Baghdad in order to explain their competence); Signes Codoñer 2014: 423–448 (which severely criticizes certain aspects of Jokisch's version of the argument); Sypiański 2011; 2012.

11 On the influence of Gibbon on the modern historiography about the Middle Ages, see contributions to the collected volume by McKitterick and Quinault 1997; see also the discussion in Mavroudi 2015b: 30–34.

12 For a summary of the religious issues at stake, and an argument that the enterprise of science cannot be viewed as separate from (or inherently inimical to) religion because natural scientists were no more or less religious than their contemporaries, see Henry 2010.

13 On the modern denial that Byzantine philosophy existed, see Mavroudi 2021.

14 A summary of the modern political and ideological issues involved in the denial for the existence of science overall, can be found in Numbers and Kampourakis 2015, and especially the essays by Harrison

itive evaluation of Western medieval philosophy and science began soon after World War II. Byzantine studies started catching up with this trend only a decade or two ago.[15] This ongoing investigation has benefited (and can further benefit) from broader historiographical trends. Specifically, under the influence of post-modernism during the second half of the twentieth century, various disciplines within the social sciences and the humanities have increasingly recognized that distinctions between "science", "religion", and "magic" are neither universal nor stable over time because they are socially constructed.[16] In addition, if one abandons the outdated but still pervasive understanding of "religion" as antithetical to "science", one can revisit the extensive Christianization of alchemy and of the disciplines related to it and draw new conclusions.[17]

Within Byzantine studies, the image of Byzantium as intellectually inconsequential for anything that happened beyond its frontiers after the sixth century is not entirely compatible with Byzantine research produced during the last several decades. As Byzantinists well know, half a century ago, Paul Lemerle highlighted detailed evidence for the continuity of Byzantine intellectual and educational traditions between the seventh and the early ninth century within the Byzantine Empire.[18] More recent scholarship has certainly confirmed this, even if the way in which Lemerle construed the concept of "humanism" and "Renaissance" is under reconsideration.[19]

Consistently with this trend, several articles published within the last few years explained that the Greek-into-Arabic translators and their patrons did not have to rely on Greek manuscripts produced before the Muslim conquests of the seventh century and preserved in monastic outposts under Muslim control but had access to their contemporary world of Byzantine learning.[20] And that, when approaching the Greek literary heritage, Arabic readers were not searching for what modernity saw in it – for

(2015) and Shank (2015). Unsurprisingly, Byzantium cannot be found in the volume's index. Likewise, the second edition of Ferngren (2017) has, compared with the first edition, new chapters in Islam, but no chapter on Byzantium. A thorough discussion of the modern denial that Byzantine science really existed is long and cannot be undertaken here.

15 For details, see Mavroudi 2015b; on philosophy, see Mavroudi 2021.

16 For a reflection on definitions of science in the pre-modern and, by extension, the Byzantine world, as opposed to modern approaches to science, see Pérez Martín and Manolova 2020.

17 See, for example, Principe 2015.

18 Lemerle 1971.

19 Flusin and Cheynet 2017. Twenty-first century objections to Lemerle's mid-twentieth-century concept of a Renaissance that are explained in this volume include a different understanding of what a Renaissance is, a decline in the conviction that Greek antiquity counts among the cornerstones of Western modernity, a new conceptualization of what Byzantine encyclopedism was, and a greater attention to the Byzantine peripheries instead of a primary focus on the imperial center.

20 For evidence that the Muslim world was in dialogue with its contemporary eighth- and ninth-century Byzantine science, see Mavroudi 2012; 2014; 2015; 2024a; Forthcoming a.

example, "democratic" or "secular" (i.e., non-Christian) ideals.[21] Rather, they were interested in elements of their contemporary Byzantine culture because it was monotheistic and imperial, and therefore its ideological priorities coincided with those of the Islamicate world.[22] The Greek and Arabic evidence on the cultivation of alchemy can further support this picture. Accordingly, the present paper highlights commonalities between Byzantine and Arabic alchemy from the seventh into the tenth century. In both the Byzantine and the Arabic context alchemy qualifies as "wisdom" rather than "science" because it provides an understanding of the universe as an interconnected whole, within which the connection with God is the most important relationship. In a tradition going back to Philo of Alexandria, and shared by Judaism, Christianity, and Islam, wisdom (and therefore alchemy) is inspired by God and for this reason is compatible with revealed scripture (whether the Bible or the Qur'ān).[23] The paper also compares the Greek and Arabic disciplines related to alchemy and proposes an alternative to the linear narrative on the transmission of philosophy and science originally received from the sixteenth century.[24]

2 Alchemy as Philosophy

The alchemical texts beg the question of genre, or intellectual category, to which they belong. This touches upon the larger problem of how each era and society classifies knowledge – in other words, how it defines and labels each of its branches according to the purpose it expects them to serve. In the second half of the nineteenth century, when the history of alchemy took shape as a modern academic discipline, the ancient and medieval sciences were approached according to nineteenth-century categories and were styled as ancestral to modern scientific disciplines. Accordingly, ancient and

---

21 On how the nineteenth- and early-twentieth-century pursuit of secularism influenced the modern study of Islamic and Byzantine philosophy, see Mavroudi 2021, where the use of ancient "paganism" as a metaphor for modern "secularism" is also explained.

22 This argument is made in Mavroudi 2015b. On the importance of Greek as an imperial language in the early Islamic context, see Stroumsa 2008; also 2015. On the fact that paying attention not to ancient but to eighth- and ninth-century Greek grammatical treatises can help understand the emergence of Arabic grammar as an organized discipline, see Mavroudi 2015a: 328–329; 2026.

23 For a summary discussion of Philo's take on Greek philosophy and the concept of wisdom, see Lévy 2018; see esp. 2.2.3 "'Barbarian' Philosophers" and Supplement "The Meanings of Philosophy in Philo of Alexandria". The term is also found in the New Testament (1 Tim. 3:7, which discusses the distinction between false teachings and true Christianity according to the Gospels) and was further elaborated upon by the Church Fathers; see Calzolari 2009. On the Qur'ānic concept of *ḥikma* ("wisdom"), see D'Ancona 2019.

24 For an analysis of the same with astronomy, and a proposition on what to replace this old narrative with, see Mavroudi 2023: 232–238 and 281–288.

medieval alchemy was supposed to have "practical" aspects (teleologically leading to the birth of modern chemistry as a "rational" science) and "theoretical" or "mystical" strands (deemed irrelevant or obstructive to such a process).[25] The latter were picked up by figures like Carl Jung and Mircea Eliade and interpreted through the lens of psychology and anthropology as part of a universal heritage of myth and ritual. This interpretative line has its own career among modern audiences that search for alternatives to traditional religion and are instead attracted to "spirituality", especially its non-Western varieties.[26]

As for the study of pre-modern alchemy's "practical" aspects, since the end of the nineteenth century, it was driven by an assumption, prevalent into the 1950s, that "science" as an enterprise had, across time and space, many of the same "rational" goals. This spirit is detectable in the ground-breaking publications on Greek alchemy by the chemist and politician Marcellin Berthelot, who approached it as ancestral to modern chemistry.[27] It can also be found in older and ground-breaking scholarly approaches to Arabic alchemy, such as Julius Ruska's *Arabische Alchemisten* (1924), which dismisses alchemy's association with the Imam Ja'far al-Ṣādiq (a figure prominent in later Islamic mysticism) in spite of its prolific attestation in Arabic literary culture. From the second half of the twentieth century onwards, science of any historical period is increasingly viewed as the product of cultures specific within time and space and influenced by changing political and social realities. Accordingly, the ancient and medieval sciences are no longer understood as early and clumsy attempts to reach truths attained only in the modern period and are increasingly studied within the concrete historical context in which they were pursued.[28] In addition, fundamental and continuous changes in modern scientific disciplines (physics, mathematics, etc.) since the nineteenth century are constantly influencing historians of science. This results in significant, and ongoing,

---

25 On this bifurcated course taken by the modern study of alchemy, see Viano 2018: 468–470. The identification of this bifurcation in historiography was pioneered by two co-authored essays, Newman and Principe 1998; Principe and Newman 2001. On the importance and impact of these essays, see Moran 2011. For a useful historiography of the modern study of alchemy, see Matus 2012. See also Newman 2011, explaining that recent developments in the study of medieval and early modern alchemy show that, contrary to what historians of the Scientific Revolution have argued, alchemy was neither marginal nor antithetical to experiment (by virtue of being too "theoretical" and "mystical").

26 For an outline of how Sufism (for which alchemy is an important tool) was received by modern Western audiences in Europe and the United States since the early modern period, but especially since the nineteenth century, see Sedgwick 2017. Sedgwick does not discuss alchemy, but a number of figures that he mentions (Ibn ʿArabī, the Brethren of Purity, etc.) were its adepts.

27 On the works of Berthelot on Greek alchemy, see Viano 2018: 469. On how the same concerns influenced the study of ancient astronomy, see Rochberg 2016.

28 For the history of dream interpretation, see Price 1986. For the history of ancient astronomy, see Rochberg 2016, and especially the comments on David Pingree's work (pp. 10–13).

reconsiderations regarding the definition, methods, and goals of the ancient and medieval sciences.[29]

For alchemy, a scholarly trend in evidence since the 1990s is again bringing together and examining in tandem its "practical" and "theoretical" or "mystical" aspects.[30] Indeed, if one focuses on alchemy as a "practical" pursuit, one may narrowly define it as the effort to transmute metals or other substances. However, in the quest for ways to influence matter, alchemy broadens in scope and addresses a whole range of relationships between the macrocosm and the microcosm.[31] The relationship of alchemy with philosophy is established by the fact that both respond to some of the same questions. Already Zosimus, author of the earliest alchemical writings that survive in Greek at any considerable length, attempted to answer questions regarding the provenance of divine/cosmic evil and the possibility of eternal life.[32] In addition, alchemy is associated with philosophy in works on broader subjects. For example, in the sixth century, John the Lydian talked about the Neoplatonist/Pythagorean aversion to beans (κύαμοι) and explained it through the transformation of beans into human forms if treated with alchemical procedures.[33]

Further, an explicit connection between alchemy and Christian spirituality is made in four alchemical poems attributed to Heliodorus, Theophrastus, Hierotheus, and Archelaus, and dated to the seventh or early eighth century.[34] At around the same time,

29 See Crowe 1990. This article correlates the views of Pierre Duhem on the history and philosophy of science with various profound changes in the study of mathematics and physics (e.g., the movement away from Euclidian geometry) that took place in the nineteenth and early twentieth centuries, and especially the influence of his contemporary Henri Poincaré. For a brief outline of how Duhem's views influenced the study of ancient astronomy, effectively making ancient astronomers look like nineteenth-century physicists influenced by positivism, see Evans 2000. For more details, see Lloyd 1978.

30 A recent book-length contribution along these lines is Dufault 2019. An equivalent trend is palpable also in the study of the mathematical sciences. For example, modern scholarship has interpreted Platonic astronomy as a "theoretical" and non-practiced endeavor. Against this earlier interpretation goes, for example, the recent explication by Paul Kalligas that the Antikythera mechanism (found in the late nineteenth century and styled as the earliest known "computer") was an astronomical model of the universe along the lines of Platonic astronomy; see Kalligas 2016.

31 See the explanations and further bibliography in Lory 2008.

32 See Dufault 2019: 8 on alchemy as a response to cosmic questions: "where does divine/cosmic evil come from? How can it be avoided? Are traditional sacrifices a solution? How an eternal life, with or without the body, be acquired?". The *Four Books* by Pseudo-Democritus, dating from the first century CE, are older than Zosimus, but survive in Greek only fragmentarily. Their Syriac version (more extensive than the surviving Greek fragments) was made after they had been epitomized, perhaps in the eighth century. See Letrouit 1995: 85 and Martelli 2013: 29.

33 Ioannes Lydus, *De mensibus* 4.42 (ed. Wünsch 1898: 99–100) (on the month of March). The overall presence of alchemy in John's work and his familiarity with the Hermetic corpus is a large topic that cannot be discussed here.

34 Reitzenstein 1919 considered these poems the product of the same author, who dedicated them to Emperor Theodosius III. On this basis, Reitzenstein dated them to the early eighth century. See also Browne 1946; commentary and discussion in Browne 1948. Browne (1946: 130) considers the mention of

Stephanus (a figure of uncertain identification) is counted by modern scholars as the earliest purely Christian alchemical author.[35]

Such an approach to these philosophically based disciplines reflects an older, pre-Christian tradition recently discussed by Matteo Martelli:[36] the exercise of a *technē* is a means of assimilation with God (*theōsis*). Assimilation with God is one of the six standard definitions of philosophy that recur among pagan and Christian authors writing between the third and the eighth century (e.g., Ammonius, David, and John of Damascus).[37] It ultimately allows one to identify Christianity as a philosophy, and a Christian monk as a philosopher. This is consistent with the ideal practitioner of alchemy as described by Archelaus, who must be pure, fast, avoid worldly concerns, embrace solitude (μένων μόνος), pray, offer adoration to God, and expect that knowledge (γνῶσις) will come from above through divine grace.[38] Although these are not exclusively Christian ideals (they recur in pagan Neoplatonism, Jewish contemplative practice, and Muslim Sufism), they certainly coincide with the pursuits of Christian monasticism. The identification of alchemy with philosophy has been repeatedly noted in recent scholarship.[39] If alchemy can lead to assimilation with God, then it can, like Christianity, be defined as a form of philosophy and be deemed an appropriate and even vital Christian pursuit.[40] The identification of Christianity and alchemy as philosophy becomes more evident when one conceives of philosophy as a way of life – in the manner explicated

Theodosius to be a possible later insertion. Letrouit (1995: 82–83, 88) dated them to the seventh century. Recently, Lauxtermann (2019: 205–207) has redated them to the fifth or early sixth century. On the four alchemical poets, see also Maria Papathanassiou's contribution to this volume.

35 For the identification of Stephanus as the earliest purely Christian alchemical author, see Carlotta 2022. Carlotta accepts a seventh-century date for Stephanus. Earlier scholarship identified "Stephanus" (of Alexandria or of Athens) as a single figure who wrote on philosophy, astrology, alchemy, and medicine. More recent scholarship separates this once unitary Stephanus into four separate authors: Stephanus the astrologer and alchemist, Stephanus the commentator on Aristotle's *Rhetoric*, Stephanus of Alexandria who commented on other Aristotelian works, and Stephanus the medical commentator. See Martelli 2016; Goulet 2016; Searby 2016; and Boudon-Millot 2016. Goulet 2016 uses a reference to the *Suda* to place the *terminus ante quem* for Stephanus, the commentator of Aristotle's *Rhetoric*, to the twelfth century. Goulet appears to use the date of the *Suda*'s earliest surviving manuscript (MS Leiden, *Vossianus graecus* F 2), not its generally accepted date of compilation during the second half of the tenth century.

36 Martelli 2014a, which does not mention alchemy explicitly but explains this within the pagan context.

37 These six standard definitions of philosophy were inherited from Plato and his later interpreters, Aristotle, and the Stoics; see Trizio 2007: 275. The six definitions are described in Ierodiakonou 2012: 5: "(i) knowledge of being as such; (ii) assimilation to God as far as humanly possible; (iii) knowledge of divine and human things; (iv) preparation for death; (v) art of the arts and science of the sciences; (vi) love of wisdom".

38 [Archelaus], *De arte sacra*, vv. 41–48 (ed. Goldschmidt 1923).

39 See Koutalis et al. 2018: 31–37; Merianos 2017: 243; 2022: 285–293; Dufault 2019: 141.

40 The Christian aspects of alchemy have been repeatedly drawn out for the late medieval and early modern period in Western Europe. On Latin alchemy from the twelfth century onwards, see, for example, Matus 2012; Nummedal 2013 (offering also a useful review of earlier literature).

by Pierre Hadot – rather than a type of discourse, which has dominated histories of philosophy from the eighteenth into the late twentieth century under the influence of giants like Johann Jakob Brucker and Eduard Zeller.[41]

All six standard definitions of philosophy that were formulated in antiquity recur prolifically in Greek literary culture throughout the Byzantine millennium.[42] They can also be found in the work of the first self-identified philosopher in the Arabic tradition, al-Kindī.[43] Further, they are echoed in the texts that (reputedly or really) transmit the thought of his older contemporary, Dhū al-Nūn al-Miṣrī (ca. 772–ca. 860), an important figure during the formative period of Sufism and reportedly also an alchemist. There, the longing to see God leads to love of death (*ḥubb al-mawt, istildhādh al-mawt*).[44] This calls to mind the Platonic *meditatio mortis* (*Phaedo* 81a) which is repeated by Byzantine authors writing around the early Islamic period, like John of the Ladder (sixth/seventh century) and John of Damascus (seventh/eighth century).[45]

As in Greek, so in Arabic, alchemy is called "the divine art" (*hiera kai theia technē* = *al-ṣinʿa al-ilāhiyya*).[46] As in Greek, so in Arabic, the term *chēmeia/kīmiyāʾ* (primarily "the art of alloying metals") appears to have had, at least at the beginning, a pejorative meaning connoting trickery.[47] The reason for this must be that several alchemical

---

41 On how these two different conceptualizations of philosophy were introduced to modern scholarship and how they influenced the study of Byzantine and Islamic philosophy, see Mavroudi 2021.

42 See Trizio 2007: 274–275.

43 On al-Kindī's definition of philosophy as assimilation to God, see Gannagé (Forthcoming). Six definitions of philosophy are found in al-Kindī's *Definitions* (a work that may not be from his pen but certainly authored in his circle). On them, see Atiyeh 1966: 17–19. On the reception of the six standard definitions of philosophy in Byzantine literary culture, see Trizio 2007: 274–275.

44 Ebstein 2014: 573. This article provides a thorough exposition of the primary sources that provide information on the thought and biography of Dhū al-Nūn. Like Socrates and Plotinus, Dhū al-Nūn did not write anything. Contrary to them, he did not have disciples who recorded his thought in writing. Information on his teaching can be gleaned from a variety of medieval Islamic sources dating from the tenth century and later – that is, considerably later than his lifetime. Ebstein emphasizes the need for a thorough philological investigation of these sources in order to retrieve his authentic thought. According to the medieval Arabic sources, he was born in Nubia or Akhmīm (Panopolis) in Upper Egypt to Christian parents who had been enslaved by Muslims. He is said to have spent time at the ancient Egyptian temple of Akhmīm when the famous White Monastery (at a distance of twenty minutes by car from modern-day Akhmīm) was the site of lively book production. He also traveled extensively and met several ascetics, both men and women, with whom he engaged in spiritual conversations. There is no reason to doubt that many of these ascetics were Christian.

45 See Trizio 2007: 274. On the early Christian career of this Platonic idea, see Munkholt Christensen 2021.

46 See Lory 2008.

47 For the Arabic evidence, see Lory 2008: "This science was generally called *al-ṣanʿa al-elāhiyya*, the 'divine art'. The name *kimiā* (Gr. *khēmeía*, the art of alloying the metals) seems to have been used at first in a rather pejorative way (meaning something like 'trickery')"; cf. also Sezgin 1967–1984: 4.3–7. Strohmaier 2016: 423–434 argues that the Arabic term *al-kīmiyāʾ* is not a word of Greek origin due to the final glottal stop. This does not cancel the semantic analogy between χυμεία/χημεία and *al-kīmiyāʾ*. In fact, the Greek term χυμεία/χημεία (unstable in its orthography, which indicates a foreign origin)

texts – including some of the earliest surviving in Greek, such as the Stockholm and Leiden papyri and the fragments of the *Four Books* by Pseudo-Democritus collected and studied by Martelli – explain how to confect substitutes for gold, silver, precious stones, and purple dye.[48] This must be the kind of alchemy that al-Kindī condemns by arguing that "man cannot do what nature alone can and that the practices of the alchemists are deceitful".[49] Arabic literary culture recognized a number of philosophically based disciplines as interrelated and effectively representing a form of practical philosophy (an equivalent to what Martelli explicated in the thought of ancient Greek authors, beginning with Galen, who had a considerable reception in Arabic). Evidence of this recognition is the fact that, by the ninth century, the word *ṣan'a* and *ṣinā'a* ("art" or "craft") is applied not only to alchemy but also the sciences of the stars (astronomy and astrology).[50]

Further equivalences between Byzantine and Arabic alchemical concepts and terminology have also been identified by scholarship. An important example is the old Greek alchemical formula *hen to pan* ("the all is one") which is based on a Platonic/Stoic conception of the universe as immanent. By the time it occurs in the seventh-century alchemical text by Stephanus, it has become "a nearly liturgical chant".[51] As was observed long ago, the formula is repeated in the tenth-century alchemical work by Muḥammad b. Umayl al-Tamīmī titled *al-ma' al-waraqī wa al-arḍ al-najmīya* (The Silvery Water and the Starry Earth).[52] Centuries later, it was reflected in the Sufi term *waḥdat al-wujūd* ("unity of existence").[53]

appears to reflect the ancient Egyptian word for Egypt (Χημία, according to Plutarch). See Merianos and Sakorrafou 2013: 48.

48 The word *chēmeutēs* is rare in Greek texts and appears to have negative connotations because it seems to have "referred primarily to artisans who imitated or counterfeited gold or jewelry"; see Dufault 2019: 95 and in general pp. 94–97. See also the comments in Merianos and Sakorrafou 2013: 51–52 on the important role of color in conveying the appearance of genuineness, whence the potential for trickery; see also Koutalis et al. 2018: 12.

49 See Levey 1962: 12.

50 See note 47 above. *Ṣan'a* can also mean "work" (that is, it can be construed as closer to the Greek word ἔργον = "divine work" rather than τέχνη). Al-Kindī called astronomy *al-ṣinā'a al-'uẓma* ("the greatest craft"); see Ya'qūb b. Isḥāq al-Kindī, *Fī al-ṣinā'a al-'uẓmā* (ed. Aḥmad 1987). Astrology is called *ṣinā'a* by Abū Ma'shar, a disciple of al-Kindī; see Saif 2015: 13 and n31.

51 See Plass 1982.

52 Plass 1982: 72; Stapleton et al. 1949: 76, 88.

53 This term is associated with Ibn 'Arabī, but it was used for the first time by Ibn Taymiyya to pejoratively describe Ibn 'Arabī's thought. Dhū al-Nūn (or the later authors who transmit his thought) used the cognate verb *wajada* in a similar technical sense; see Aladdin 2012.

3 The Byzantine and Arabic Reception of Hermetism

Alchemical works mention and quote from texts attributed to Hermes – an overlap that modern scholarship noted several decades ago.[54] In addition, Christian approaches to both alchemy and Hermetism are in evidence since the early Christian centuries.[55] This means that, by the time Islam appeared in the seventh century, alchemy and Hermetism had become effectively Christian disciplines cultivated, in their Christian version, throughout the so-called "dark centuries" and into the tenth, when the earliest surviving Greek alchemical manuscript (*Marcianus graecus* 299 [coll. 584]), was created as a reflection of contemporary Byzantine interest in alchemy rather than an antiquarian exercise.[56]

Regarding the Hermetic corpus, modern scholarship has contemplated two possible ways leading to its Arabic transmission: either directly from pagan antiquity (as if placed in a time-capsule) through the Sabeans of Ḥarrān, a pagan community living in the Islamicate world;[57] or, as David Pingree briefly discussed and Kevin van Bladel elaborated in detail, via its earlier translations into Middle Persian.[58] More recently, Emily Cottrell argued against the centrality of the Middle Persian translations for the transmission of Hermetism (and, with Micah Ross, also of Greek astrology) to the Arabic-speaking world.[59] What is at stake for Cottrell and Ross is less the transmission of the "classical tradition" to the Arabs and more the conveyance of the Zoroastrian writings from the Sassanian to the Abbasid period, including an important open question: whether the Neoplatonist elements detected in them are additions of the Islamic period.[60] For Cottrell and Ross, the transmission of Hermetism, Neoplatonism, and Greek astronomy and astrology to the Iranian and Islamic world is a complex and multi-directional, rather than linear process. An exploration and appropriate description of this complexity is better achieved if one takes into consideration the fact that, from antiquity into the sixteenth century, both Eastern and Western Christian authors

54 For the history of the scholarship on this, see Moreschini 2011: 1–26.

55 For documentation, see Moreschini 2011.

56 On the Christianization of alchemy, see note 40 above (Matus, Nummedal), as well as Carlotta 2022 and Merianos 2024. On MS *Marcianus graecus* 299 (coll. 584) as a product of tenth-century Byzantine alchemy, see Roberts 2019.

57 On the scholarly history of this argument, a discussion of its merits, and an ultimate rejection, see van Bladel 2009: 64–114.

58 See van Bladel 2009. The discussion on the importance of the Middle Persian translations for the transmission of originally Greek materials into Arabic is much older. An earlier influence on Pingree was Carlo Alfonso Nallino (1872–1938). Gutas 1998 places the Middle Persian translations of Greek material center-stage among the triggers of the Greek-to-Arabic translation movement.

59 See the review article by Cottrell 2015; and Cottrell and Ross 2019.

60 On this problem, see König 2018.

identified convergencies between Christian and Hermetic concepts and many considered Hermetism a vehicle of quasi-divine revelations.[61]

4 Byzantine and Arabic "Science of Letters"

A Byzantine text with concerns and goals similar to those addressed by alchemy and Hermetism – and therefore one belonging to the same broader genre because it deals with questions about the universe and human salvation in similar ways – is the treatise titled *On the Mystery of the Letters* (*Περὶ τοῦ μυστηρίου τῶν γραμμάτων*). Scholars have generally dated it towards the end of the sixth century, but it may have been written later.[62] As is the case with alchemical and Hermetic treatises, it is addressed by a master (Abbot Saba) to a disciple, identified as a Canaanite who cannot speak his ancestral tongue.[63] It offers an interpretation of the letters of the Greek alphabet as symbols of the Christian universe and the salvation promised by Christianity. Although at the center of its symbolism is the Greek alphabet, it is extensively correlated with the shapes and sounds of the letters in the Syriac and Hebrew alphabets – two languages that the author appears to know well. The *On the Mystery of the Letters* is a late antique equivalent (expressed in Greek) to later Arabic/Islamic works on *ʿilm al-ḥurūf* ("science of letters"), a discipline closely related to alchemy and used as an instrument for divination, prophecy, and apocalypticism.[64] In an impressive recent study, Juan Acevedo traced the roots of this discipline to the conceptualization of *stoicheion* by Aristotle, Plato, and the Greek grammarians, highlighted the Pythagorean correlation of *stoicheion* to number, and followed the absorption of this correlation into the three Abrahamic traditions through the key intervention of Philo of Alexandria. This resulted in "a fully shared intellectual tradition among scholars of the three faith communities".[65]

The cosmopolitanism of the sources adduced in *On the Mystery of the Greek Letters* calls to mind a treatise described in the ninth century by Photius in his *Bibliotheca* (codex 170).[66] It compiled testimonies and quotations by the third/fourth-century alchemist Zosimus of Panopolis and a number of Greek, Persian, Thracian, Egyptian, Babylonian,

---

61 For a book-length documentation of this phenomenon, see Moreschini 2011.

62 See Bandt 2007. Elements in the text that can help contextualize it include its mention of the Turks; the Arabs have multiple dialects (that is, the author is familiar with many kinds of Arabs); paganism appears as a living intellectual tradition; letter *iota* has a *trēmma* and letter *sigma* is nunated, but paleographically this does not take us very far because both are attested in Greek handwriting from the papyri into the Middle Byzantine period and beyond.

63 *Mysteria litterarum* 37 (ed. Bandt 2007: 184, 26–29).

64 Famous examples from the late medieval and early modern period include Astarabadi, on whom see Algar 1987; and al-Bistami, on whom see Fleischer 2015, a videotaped but unpublished lecture.

65 See Acevedo 2020: 200.

66 Photius, *Bibliotheca*, cod. 170 (ed. Henry 1959–1977: 2.162–165).

Chaldean, and Roman authors on various subjects. Its goal was to demonstrate the truths of Christianity, albeit in ways that Photius disapproved of. Its considerable volume and the variegated provenience of its source material as reported by Photius suggest that Constantinople's libraries during the so-called "dark centuries" were richer than modern scholarship usually grants.[67] Modern scholarship has identified this text as the so-called "Tübingen Theosophy", belonging to a kind of writing that is also closely associated with alchemy.

5 Theosophy and the Encounter of Christianity with Islam

Knowledge of the affinities and correspondences between letters, numbers, the cosmogonic elements, and the heavenly bodies (which was necessary to an alchemist), could be used to influence the physical world and foretell the future.[68] The ability to divine and prophesy through being an adept in the principles needed to understand and practice alchemy brings up the connection of alchemy with theosophy (literally, "wisdom regarding things divine"). The earliest surviving attestations of the word occur within pagan and Christian writers of the third and fourth century: Iamblichus, Eusebius, Emperor Julian the Apostate.[69] From these references it is clear that, through accessing "indigenous" as well as "alien" wisdom, alchemy and theosophy address some of the same fundamental questions: the constitution of the universe and the course to salvation. For example, Iamblichus, in *On the Mysteries*, outlines how the "theosophical muse" can help a "Hellene" understand Egyptian theology as an "intellectual interpretation of symbols" embedded in the elements of the universe and their powers.[70] The constitution of the universe, the course to salvation, and the use of "alien wisdom" in order to access their truths, are major themes that alchemy and theosophy share with

---

67 For a particularly pessimistic view, see Mango 1975: 39–43. This article's points need to be reconsidered one-by-one, by taking on board several details in the manuscript tradition of technical disciplines (the genre of writing that Mango projects as the most scarce in Byzantium at the time). However, this is a lengthy task and cannot be undertaken within the confines of the present contribution.

68 For an example of the connection between the "science of letters" and apocalypticism, see the analysis of the Ottoman case by Fleischer 2018: 18–90.

69 Iamblichus and Julian mention the word (or a cognate) only once. See Iamblichus, *De Mysteriis* 7.1 (ed. des Places 1966); Iulianus Imperator, *Epistulae* 12.14 (ed. Bidez 1960). This passage makes clear that Julian is aware of Iamblichus' understanding of theosophy. There are multiple mentions of the word in Eusebius. The Christian tradition of adducing non-Christian texts in order to prove the truth of Christianity is, of course, much earlier than Eusebius.

70 Iamblichus, *De Mysteriis* 7.1–5 (ed. des Places 1966); Engl. transl. Clarke et al. 2003: 290–303 (which also prints the Greek text as established by des Places).

the Hermetic corpus.[71] And if pagan authors in these genres highlight "Egyptian" wisdom as an "alien" source from which a "Hellene" can be enlightened, Christian authors construe non-Christian wisdom as "alien". Accordingly, Christian theosophical writing explicates how "pagan" or "alien" (i.e., non-Christian) wisdom effectively agreed with – and, therefore, announced to those who did not have the privilege of first-hand contact with Christianity – the tenets of Christian and biblical revelation.[72]

The theosophic character of alchemy is evident also from the attributions of the aforementioned alchemical poems: Archelaus was the teacher of Socrates and Theophrastus the student of Aristotle. As such, they count among the ancient pagan thinkers who, according to a robust and continuous Byzantine tradition beginning with Justin Martyr and Clement of Alexandria and extending into the post-Byzantine period, had conquered parts of the Christian truth before the Christian revelation. Heliodorus can be identified as the author of the novel *Aethiopica*, set in a pagan historical context but interpreted by Byzantine readers as a Christian allegory. Hierotheus is likely Hierotheus the Thesmothete, a learned Athenian converted to Christianity by Saint Paul and, according to some traditions, the teacher of Dionysius the Areopagite.[73] In other words, while Archelaus and Theophrastus are pre-Christian pagans, Heliodorus and Hierotheus stand at the cusp of the transition from paganism to Christianity. It is easy to see why all four figures can be identified as non-Christians who recognized and conveyed in their writings the truths of Christianity, either by intuition or through early contact with Christian teachings.

The composition of theosophic texts, adapting to Christian purposes originally non-Christian material, is generally dated by modern scholars to Late Antiquity and considered the product of the struggle between ancient paganism and Christianity. A case can be made that such texts, which are closely related to alchemy, circulated (and were perhaps even written) during the "dark centuries" and were used in the Christian encounter with Islam.

Perhaps the theosophical text most famous in modernity are the *Chaldean Oracles*. Their Christian use is clear already in the early Christian period.[74] Their circulation in Greek during the "dark centuries" is suggested by the fact that they were known to Bishop Arethas of Caesarea in the late ninth century.[75] For the time being, their

---

71 The intellectual characteristics shared between alchemy and Hermetism were analyzed above. On salvation in the Hermetic corpus, see Luck 1991.

72 This is the approach to pagan revelatory and divinatory texts in Eusebius' *Praeparatio evangelica* (ed. Mras 1954–1956).

73 On these attributions, including previous scholarship on them, see Roberts 2022. Particularly on Heliodorus and his identification with the namesake author of the *Aethiopica*, see Gerasimos Merianos' contribution to this volume.

74 See Majercik 1998.

75 Lemerle 1971: 231. The modern narrative on the history of Byzantine philosophy sees a disruption between John of Damascus (at the latest) and Michael Psellus, recognized by modernity as the first "true"

only known Arabic/Muslim reception occurred in the fifteenth century at the court of Mehmet the Conqueror – a time rife with apocalypse and therefore receptive to the theosophic message of this text.[76]

A second theosophic text, this one certainly circulating during the "dark centuries", is the aforementioned "Tübingen Theosophy". It uses pagan Greek, Egyptian, Persian, and Chaldean wisdom to prove the truths of Christianity. It appears to have been written around the year 500 and to have been epitomized some time between the seventh and the thirteenth century, when it certainly was read in Constantinople.[77] Today it survives only in fragments but, as already indicated, modern scholarship has identified its full version as the text described by Photius (codex 170).[78]

The evidence from Arethas, Photius, and various theosophic fragments edited during the last century-and-a-half indicate that theosophy (a genre akin to alchemical writing, as explained earlier) continued to be read and commented upon, quoted, excerpted (that is, written about) in Byzantium considerably past the sixth century and into the Late Byzantine period. The reason for this recirculation was not Byzantine antiquarianism but the ability of these texts to address concerns of the living present, including the encounter of Byzantine with Islamic culture. This suggests that the theosophic genre may have been created in response to Hellenic paganism as a living tradition but was certainly used in response to other kinds of unbelief past the seventh century, including Islam. The later Byzantine reception of late antique theosophic texts was, like the reception of alchemical texts, not an exercise in antiquarianism but a use of earlier tradition in order to address the problems of a living present.

This is clear if one focuses on Question 136 among the collection of *Questions and Answers to Antiochus*, attributed to the Church Father Athanasius of Alexandria but in

---

philosopher writing in Greek since Late Antiquity; on this narrative, see Ierodiakonou and Bydén 2018. Part of why Psellus counts for modernity as a philosopher is his edition and commentary of the *Chaldean Oracles*, a text received in Renaissance Europe. The acquaintance of Arethas with the *Chaldean Oracles* belies this narrative.

76 See Mavroudi 2013.

77 On this dating and localization, see Carrara 2019.

78 See note 65 above; Lewy 2011: 16. This identification is largely the product of philological methodology current from the nineteenth into the twenty-first century, whereby Byzantine authors (with Photius' *Bibliotheca* at the forefront) are used as a warehouse from which to retrieve fragments of ancient authors. Several modern editions of ancient texts (the *Chaldean Oracles* are an example) are collections of fragments selected and subjectively ordered by modern editors. Accordingly, Beatrice 2001: LIX collects theosophic references found in Greek texts or manuscripts produced between the seventh and the ninth centuries and considers them as evidence that could help an editor piece together a single text. However, these references may well be to different works within the same profuse genre – in other words, they may be excerpts from different texts that deal with the same topics in similar ways. This is suggested by the variegated provenience of the theosophic fragments edited in Erbse 1941 and new edition in Erbse 1995. On the history of these editions, see Beatrice 2001: XI–XVIII. For criticism of the subjectivity of Beatrice's edition, see its review by Van Deun 2003.

fact composed some time before the middle of the eighth century.[79] This brief text was recently edited by Caroline Macé and Ilse De Vos out of three manuscripts. The two oldest were copied in times and places characterized by close contact between Muslims and Christians: tenth-century Sicily and early eleventh-century Egypt (by a cleric named Leo).[80] This suggests that the scribes and intended readers of these manuscripts understood the "pagans" named in Question 136 not as followers of the Graeco-Roman pantheon (which was irrelevant in their time) but as their contemporary Muslims. This impression is strengthened when we consider that part of what Question 136 addresses is thunder. The passage seems to be a response to the cosmological and other concerns laid out in Chapter 13 of the Qur'ān ("Chapter of Thunder"). As for the prophecies on Christ in Question 136, they pertain to his birth by the Virgin Mary, his hanging from the cross, and his ability to cure. Christ's virgin birth and curing miracles are mentioned and accepted as true in the Qur'ān. However, the Qur'ān does not accept his crucifixion. Instead, he is said to have been miraculously saved by God and to have ascended to Heaven. Insisting on Christ's death on the cross means insisting on the fact of his resurrection, which is key to the salvific message of Christianity. Accordingly, for Christians living among Muslims, Question 136 can serve apologetic purposes. Significantly, the prophecy of Question 136 on Christ's crucifixion is repeated in a further text that was probably written under Arab rule: the *Homily on the Nativity of Christ* attributed (with some likelihood) to John of Damascus.[81]

6 Byzantine and Arabic Magic

The manipulation of the physical world possible through knowledge of alchemy mentioned earlier is, of course, the pursuit of magic. With the exception of the excerpts from Julius Africanus' *Cestoi* that survive in tenth- and eleventh-century codices on military tactics (MSS Vatican, *Barberinianus graecus* 276; Florence, *Laurentianus Pluteus* 55.4; Escorial, Y.III.11), Greek magical texts came down to us in manuscripts from the fifteenth century and later.[82] This creates considerable difficulty in retrieving the history

79 On dating and other problems surrounding this text, see Macé and De Vos 2013: 320. The article states that QA 136 looks "a bit out of place in the collection, since the preceding questions are dealing with sins, death and the human soul, while QA 136, as we will see, is concerned with the question of how to convert Pagans" (p. 320). In fact, these topics fall squarely within a larger and important issue that alchemical, Hermetic, and theosophic texts address, namely the course to human salvation (as explained earlier). At the same time, the suggestion that "the collection of 137 QA that we now have may be the result of successive additions and interpolations" (p. 320) is reasonable.

80 See Macé and De Vos 2013: 321–322.

81 Macé and De Vos 2013: 332.

82 Key manuscripts for the transmission of Greek magic are MS *Bononiensis* 3632 of the fifteenth century; MS *Atheniensis* 1265, a miscellany copied in the sixteenth and seventeenth centuries; see also the

and circulation of magical texts in the Early and Middle Byzantine period, including the "dark centuries". The most explicit textual evidence for the eighth-century pursuit of magic as a sophisticated, philosophically based discipline appropriate for the highest echelons of Byzantine society, is the *Parastaseis Syntomoi Chronikai* (Brief Historical Notes). This compilation discusses the monuments of Constantinople, including its statuary.[83] It is not a text on philosophy or natural science but does reflect the uses of these disciplines at the time when it was written. In many ways, it represents a medieval continuation of the ancient Roman tradition of state augury that Byzantine administrators needed to be familiar with.[84] An earlier sample of works servicing this goal are the compositions of John the Lydian, written in the sixth century. The fact that John and the author of the *Parastaseis* are participants to the same administrative lore is demonstrated by demonstrated by their shared familiarity with the figures of earlier important state officials, such as Demosthenes in *Parastaseis*, who is definitely not the ancient Athenian orator but Demosthenes the eparch, also mentioned by John the Lydian in *De magistratibus* (On Powers).[85] Perhaps Demophilus in *Parastaseis* (a general who likes idols) can be identified with the person by the same name mentioned by John the Lydian in *De mensibus* (On the Months).[86] These individuals that serve as shared points of reference indicate a continuous knowledge of the past from the sixth into the eighth century. A number of philosophers appear in the *Parastaseis*. Some are pagan, others Christian. They are portrayed as advisors to kings, but their most important function is to impart magical powers to statuary (στοιχειώσασθαι). Like alchemy, Hermetism, and the "science of letters", the *stoicheiosis* was also Christianized in Late Antiquity.[87] The administrators who compiled and read the *Parastaseis* would have identified with these philosophers by virtue of their education, social position, and professional role, which included manipulating for talismanic purposes the considerable ancient statuary found throughout eighth-century Constantinople. Indeed, the talismanic potency of ancient statues is amply recorded in both Byzantine and Arabic sources from around the ninth century.[88] This suggests that their accurate identification with the help of mythology

---

fifteenth- and sixteenth-century dates of the earliest manuscripts in which the *Testament of Solomon* survives in McCown 1922: 20–28.

83 The *Parastaseis* is dated by Cameron and Herrin to the early eighth century, late eighth or early ninth by Kazhdan. See Odorico 2014, which offers a history of this text in scholarship and outlines the range of dates ascribed to it by various scholars (from the early eighth to the ninth century at large).

84 On the connection of the *Parastaseis* with administrative circles, see Anderson 2011: 1. The text survives in two manuscripts, one of the eleventh century (on which the edition by Preger is based) and one of the tenth, published separately and known as "anonymous Treu". See Odorico 2014.

85 *Parastaseis Syntomoi Chronikai* 28 (ed. Preger 1901: 36, 18–22); Engl. transl. Cameron and Herrin 1984: 91. Ioannes Lydus, *De magistratibus populi Romani* 3.42 (ed. Bandy 1983: 198, 20).

86 *Parastaseis Syntomoi Chronikai* 49 (ed. Preger 1901: 54, 3–7); Engl. transl. Cameron and Herrin 1984: 127. Ioannes Lydus, *De mensibus* 4.2 (ed. Wünsch 1898: 66, 8).

87 See Raggetti 2019.

88 Magdalino 2006; El-Cheikh 2004: 147–149.

and its cosmic (that is, alchemical) connotations was an important component of magic as a form of natural science.[89]

In modern times, John the Lydian's extensive references to pagan mythology and explanations on the correspondence of ancient gods to elements of the universe were understood as antiquarianism. Such an interpretation misses the function of this information as a guide to the talismanic manipulation of statues representing the ancient gods, which served important civic and dynastic purposes.[90]

The earliest known Arabic works on talismans were written a few decades later than the *Parastaseis* by al-Kindī and Thābit b. Qurrah (ca. 826–ca. 901). The most famous talisman maker of Graeco-Roman antiquity, Apollonius of Tyana, also enjoyed an Arabic reception.[91] Although these texts do not explicitly mention ancient statues when they discuss effigies, the correct identification of statues representing figures from Greek mythology seems to have played a role in Arabic talismanic magic. Evidence to this effect comes from the the *Kitāb al-aghānī* (a voluminous tenth-century collection of songs and poems dating from the pre-Islamic period to the end of the ninth century). This text describes how a young man from Edessa became sick with love for the statue of a beautiful woman, Haylāna (Helena). He recovered when his father decapitated it.

7 Allegory in Byzantine and Arabic Literary Culture

The allegorical interpretation of Greek mythology was important not only for talisman-making but for all kinds of natural science, including alchemy. Perhaps the most diffuse type of alchemical allegory is based on the well-known correspondences between planets named after the ancient gods and metals. These are first attested in the ancient Near East. Their adaptation to the Graeco-Roman pantheon is evident in Greek astrological texts.[92] It is also evident in the seventh-century writings of the alchemist Stephanus.

Within the Byzantine intellectual universe, it follows an intellectual tradition rooted in pagan antiquity and further developed after the appearance of Christianity.[93]

89 On the alchemical connotations of ancient statuary, see below the discussion on the correspondences between the ancient gods and elements of nature.

90 For descriptions of the talismanic manipulation of statues in Constantinople during the Middle Byzantine period for civic and dynastic purposes, see Magdalino 2006.

91 Raggetti 2019.

92 See Bouché Leclercq 1899: 313–317.

93 As a broader method of understanding texts, allegorical interpretation has pagan, Jewish, and Christian roots. Allegory is attested already in the fifth century BCE and picked up by Stoicism and Platonism. See Long 1992: 44. In the first century BCE–first century CE, Philo of Alexandria (whose influence on early Christian thinkers was considerable) used allegory to interpret the Old Testament. Many (though not all) Byzantine thinkers considered allegory an acceptable method of interpretation.

The allegorical interpretation of the ancient gods identified them as elements of nature (fire, air, water) or as the moon, sun, stars, and planets. It also drew correspondences between the gods and the powers and passions of the soul and intellect. Alternatively, and especially within the tradition of the Christian world chronicle, the ancient gods could be understood as historical figures (heroes, kings, or wise men).[94] Examples for the Christian allegorizing of ancient mythology include various texts by the fourth-century Church Father Cyril of Alexandria, who interpreted the ancient gods as representations of the natural elements or moral qualities of the soul.[95] Closer to the discipline of alchemy, John of Antioch, a seventh-century chronicler writing in Greek, provided an alchemical allegorical interpretation of the myth of the Argonauts, and particularly of the Golden Fleece.[96]

The allegorical interpretation of texts explicating pagan mythology opened the way for their incorporation to the reading lists of technical literature in the natural sciences, to which alchemy belongs. For example, both in the ancient and the medieval world the Homeric epics were used to explicate the *Phaenomena* by Aratus of Soli, which described the constellations and introduced students to the study of astronomy. This meant that, within the classroom, the allegorical interpretations of ancient mythology could be used to explain cosmology in both pagan and Judaeo-Christian terms. This intellectual practice is attested in Greek from the Church Fathers to the end of the Byzantine period and clearly influenced its contemporary Syriac and Arabic literary culture.[97] Accordingly, the *Phaenomena* were translated into Arabic and offered an avenue through which Arabic readers could become acquainted with allegory in the course of technical instruction in astronomy and astrology.[98] A further example of such an approach may be a text, no longer extant, by Theophilus of Edessa, the Christian astrologer to Caliph al-Mahdī. He reportedly rendered from Greek into Syriac "the two books on the conquest of the city of Ilion in ancient times by the poet Homer".[99] This may be not a translation of the two Homeric epics in their entirety (an almost impossibly laborious task) but a recasting of the two poems informed by Theophilus' astrological, astronomical, and historical knowledge – an equivalent to John Tzetzes' approach

94 On these methods of interpretation, see Hunger 1954: 47–55; 1955: 4; Vasilikopoulou-Ioannidou 1971: 69. On allegory and world chronicles, see Mavroudi, 2019: 447–455.

95 For concrete examples, see Bartelink 1983.

96 Martelli 2013: 60. See also the contributions to this volume by Alexandre Roberts and Matteo Martelli.

97 Cosmological readings of Homer based on allegory can be found in the earliest surviving Greek manuscripts with Homeric commentaries (which are earlier than the ninth century), the well-known twelfth-century commentaries by John Tzetzes and Eustathius of Thessaloniki, and the fifteenth-century philosophical writings by Plethon. For references, see Mavroudi 2020: 456. See also Pontani 2014.

98 This translation is no longer extant but is mentioned in the eleventh-century description of India by the astronomer al-Bīrūnī and the tenth-century world chronicle by Agapius, bishop of Manbij.

99 This is reported in the thirteenth century by Bar Hebraeus. For details, see Kraemer 1956: 261; Hilkens 2013: 288.

to Homer a few centuries later.[100] In the ninth century, Ḥunayn b. Isḥāq (809–873 CE), a physician famous for his voluminous translations of Galen from Greek and Syriac into Arabic, frequently rendered the ancient gods as their homonymous planets, or as angels.[101] The identification of the ancient gods with angels has its roots in Philo of Alexandria and early Christian literature in Greek (e.g., Justin Martyr). Hans Daiber argued that its repeated manifestation in Ḥunayn's translations probably reflects a Christian reception of the pseudo-Aristotelian *De mundo*, which is also evident in Syriac and Arabic authors active somewhat earlier or around the same time as Ḥunayn.[102] The *De mundo* mentions correspondences between the elements of nature and several gods and heroes from Greek mythology. For example, it explains which god or hero corresponds to each of the seven celestial spheres and extensively describes the benefic influence of Zeus as the planet Jupiter and as an omnipotent and timeless god in terms very compatible with Abrahamic monotheism.[103] Additional ninth-century Arabic texts of Greek inspiration indicate that the intellectual circles involved in the Greek-to-Arabic translation movement were familiar with the allegorical interpretation of Homer.[104] Correspondences between figures from Greek mythology and elements of nature were embedded in fundamental concepts of astrology and were further echoed in widely influential works such as the *Epistles of the Brethren of Purity* (*Ikhwān al-Ṣafā'*).[105] These Syriac and Arabic literary phenomena are perfectly paralleled in Byzantine Greek: several ancient and Byzantine commentaries on Homer also identify the ancient gods with celestial bodies. Further, Homeric scholia use the term *angelos* for a number of gods. In Byzantine Greek, the kind that Ḥunayn would have been familiar with as a living language, *angelos* means both "messenger" and "angel". Mythological allegory is featured in the poetry of the prominent ninth-century Byzantine scientist Leo the Mathematician.[106]

The above suggests that Homer's portrayal as an alchemist within Syriac and Arabic texts – discussed in the papers by Matteo Martelli and Bojidar Dimitrov in the present volume – must be inscribed within a wider, multi-lingual, and multi-confessional intellectual phenomenon. The manifestations of this phenomenon in the translations of Ḥunayn b. Isḥāq and the *Epistles of the Brethren of Purity* resulted from the Christianization of the technical disciplines (including alchemy) that began in Late Antiquity and continued in the seventh-, eighth-, and ninth-century. This implies a dialogue between Syriac, Arabic, and Greek literary cultures as contemporary medieval

100 See Mavroudi 2020: 457–458.
101 Strohmaier 1968.
102 Daiber 2014: 171–178.
103 The final chapter in *De mundo* is dedicated to explicating Zeus as a benefic presence.
104 For details, see Mavroudi 2020: 459–460.
105 See Ikhwān al-Ṣafā', *De astronomia* (*Ep.* 3), Chapter 1 (ed. Ragep and Mimura 2015: 26–27) on the correspondences between celestial spheres and planets/gods), and Chapter 6 on Jupiter as benefic.
106 Pontani 2005: 142–143.

intellectual phenomena.[107] A place to begin in order to understand the Byzantine contribution to this dialogue would be the Byzantine reception of the *De mundo*: if we follow Daiber's suggestion for its Syriac and Arabic reception and grant that it was a key to the monotheistic reinvention of ancient mythology for the purposes of the natural sciences, then we must also map out its Byzantine reception after the sixth century. The manuscript tradition of the *De mundo* in Greek suggests its potential utility for Christianized alchemy: its earliest surviving manuscript, *Parisinus graecus* 1102, belongs to the eleventh century and is therefore reasonably close in date to the rising popularity of allegory in Arabic literary culture.[108] It copies the *De mundo* together with texts on monastic and Christian spirituality. This suggests that, in the Middle Byzantine intellectual universe, the *De mundo* was understood as a text compatible with Christianity. Indeed, Chapter 6 of the *De mundo*, which discusses God in relationship to the cosmos, states that "all things have come into existence from god and because of god, and that no thing by itself is self-sufficient, if deprived of the preservation deriving from him [...] For god is really the preserver (*sōtēr*) of all things and the begetter (*genētōr*) of everything however it is brought about in this cosmos [...]".[109] This agrees with the Christian (but also Jewish and Muslim) idea of a timeless and providential god who created the universe within time, as opposed to the Aristotelian concept of a timeless, uncreated universe, and a divinity that does not interfere in the cosmos or human affairs. As for the alchemical connotations of the *De mundo*, Chapter 5 briefly explains that cosmic harmony results from bringing together opposites (for example dry and wet, cold and warm) and compares this with how painting, music, and grammar produce harmony out of different colors, sounds, and letters. This calls to mind the goal of alchemy and alchemically inspired approaches to music and grammar (to be further discussed below), which is to discover and take advantage of the way in which all kinds of opposites come together to form the universe as a single harmonious whole. Paying attention to the multiple eleventh-century marginalia on the *De mundo* in MS *Parisinus graecus* 1102 can help further illuminate the Middle Byzantine understanding of this text and its relationship with eleventh-century philosophy and other disciplines.[110]

107 The Syriac and Arabic reception of the *De mundo* has been repeatedly discussed in scholarship. See, for example, Raven 2003. When it comes to the Greek text, all efforts have been expended on understanding the intellectual and historical context that produced it, never its later Greek reception. It is telling that Thom 2014 includes three articles on the Syriac and Arabic versions and none on the Byzantine reception of the text.

108 The *De mundo* survives in several tens of Greek manuscripts, most of them written in the post-Byzantine period. The second oldest manuscript was copied in the thirteenth century. It was translated into Latin in the twelfth century, which suggests a considerable Byzantine reception at that time.

109 Transl. Thom 2014: 43.

110 Even a cursory examination of *Parisinus graecus* 1102 on https://gallica.bnf.fr/ark:/12148/btv1b10723486r/f222.item alerts the researcher to the existence of marginalia to the *De mundo* contemporary to the main text. From f. 225 recto onwards, the text is copied in a tidy and clear Palaeologan hand.

8 Christian Spirituality, Early Sufism, and the Beginnings of Arabic Alchemy

Arabic literary culture recognizes a connection of Christian spirituality with alchemy. A prime example is the assertion, found repeatedly within Islamic Arabic texts, that the Byzantine monk Marianus (Lat. Morienus) taught alchemy to the Umayyad prince Khālid b. Yazīd (d. 704).[111] Its veracity was frequently doubted by modern scholars. Part of the reason was the fact that a text describing the meeting and dialogue between the two was for a while known only in Latin and was considered an original composition in this language rather than a translation from Arabic.[112] However, twenty years ago Ahmad Y. Al-Hassan identified a comparable Arabic text.[113] More recently, Marion Dapsens called attention to the multiplicity of the Latin and Arabic versions and inventoried several Latin and Arabic manuscripts.[114] She also highlighted the numerous alchemical texts circulating in Arabic under the name of Khālid.[115] Overall, the more

111 The instances were collected by Ruska 1924, a publication that rejects the veracity of the Arabic tradition.

112 This is the *Liber de compositione alchemiae*. Translation of the Latin text https://rexresearch1.com/AlchemyArchives/zalchhtm/MorienusBook.htm. A comparison of the Arabic with the Latin text (presented in English translation) is available in Al-Hassan 2004. On the basis of this comparison, one of the discrepancies is the place from which Marianus came: in the Latin text, he is a "Roman" (*Romanus*) monk from Jerusalem. In the Arabic, he is a Byzantine (*rūmī*) monk from Dayr al-Murrān near Damascus (a monastery famous as caliphal retreat in the Umayyad and early Abbasid period). The name "Murrān" is garbled in the Arabic and Latin manuscripts (evidently because someone away from the geography of Damascus had a hard time recognizing it). The translation was made by Robert of Chester, an Englishman who studied Arabic in Spain in the twelfth century and is also translator of one of the earliest versions of the Qur'ān in Latin. To a twelfth-century Englishman, the focus on Jerusalem is Crusader (it highlights the "Romanness" of Crusader Jerusalem). But in the multilingual and age-old Christian apocalyptic tradition, Jerusalem is also a site fundamental to the end of times. It makes sense for Robert of Chester to have encountered a text dealing with an Umayyad prince in Spain (since the Umayyad dynasty lasted there until 1030).

113 See Al-Hassan 2004.

114 Dapsens 2016 indicates a variety of Latin versions (not yet collated). The Latin translation is central, because scholarship considers it as the earliest alchemical text in Latin which launched a Western interest in the discipline (Dapsens 2016: 121). The Arabic manuscripts are late; the earliest date from AH 900 (i.e., towards the end of the fifteenth century). The article provides more information on the individual manuscripts.

115 The alchemical works attributed to Khālid (both in prose and in verse) are presented in Dapsens 2021. See also the edition and translation into English of the Arabic and Latin version of a text attributed to Khālid in Dapsens and Moureau 2021. It is interesting that, based on the inventories of MSS in Dapsens 2016 and Dapsens and Moureau 2021, both Arabic and Latin MSS appear to be late (thirteenth century at the earliest, with most around the fifteenth and sixteenth centuries). This coincides with the fact that the Greek manuscript tradition is also late (with the eminent exception of MS *Marcianus graecus* 299 (coll. 584), which dates to the late tenth century).

recent scholarly trend is to accept, if not the historical accuracy of Khālid's apprenticeship under Marianus, at least its veracity as a metaphor for a Byzantine Christian contribution to the beginnings of Arabic alchemy.[116] In addition, according to the tenth-century geographer and historian Ibn al-Faqīh al-Hamadhānī, the engagement of the eight-century emperor Constantine V with alchemy was the motivation for Caliph al-Manṣūr to experiment with alchemy.[117]

Numerous authors cited in the Arabic alchemical literature are Greek or occur in the Greek Alchemical Corpus. Examples include Plato, Aristotle, Hermes, Apollonius of Tyana, and Ostanes (one of the "Hellenized magians", descendants of Zoroaster, who entered the Greek literary tradition in the Alexandrian period).[118] Modern scholarship has also detected genuine Iranian elements, likely due to the influence of the Sassanian medical schools on the development of early Arabic alchemy.[119] Other alchemists that appear in the Greek and Arabic alchemical corpus, such as Zosimus, Maria the Copt, and Cleopatra, flourished in Egypt – a formerly Byzantine territory where both Greek and Coptic were spoken. The Coptic contribution to alchemy is acknowledged by tenth-century Arabic authors like the alchemist Ibn Umayl and the historian and geographer al-Mas'ūdī.[120] The Egyptian Panopolis/Akhmīm, famous as a site of alchemical activity during the Byzantine period, continued to enjoy this reputation during the early Islamic period.[121] The Coptic engagement with alchemy is confirmed by modern papyrology.[122]

The Arabic literary tradition describes alchemy as coming from multiple languages and cultural spheres but also associates it with two figures seminal for Sunni and Shī'ī Islam (which is significant, in view of alchemy's importance within Ismailism): 'Alī b. Abī Ṭālib, the fourth caliph and son-in-law of the Prophet, and the Imam Ja'far al-Ṣādiq.[123] This association thoroughly Islamicizes an imported discipline – a move equivalent to the Christianization of alchemy discussed earlier.

Ja'far al-Ṣādiq (702–765) was allegedly the teacher of Jābir b. Ḥayyān, the purported author of approximately 215 works on alchemy that survive today.[124] About four hun-

116 Scholars who accept the early contact between Byzantine and Arabic alchemy are listed in Dapsens and Moureau 2021: 559 and include Burnett 1992; Bacchi and Martelli 2009; Martelli 2014b.

117 For a discussion of this narrative (with references to primary sources and earlier scholarship), see Strohmaier 2016; also Strohmaier 1991. On the portrayal of Constantine V as hungry for gold in the Byzantine sources, see Merianos 2022: 307.

118 The Greek Zoroastrian pseudepigrapha were collected by Bidez and Cumont 1938. See also Smith 2002.

119 See Ruska 1923; Lory 2008.

120 For references to primary sources, see El Daly 2005: 51–55 and 169 (for Ibn Umayl), 133 (al-Mas'ūdī).

121 El Daly 2005: 112.

122 On alchemy in Coptic, and the importance of the Egyptian Panopolis, see Richter 2015.

123 For a history of the scholarly investigation on alchemy's shī'ī and ismā'īlī connections, beginning with Kraus, see Lory 2003: 47–51.

124 For a list of titles belonging to the Jābirian corpus, see Kraus 1942–1943: 1.203–210. On the 215 surviving treatises, see Lory 1983: 51.

dred additional treatises, no longer extant but known by title, are also attributed to Jābir.[125] The Jābirian corpus poses serious problems of authorship and dating.[126] The association of Jābir with Ja'far al-Ṣādiq places it to the eighth century. In the 1940s, Paul Kraus argued that the Jābirian corpus is characterized by an incoherence of content; a dizzying multiplicity of texts that could not have been written by the same author; and a number of historical anachronisms, including an awareness of concepts and terms originating in Greek philosophy and introduced to Arabic during the Greek-to-Arabic translation movement of the ninth and tenth centuries. Accordingly, he dated it to the ninth/tenth century. In the 1980s, Pierre Lory suggested that the Jābirian corpus may have grown out of an authentic eight-century core that was rewritten a century later, when more texts attributed to Jābir were also composed.[127] In the early 1990s, Syed Nomanul Haq insisted on the reliability of an eighth-century date. He argued that the Jābirian corpus is internally consistent and itself provides the key to comprehending it: Jābir articulates

> a comprehensive theory of knowledge, language, and music. Thus, under one central principle, namely *Mīzān al-Ḥurūf* ["Balance of Letters"] Jābir attempts a grand philosophical synthesis in which the four natures systematically and coherently relate to phonetics and morphology on the one hand, and to prosody and musical harmony on the other. From all this emerges the ontological counterpart of his reductionist thesis: we have seen that in Jābir's physics all explanations of the natural world were reduced to an explanation of the four natures – now he proposes an ontological equivalence between the natures and the letters of the alphabet. The *Kitāb al-Aḥjār* ["Book of Stones"] is devoted to an elaboration of this very thesis.[128]

Nomanul Haq identified a portion of Aristotle's *Categories* embedded in the *Book of Stones*. This version of the text is completely unrelated to its ninth-century translation by Ḥunayn b. Isḥāq, employs a cruder terminology, and was apparently created for readers not familiar with the Greek logical tradition.[129]

Nomanul Haq's argumentation means that Greek philosophy had been introduced to Arabic about a generation or two earlier than al-Kindī, although he is known as the first self-identified philosopher in the Arabic tradition. This precedent underscores that the Greek-to-Arabic translation movement of the ninth and tenth centuries took place against an intellectual background whereby Muslim speakers of Arabic, who did not

---

125 Kraus, an important contributor to the study of the Jābirian corpus, whose primary objective was to date it to the ninth and tenth centuries (during or after the Baghdadi translation movement from Greek into Arabic), counted 3,000 titles and insisted on the incoherence of their doctrine. This was used as evidence for the pseudepigraphy of the corpus and its dating to the ninth and tenth centuries. For criticism of Kraus, see Nomanul Haq 1994: 11–15.

126 The literature is summarized in Nomanul Haq 2011.

127 Lory 1983: 62–79.

128 Nomanul Haq 1994, based on a thesis submitted to the University of London in 1990; Nomanul Haq 1990: 90–91.

129 Nomanul Haq 1990: 30.

understand Greek or Syriac, had already absorbed concepts and processes of thought originating in Greek philosophy. Although this realization is not new, it does shift the way that modern scholarship evaluated the ninth/tenth-century translation movement. Nineteenth- and twentieth-century scholarship viewed it as a watershed moment after sporadic and unsystematic contact with Greek ideas and translations of Greek material into Arabic that was taking place since the Umayyad period. Given the extent and complexity of the Jābirian corpus, accepting its eighth-century date questions the watershed quality of the translation movement. Instead, it suggests an intellectual continuum from the pre-Islamic through the Umayyad and into the early Abbasid period. This continuum cannot be identified by using religion and language as markers because it adapts to new religious and linguistic environments without losing its recognizability. This transfer of ideas depended not on translated texts but on the cultural transfers enabled by recent converts to Islam, well versed in the Judaeo-Christian intellectual traditions received from Late Antiquity, and able to express them in Arabic and adapt them to the ideological and practical requirements of the new religion.[130] Such a description of intellectual history seriously undermines the linear and teleological narrative on the transmission of philosophy and science outlined at the beginning of this paper.

9 The Connection of Byzantine and Arabic Alchemy with Medicine, Astrology, and the Disciplines of the *Quadrivium*

A modern scholar trying to research how Byzantine and Arabic alchemy were connected with other disciplines and each other from the seventh until the ninth century stumbles upon a practical obstacle that recurs across languages and disciplines: the surviving texts present serious problems of authorship and dating. In addition, they are relatively few, with two notable exceptions: medical writing in Greek and the Jābirian corpus in Arabic. The convergence must be due to comparable historical circumstances that affected the survival of literary production. For Arabists, Islamic culture as we know it was systematized – through imperial intervention contributing to the canonization of its texts and disciplines – in the ninth and tenth centuries. This means that the surviving texts represent this systematization and that the intellectual and educational landscape of the seventh and eighth centuries can be described only vaguely. For Byzantinists, the decline of urban centers (the late antique hubs of education) during

130 Further such instances of cultural transfer have been identified by scholarship. An example is the similarity of early Muslim discussions on free will which call to mind the writings of John of Damascus. For a discussion, see Mavroudi (Forthcoming b).

this period is supposed to have caused the collapse of "secular" education. Literary activity is supposed to have continued in monasteries, which caused the content of education to become more "religious".[131] The concept of a clear division between "secular" and "religious" education in the pre-modern period is, of course, problematic for cultures that did not apply these categories in the way modernity understands them, and Byzantinists need to revisit it.[132] In addition, archeology suggests that some urban centers declined but others did not.[133] Due attention to both disruptions and continuities in urban life during this period can help scholars rethink the conditions under which education was dispensed between the seventh and the ninth centuries. A further explanation for the texts lost in both Greek and Arabic is the fact that, in the course of the ninth century, both the Byzantine and the Islamic world were dealing with the same broader problem that many philosophically based disciplines could help resolve: the relationship of matter with the non-physical realm. In Byzantium, this problem was related to defining what icons were and whether they could play a role in approaching the divine. This was an ongoing debate during the two phases of Iconoclasm (726–787 and 815–843). In the Muslim caliphate, an equivalent problem that climaxed at the very beginning of the ninth century was the createdness or non-createdness of the Qur'ān. In both cases, what was initially accepted as "orthodox" was eventually discredited as unorthodox. The switch must have led to the eventual disappearance of texts that systematized the wrong intellectual position. This common-sense conclusion is confirmed by a palpable fact: modern researchers of the Greek, Arabic, and Syriac literary traditions agree that significant changes in the selection of texts chosen for reading (and therefore copying) were taking place in all three languages during the ninth century.[134]

---

131 For a summary of this historiographical trend, see Chrysos 2002. On monasteries and literary production, see also Hatlie 2007: 262.

132 The problems are outlined in Mavroudi 2021.

133 For example, the decline of Ephesus on the West Coast of Asia Minor is countered by the continued vitality of Thessaloniki, the largest city in the European territories of the Byzantine Empire throughout the Byzantine millennium. It recently became possible to explore the urban fabric of Thessaloniki (which remained one of the largest cities in the empire throughout Byzantine history) from Late Antiquity into the Komnenian period thanks to the spectacularly well-preserved archeological findings during the construction of the city's underground transportation system (effectively begun in 2013). Unfortunately, important opportunities for the preservation and study of these findings were lost. Still, whatever was possible to record in the process of digging for the metro will certainly shift our perspective on the history of Byzantine urban centers within the next two decades as it gradually gets absorbed in the scholarship on Byzantine urban life.

134 For Syriac, on the break in the choice of texts selected for copying around the end of the ninth century, see Brock 2004. Scholars of Arabic literature are well aware of the fact that the handful of texts that survive from earlier than the ninth century (pre-Islamic poetry, works connected with the Islamic revelation such as collections of hadith and Quranic exegesis, the *Sīra* and *maghāzī* literature, etc.) all came down to us in systematizations of the Abbasid period (post 750 CE). In Greek, in addition to the end of Iconoclasm, the introduction of the minuscule is believed to have accentuated the break; see Mavroudi 2015a: 319–321.

If we look for strictly alchemical texts, Jean Letrouit's effort to revisit the chronology of the Greek Alchemical Corpus resulted in placing the alchemist Stephanus and the four poems attributed to Heliodorus, Theophrastus, Hierotheus, and Archelaus to the seventh century,[135] Christianus to the seventh or eighth century,[136] and two anonymous philosophers to the eighth or ninth centuries.[137] More recently, the sixth century resurfaced as a possible date for the alchemical poems and Christianus.[138] In addition, the Christian alchemical oath attributed to Pappus – a name perhaps meant to evoke the fourth-century mathematician Pappus of Alexandria – can be assigned to a time between the sixth and the eighth centuries.[139] Whatever the exact date of these authors, their inclusion in *Marcianus graecus* 299 (coll. 584) guarantees their composition earlier than the tenth century.[140] The continuity of the alchemical tradition in Byzantium between the seventh and the ninth centuries is also supported by the information on its texts and practitioners provided by George Syncellus and Patriarch Photius, who wrote between the late eighth and the late ninth century.[141]

If we compare this Greek evidence with the number of texts in the Jābirian corpus, even if we accept, with Lory, that only some of them actually date from the eighth century, the Greek yield is small: it totals approximately 30,000 words, about half of which are taken up by the nine lectures (*praxeis*) circulating under the name of Stephanus the alchemist.[142] Scholarship has entertained sixth-century dates for all of the aforementioned Greek alchemical authors except for Stephanus, which means that his date within the seventh-to-ninth-century period is the most secure. This implies that the total number of surviving alchemical treatises produced during the "dark centuries" may be limited to what circulates under Stephanus' name. Given the modern reputation of the

135 Letrouit 1995: 63–65. See also Martelli 2013: 1n5.

136 See also the discussion of Christianus' date in Merianos 2022: Section 5.

137 Letrouit 1995.

138 Merianos 2022 entertains the late-sixth century date for Christianus. On the sixth-century date for the alchemical poems, see Lauxtermann 2019: 205–207. Lauxtermann's dating relies on the preferences in the construction of the poetic meter in which they are composed.

139 On the date of Pappus, his Chistian identity, the problem of whether the oath is an earlier text rewritten to include Christianity, and a discussion of its content, see Merianos 2022.

140 Scholars since the eighteenth century have used paleography to date this manuscript; for a list of their opinions, see Roberts 2019: 75n46. For an extensive discussion of its paleography, including photographs of several comparable samples of script, see Roberts 2020: 11–25, arguing in favor of a date in the second half of the tenth century.

141 Merianos and Sakorrafou 2013; Merianos 2017.

142 The specific breakdown, arrived at on the basis of number of words reported for each text on the *Thesaurus Linguae Graecae* (https://stephanus.tlg.uci.edu/; hereafter *TLG*) is approximately as follows: Heliodorus 1,500, Theophrastus 1,500, Hierotheus 1,200, and Archelaus 2,000 (for a total of 6,200 words); Christianus is 5,000 words; the fragments of the two anonymous philosophers (some modern scholars consider them as a single author) total 3,500 words. This totals 14,700 words. Stephanus is 16,500 words. The new edition of Stephanus the alchemist's *De magna et sacra arte* (it now replaces Ideler, whose text is searchable in the electronic *TLG*) is Papathanassiou 2017.

seventh-to-ninth centuries as intellectually impoverished, a modern researcher must assess whether this reflects a low number of alchemical texts produced during this period or results from accidents of survival. The latter is supported by two facts: first, our knowledge of the entire Greek alchemical tradition effectively depends on three or four manuscripts: *Marcianus graecus* 299 (coll. 584) of the tenth century, *Parisinus graecus* 2325 of the thirteenth, *Parisinus graecus* 2327 of the fifteenth, and perhaps also *Laurentianus Pluteus* 86.16, also of the fifteenth century.[143] Second, the surviving Greek technical manuscripts (alchemical, astrological, medical, musical, philosophical) dating earlier than the twelfth century represent only 5% of the total number of manuscripts that we have in these disciplines. This means that what survives does not reflect the state of the Byzantine technical disciplines (including alchemy) during the whole Byzantine millennium (as is sometimes presumed by scholars). Rather, it effectively represents the selection of texts and interpretative approaches to these disciplines that prevailed during the Palaeologan period, from which three of the four key alchemical manuscripts also date.[144]

Compared with the Greek, the world of Arabic manuscripts is considerably less explored in modern scholarship. It is therefore harder to quantify their patterns of survival. Still, Kraus' census of Jābirian manuscripts conveys the impression that they mostly date from the Ottoman period.[145] This suggests that the role of alchemy for Sufism

143 These are the most important manuscripts for the two most extensively surviving Greek alchemical authors, Zosimus and Stephanus. This situation is not particular to alchemy. Our knowledge of the Greek magical tradition also depends on a handful of manuscripts, as outlined earlier, note 82 (MSS *Bononiensis* 3632, *Atheniensis* 1265, and the manuscripts with the *Testament of Solomon*).

144 More detailed discussion and concrete figures in Mavroudi 2023: 253–254 and 269–270; 2024a.

145 See Kraus 1942–1943: 1.176–188. The evidence on which this impression is based is not thorough: Kraus sometimes does not list individual manuscripts but sends the reader to library catalogues. And even when individual manuscripts are listed, Kraus does not always provide dates. When he provides dates, he does not always specify whether they correspond to the Christian or the Hijrī calendar. The reluctance to date is understandable for reasons inherent to the study of Arabic paleography. Greek paleography was inaugurated as a modern discipline with Bernard de Montfaucon, *Palaeographia graeca* (1705), while the study of Arabic paleography effectively begun in the late nineteenth century, through encountering the scripts of Arabic papyri. The first reference books on the study of Arabic manuscripts appeared only at the beginning of the twenty-first century. As a result, attempting to date Arabic manuscripts through paleographic observations is an exercise in which Arabists engage much less frequently than Byzantinists or Hellenists. For example, while catalogues of Greek manuscripts always provide a date for the manuscript described (even if it is suggested only by paleography), many Arabic manuscript catalogues do not attempt to date the described manuscripts unless there is a colophon or other note that helps such dating. I took all of Kraus' references to a date as corresponding to the Hijrī calendar. The impression of late manuscripts from Kraus is confirmed by the following: three of the four manuscripts that contain the Jābirian *Book of the Stones*, listed by Nomanul Haq 1990: 120–121, belong to the seventeenth century (I have been unable to verify the date of the Cairo manuscript). Likewise, the inventory of manuscripts in Dapsens 2016: 124–126, when their date is indicated, are all from the Ottoman period. In addition, among the twelve manuscripts listed in Dapsens and Moureau 2021, one

and the importance of Sufism as a living tradition in the Ottoman and post-Ottoman world played a role in the production and preservation of these manuscripts. There is no equivalent social role for Greek alchemy in the late medieval or early modern context, which helps explain why Greek alchemical manuscripts survive in considerably fewer numbers than Arabic ones.

Astrology and medicine are disciplines related to alchemy. Although the quantity of texts that survive in these disciplines in Greek and Arabic are not equivalent, they share with alchemy an important feature: statements from within the Arabic literary tradition claim that knowledge of alchemy, astrology, and medicine was transferred from Greek into Arabic earlier than the ninth century – that is, earlier than the Greek-to-Arabic translation movement.

Greek astrology is characterized by a paucity of surviving texts and authors securely dateable to the seventh, eighth, or ninth centuries. Two are the known authors of this period: Stephanus (once thought to be the same person writing in multiple disciplines, including alchemy but now considered as four different authors);[146] and the Graeco-Syriac Theophilus of Edessa, who served Caliph al-Mahdī.[147] The profile of Theophilus indicates a close relationship between Greek and Arabic astrology during this time. Indeed, both Stephanus and Theophilus are mentioned in the *Discussions* (*Mudhakarāt*) between the famous ninth-century astrologer Abū Maʿshar al-Balkhī (a member of al-Kindī's circle) and Abū Saʿīd Shādhān b. Baḥr.[148] This and other texts written in the tenth century and later, and especially the Arabic bio-bibliographical tradition, record a robust astrological literature in Arabic during the eighth and ninth centuries, enriched from Greek, Syriac, Middle Persian, and Indian sources. Within the Arabic literary tradition, it is possible to find claims that the importation of Greek astrology into Arabic started during the Umayyad period: according to al-Bīrūnī's eleventh-century testimony, the first translations of Greek astrological works into Arabic were made for the alchemist prince Khālid b. Yazīd. In addition, a Greek astrological work attributed to Hermes was rendered into Arabic in the year 743, seven years before the fall of the Umayyads.[149] The earliest surviving astrological work composed in Arabic is probably the *Epistle on the Secrets of Astrology* by Abū Sahl b. Nawbakht, written in the second

is from the thirteenth century (unusually early), one from the fifteenth, and the rest from the sixteenth until the twentieth century.

146 On the identity of Stephanus, see earlier, note 35. The historiography on Stephanus' identity is long and cannot be outlined here.

147 On the problem of reconstructing Theophilus' profile and works, see Mavroudi 2006: 87–88n148. For a biography, see Van Rompay 2011.

148 Rosenthal 1963: 454–456. The name of Stephanus is omitted from the Greek and the Latin translation of the work; Rosenthal 1963: 455n3.

149 See Sezgin 1967–1984: vol. 7, Part I.II (*Quellen*).

half of the eighth century, before the inception of the translation movement.[150] Several astrological treatises written in Arabic during the ninth century survive.

The considerable discrepancy between the Greek and Arabic literary record, in terms of both surviving texts and names of known authors whose writings no longer exist, can be explained by a considerable difference in patterns of preservation. The most important avenue for our knowledge of early Arabic astrology is the bio-bibliographical tradition – a genre that does not exist in Greek.[151] In addition, the world of Arabic manuscripts is exponentially larger than its Greek equivalent: surviving Greek manuscripts are approximately 100,000 and are mostly kept in European libraries; surviving Arabic manuscripts can be anywhere between one and two million and exist in libraries from Indonesia to Gibraltar. In addition, Greek literary culture began transitioning to the printing press in the fifteenth century, while Arabic literary culture did the same in the nineteenth.[152] A factor aggravating the survival of Greek astrological texts written between the seventh and the ninth centuries is the disarray in the manuscript tradition of the Greek astrology in its entirety – a phenomenon completely analogous to the problems inherent in the Greek alchemical tradition outlined earlier and known to everyone who has worked with Greek astrological manuscripts. As Stephan Heilen pointed out, Greek astrology came down to us in Byzantine manuscripts that are working copies of practicing astrologers. Only the *Tetrabiblos* remained relatively free of scribal and user intervention over the centuries. To illustrate the extent of the problem: David Pingree, who prepared the available critical editions of most ancient and Byzantine Greek astrological texts, accepted that, for works written in prose, what came out of the pen of the ancient author is not retrievable. The only viable alternative was to carefully distinguish and publish the earliest Byzantine versions of these texts that can be retrieved from the manuscripts, always preserving their lacunae and contradictions. This editorial approach resulted in texts that are frequently incomprehensible to modern readers.[153]

Compared with alchemy and astrology, medicine written in Greek is significantly better attested between the seventh and the ninth century. Its surviving texts are literally twenty times more abundant than those of the second runner up, alchemy.[154] Perhaps the most widely received medical author writing in Greek between the seventh

150 This is my conclusion from reading Sezgin 1967–1984: vol. 7 and Ullmann 1972: 303. On Abū Sahl b. Nawbakht, see Pingree 1983. It is impossible to address here the complicated question of originally Greek astrological works translated from Greek into Pahlavi in the Sassanian period and from Pahlavi into Greek in the Islamic period.

151 See Mavroudi 2024b.

152 For an additional comparison of Greek and Arabic manuscript survival, see Mavroudi 2024a.

153 Heilen 2010.

154 According to the word count of alchemical works above, alchemy is close to 30,000 words. As the word count of medical texts below will make clear, the total length of medical works written in this period amounts to something like 600,000 words.

and the ninth century is Paul of Aegina. Others are Paul of Nicaea, Leo the Physician, Theophanes, and Meletius the Monk.[155] Medical writings are also associated with the aforementioned Stephanus and are so many that, as in the case of alchemy under Stephanus' name, their total length surpasses that of the remaining authors of this period put together.[156] The reach of these authors beyond the Greek language is also notable: a number among them were quoted in Arabic and Latin medical writings.[157] The remarkably better survival of medical writings in Greek can perhaps be explained by the fact that, among the philosophically based disciplines, medicine would be the least affected by iconoclastic considerations.

As for medicine in Arabic, modern scholars collected the scant information that we have on its application within the pre-Islamic and early Islamic Arabian peninsula and argued for a preponderance of what modernity calls "folk" medicine.[158] This is a problematic evaluation because, during the last thirty years and longer, the social sciences are extensively revisiting the division between "folk" and "learned" traditions and insist that there was no barrier between the two.[159] In addition, the bio-bibliographies and medical writings of the Abbasid era record the names of Arab physicians from the pre-Islamic and early Islamic period whose methods had been incorporated into later Arabic medicine that followed the Greek medical tradition. An example is al-Ḥārith b. Kalada al-Thaqafī, who was born in the pre-Islamic period, studied medicine at the famous school of Gundishapur, and practiced into the Umayyad period. Part

155 An outline of medical writings after the year 700 is provided in Miller 2017. Miller adds Theophanes Chrysobalantes to this count. I exclude Theophanes here, accepting Sonderkamp's dating of his writings to the tenth century (during the reign of Constantine VII).

156 A count of words based on the *TLG* is approximately as follows: Paul of Aegina is 197,000 words; Paul of Nicaea 31,000; Meletius 40,000; Leo Medicus 7,000 + 10,000 = 17,000 words. This totals 185,000 words for all these writers together. A crude count of the different treatises attributed to Stephanus (without trying to resolve problems of authorship, dating, and overlap in content) can be added up as follows: Stephanus the medical commentator has 28,000 + 50,000 + 1,000 + 6,000 + 46,000 + 77,000 + 48,000, Stephanus the author of uroscopy has 12,000 words; Stephanus the commentator on Hippocrates' aphorisms 68,000. This adds up to 324,000 words. The count excludes the works attributed to Theophanes Chrysobalantes/Nonnus because of the complicated authorship problems and the possibility of a tenth-century date. At any rate, an exact count is not the point here. The goal is to convey an approximation of bulk.

157 See Mavroudi 2015b: 43–44.

158 This is suggested by Ullmann 1979: 1–6, which labels pre-Islamic medicine within the Arabian peninsula as "folk" and related to magic; after the Muslim conquests, the Greek medical tradition was introduced into Arabic through Greek, Syriac, Persian, and Jewish physicians already during the Umayyad period. The more recent book by Pormann and Savage-Smith 2007 rewrote Ullmann through a post-colonial lens without introducing any substantially new facts. Accordingly, Pormann and Savage-Smith 2007: 6–9 provide an account of pre-Islamic Arabic medicine by maintaining a divide between Arabic "folk" medicine and the "learned" tradition, describing the two as co-existing in creative tension to this day.

159 An example, within Byzantine studies, is Agapitos 1991.

of a medical text written in dialogue form survives under his name.[160] It bears emphasizing that, in spite of the universality of this method of exposition,[161] Byzantinists have remarked long ago that it is followed in an unusually high number of Greek technical compositions dateable between the seventh and the ninth century.[162] Many physicians active in the Umayyad period are known, mostly Christian and versed in the Greek medical tradition.[163] The earliest known translator of a medical work into Arabic is the Syriac-speaking Jewish physician Masarjawaih al-Baṣrī, who was active in the 680s. He translated into Arabic the work of Ahron, a physician and presbyter from sixth- or early seventh-century Alexandria.[164]

When it comes to the disciplines of the *quadrivium* in Greek, no author dateable to this period is known, except for Bacchius Geron (evidently a pseudonym), to whom a short treatise on music is attributed.[165] Like musically informed alchemy, this text emphasizes the connection of music with proportion and number.

If we now turn to the disciplines of the *trivium*, many of the grammatical and logical handbooks written in Greek during this period were likely composed in lands under Arab rule: Syria, Palestine, and Egypt.[166] This is significant for the present effort to draw analogies between Greek and Arabic approaches to alchemy. Acevedo has underlined the importance of grammar for Greek and Arabic alphanumerical cosmology embraced by alchemy. The connection of logic with mathematics in Greek antiquity (extending to shared technical terminology, techniques of proof, structure of mathematical demonstrations) was acknowledged by modern scholarship long ago.[167] In the ninth century, al-Kindī considered both logic and mathematics as a necessary background to all kinds of organized inquiry.[168] Unfortunately, scholarship has not yet explored the connection between logic and mathematics in the Syriac or Greek literary universe after the sixth century. Yet it is reasonable to deduce (and perhaps future research will show) that al-Kindī did not resurrect this approach directly from Greek antiquity through reading Greek texts in Arabic translation. Rather, he must have encountered it through access to his contemporary world of Christian learning, expressed in both Greek and Syriac.

---

160 References in Sezgin 1967–1984: vol. 3, Chapter 1 (*Medizin und Pharmazie*).

161 On the universality of the format, see Papadogiannakis 2013.

162 Cf. the musical treatise by Bacchius Geron below; see also Chrysos 2002: 17–18; Kazhdan 1991.

163 These are listed in Sezgin 1967–1984: vol. 3.

164 See Schulze 2006.

165 Bacchius Geron, *Isagoge artis musicae* (ed. Zanoncelli 1990: 252–284). This text consists in 3,000 words written in question-and-answer format (a popular writing choice for texts surviving from this period). The name of the purported author sounds fictional. On a shorter text extracted from this longer one, and what is known about the author, see Terzes 2023: 118n2.

166 For a discussion on grammar, see Mavroudi 2015a: 327–331; see also Mavroudi 2026.

167 A landmark was Benedict Einarson (1936) who observed similarities in the language of ancient Greek mathematics and of Aristotle's logical works. See Smith 1978.

168 Rescher 1963: 46: "Logic (and mathematics) are thus regarded as propaedeutic to all scientific inquiries".

For the time being, the liveliness of the logical tradition in Greek during the "dark centuries" is demonstrated by the composition of textbooks, many collecting definitions of terms. Mossman Roueché clarified the composition and reception of several such, explaining the rootedness of this Christian tradition in ancient logic and the manner in which it reflects contemporary Christian concerns.[169] Christophe Erisman showed the interrelatedness of this Christian logic with medicine.[170] Gerasimos Merianos recently pointed out that one of the most important authors of the second half of the seventh century, Maximus the Confessor, made extensive reference to formal logic and expounded concepts that two Greek alchemical authors writing in this period also engaged with.[171]

Collections of logical definitions were included in Greek works composed under Muslim rule during the late eighth and ninth centuries such as the *Guide* (*Hodēgos*) by Anastasius of Sinai and the *Dialectica* by John of Damascus.[172] In addition, logical definitions are included in the important anthology of patristic literature known as the *Doctrina Patrum*.[173] Its earliest surviving manuscript, the famous *Vaticanus graecus* 2200, was copied in Syria, Palestine, or Egypt towards the end of the eighth and the beginning of the ninth century.[174] This means that it circulated in Greek-speaking Christian circles under Muslim rule. Under these circumstances, it is possible to argue that the translation of Aristotle's *Categories* within the *Book of Stones* identified by Nomanul Haq, if its eighth-century date is correct, was made against the background of a very lively engagement with Greek logic in the Christian circles of the Muslim caliphate. Further, the Jābirian corpus included now lost treatises on logic (*manṭiq*).[175] If these belonged to its eighth-century core, it means that they were written during the concurrent cultivation of Greek logic in Muslim lands. The development of *kalām* in Arabic – also as a result of early ninth-century debates on the createdness or non-createdness of the Qur'ān) may help explain the loss of treatises on *manṭiq* from whatever the eighth-century core of the Jābirian corpus may have been.

A prominent feature of the Jābirian corpus that is not shared to the same degree by the Greek alchemical tradition as we now have it is its complex arithmology, referred to, within the corpus, as *ʿilm al-mīzān* ("science" or "method of the balance"). This draws numerical correspondences to the four elements (fire, air, water, earth) and their natures (hot, dry, cold, wet) as known from Greek philosophy and medicine. Every component of the universe represents an equilibrium of natures which can be expressed with numbers, and especially the sequence 1:3:5:8. These numbers reflect the musical harmony

169 Roueché 1974; 1980. A helpful summary presentation of the textbooks can be found in Louth 2002: 42–44.
170 Erismann 2017.
171 The two authors are Christianus and the Anonymous Philosopher; see Merianos 2022.
172 The logical deifinitions are included in the second chapter.
173 *Doctrina Patrum* 33 (ed. Diekamp 1907: 249–266).
174 On this manuscript, see https://spotlight.vatlib.it/greek-paleography/catalog/Vat_gr_2200.
175 For a census of these treatises, see Kraus 1942–1943: 1.162–165.

governing the heavens. Accordingly, the different numbers that express the intensity of the natures correspond to the ratios of the musical intervals.[176] Seventeen is the key number of the balance and indicates the harmony of everything in the sublunar world. Words are formed according to the image of the thing that they signify and are made up of letters that have numerical value. Accordingly, the numbers embedded in the words that express each thing correspond to the numerical value of the elements that constitute this thing.[177] Kraus identified the sources of this arithmology and cosmology as Pythagorean, Platonic, Aristotelian, Neoplatonic, and Neopythagorean.[178] This agrees with claims made within the Jābirian corpus, which professes a Greek origin for some of its key concepts, including the "method of the balance" itself. For example, the *Kitāb al-tajmīʿ* (Book of Concentration) attributes to Zosimus a *Kitāb fī al-mīzān* (Book on the Balance) which Kraus identified as a treatise mentioned in the surviving Greek alchemical fragments as *Peri stathmōn* (On Measures).[179]

Gerasimos Merianos recently drew attention to the participation of the alchemists Christianus and Stephanus to this ancient and medieval tradition of mathematical philosophical inquiry. Neither author can be dated precisely, but they wrote roughly between the seventh and the tenth century – that is, around the time that the Jābirian corpus came into being.[180] "Christianus" is not the author's proper name but an adjective that underlines his "neo-Chalcedonian" Christian faith, as opposed to the paganism of earlier alchemists whom he mentions.[181] As pointed out by Merianos, Christianus inscribes himself within the aforementioned understanding of alchemy as spiritual practice that can lead to assimilation with the divine. Mathematics is part of this spiritual path. Christianus' mathematization of alchemy is based on conceiving of matter as quadripartite, with each of matter's constituent elements symbolized by four parts of the egg. The different ways of combining various parts of the egg according to three methods of processing (with fire, without fire, mixed) and three states of compounds (dry, liquid, middle) result in 9, 36, 54, or again 36 generic classes of production. This adds up to 135 methods of producing something alchemically – a number that is taken for granted by an anonymous alchemical author writing in Greek around the eighth or ninth century.[182]

---

176 For a detailed analysis of Jābirian arithmology, see Kraus 1942–1943: 2.187–199.

177 For a brief exposition of the Jābirian system as a coherent whole, see Kraus 1942–1943: 2.308–311.

178 On the Greek sources of Jābirian arithmology, see Kraus 1942–1943: 2.199–223.

179 Kraus 1942–1943: 2.306. The *Book of Concentration* explains how to generate creatures (scorpions, snakes, humans) that obey their creator.

180 The tenth-century *terminus ante quem* is deduced on the basis of their presence in *Marcianus graecus* 299 (coll. 584), on the date of which see above, note 140.

181 Merianos 2022: 271–275; on the characterization of his faith as "neo-Chalcedonian", see Merianos 2024: 31–32.

182 For a more detailed explanation, see Merianos 2022: 295–296.

Further, as Merianos explained, Christianus approaches the alchemical matter and processes by using definitions from logic, such as “species” and “genus”, because they can help spell out what a thing is.[183] For example, he endeavors to demonstrate that “the material of science is one and only in terms of species” (a principle that he acknowledges as well-known to alchemists).[184] He also dedicates a chapter to “How Many are the Differences of Productions by Species and Genus” where he explains how substances combine through various methods of treatment and yield compounds of different states.[185] In another chapter, Christianus correlates the alchemical processes with geometrical shapes and solids (the triangle, quadrangle, semicircle, circle, pyramid, octahedron).[186] Further, Stephanus and the aforementioned anonymous philosopher in particular correlate alchemy with music.[187]

As Kraus and Merianos already observed, the elements in the Greek alchemical writings dateable between the seventh and the ninth centuries call to mind similar elements that are more extensively addressed in the Jābirian corpus. Both the Greek and the Arabic alchemical material develop concepts traceable to Greek antiquity but do so in different ways. For example, the importance of the number 17 or the analogies 1:3:5:8 cannot be found in the extant Greek alchemical writings, although it is possible to trace their origin to ancient Greek sources.[188]

The question is: what do these analogies mean for the medieval transmission of ancient philosophy and science? If we try to extract a linear narrative (who did what first and therefore which text, author, or linguistic/religious realm transmitted what where), we immediately run into the problem of precisely dating each text in the Jābirian and Christian Greek alchemical corpora, which may be unsolvable. All that can be said with certainty is that, at around the same time, alchemical authors writing in Greek and Arabic were concerned with some of the same problems and used ideas rooted in the same intellectual traditions of Graeco-Roman antiquity. In addition, the fact that both the Jābirian corpus and the Christian Greek alchemical writings explicitly acknowledge an ancient Greek origin for many of their fundamental concepts does not

183 See Siorvanes 1996: 74, quoted by Merianos 2022: 297.

184 Merianos 2022: 294.

185 Merianos 2022: 294.

186 Merianos 2022: 297–298.

187 Merianos 2022: 296–302; the treatise of the anonymous philosopher is edited in Berthelot and Ruelle 1887–1888: 2.433–441. It is, effectively, an extensive comparison of music as a mathematical enterprise with alchemical procedures. On music in Stephanus’ writings, see Papathanassiou 2017: 84, 140, 149, 151, 161. Among Stephanus’ lectures, the one most engaged with music is Lecture 2 (*De magna et sacra arte* 2 [ed. Papathanassiou 2017: 161–166]).

188 Kraus 1942–1943: 2.208–211, explains that Aristotle and Plutarch help establish the importance of number 17 for Pythagorean thought. Jābirian influence (via the Latin) can be detected in a Byzantine text composed around the year 1300; see Colinet 2000.

help answer the question of where concrete ideas originate.[189] The reason is that narratives on the transmission of disciplines found within the primary sources are motivated by the ideological and political conditions in which these sources were written. This means that they inform us on the social context rather than the factual details of intellectual transmission.[190]

10 Conclusions through the Concrete Example of Magic Squares

A circumscribed example through which to address the problem of understanding the transmission of knowledge based on fragmentary and seemingly conflicting pieces of evidence is the appearance of magic squares in Greek, Arabic, and other languages. A magic square is made up of smaller ones, each containing a number. The horizontal, vertical, and diagonal row of numbers must add up to the same total. If the universe is encoded in and represented by numbers, magic squares can serve as a means to represent, but also manipulate, its forces. If we accept an eighth-century date for the Jābirian corpus, this possibly counts as their earliest attestation in Arabic.[191] Magic squares had an enormous reception in later Arabic/Islamic magic and Sufism. Modern scholarship has debated at length whether magic squares were received in Arabic from a pre-existing Greek mathematical tradition or not. Statements within the Arabic literary culture repeatedly indicate a Greek origin. For example, a relatively simple magic square transmitted in the Jābirian treatise titled the *Kitāb al-mawāzīn al-ṣaghīr* (Small Book of Balances) is attributed to Apollonius of Tyana.[192] Later Arabic statements regarding a Greek origin can also be found. A famous example is Qazwīnī's *Wonders of the Creation* (*'Ajā'ib al-Makhlūqāt*, thirteenth century), which acknowledges Archimedes as the inventor of magic squares.[193] These Arabic testimonies conflict with the fact that, within existing Greek writings, magic squares appear for the first time in a short treatise by Manuel Moschopulus (late thirteenth–early fourteenth century). In his 1904 study of this text, Paul Tannery debated whether they were a medieval import from Arabic into Greek or were invented in the Hellenistic period without leaving any trace in the surviving Greek literary production.[194] In the 1990s, Jacques Sesiano explained

189 For example, Christianus openly acknowledges Plato as an influence. See Merianos 2022: 278–279, 294, 296–297.

190 Explanations using geomancy as an example in Mavroudi 2023: 220–227.

191 A survey of early attestations of magic squares within Arabic literary culture (edition, translation, commentary on texts) can be found in Hallum 2021.

192 Arabic text, translation, and commentary in Hallum 2021: 78–80.

193 Kraus 1942–1943: 2.73n1.

194 See the discussion in Kraus 1942–1943: 2.73n1.

that Moschopulus did not have a profound knowledge of magic squares but worked on them, as he indicates, at the instigation of Nicolaus Artabasdus Rhabdas. It bears emphasizing that Byzantinists know Rhabdas for, among other things, his treatment of a topic related to alchemy: his solution to problems pertaining to the minting of gold coins.[195] According to Sesiano, Moschopulus seems to have derived his method from Arabic sources written in the eleventh century at the latest. From this evidence, Sesiano deduced that magic squares were unknown in Byzantium and were imported from the Islamic world at the beginning of the fourteenth century thanks to the astronomical and astrological activity of Gregory Chioniades.[196]

One can point out at least three problems with this conclusion. First, it implicitly presumes that we still have approximately everything ever written in Greek and arrives at a sweeping generalization based on a single piece of evidence. The aforementioned patterns in Greek manuscript preservation and what they mean for the survival of Greek literary culture indicate the precariousness of such an approach. Second, Sesiano's conclusion is attached to a linear and overly simplistic narrative on the centrality of Chioniades for the history of the Byzantine mathematical sciences and their contact with the Islamic intellectual universe. Chioniades' image as a pioneer of Islamic ideas within the Byzantine cultural sphere was created in sixteenth-century Europe and has been repeated in various histories of the mathematical sciences ever since. Recent scholarship explained why it cannot be taken for granted.[197] Third, Sesiano's conclusion is undermined by a more recent suggestion by Nicolas Vinel.[198] According to Vinel, allusions to magic squares can be found in Iamblichus' fourth-century commentary to the *Arithmetical Introduction* by Nicomachus of Gerasa. This would be consistent with the claims of the Arabic sources that magic squares were an import from the Greek intellectual universe.

Scholarship has also discussed a probable origin of magic squares in China or India.[199] Cristian Tolsa recently examined the earliest attestations of magic squares in Arabic. Since he accepts a tenth-century date for the Jābirian corpus, he considers as the earliest Arabic attestation of a magic square the one described in the ninth-century medical encyclopedia *Firdaws al-ḥikma* (Paradise of Wisdom). There, the magic square is attributed to the author's father, Sahl Rabbān al-Ṭabarī, which dates it to the early ninth or perhaps late eighth century.[200] The instructions on using the square include the injunction to write around it two lines from the Psalms in Syriac. This is not surprising, given that the author of the *Firdaws al-ḥikma* was a convert to Islam while

195 See Stoyas 2013.

196 See Sesiano 1998.

197 On the problems of this narrative, see Mavroudi 2023: 233–238.

198 Vinel 2005.

199 See Tolsa 2020–2021: 8–11.

200 For a discussion of the date, authorship, provenience, and text of this magic square, see Tolsa 2020–2021: 9–15. See also Hallum 2021: 71–75.

his father, Sahl Rabbān, was a learned Syriac-speaking Christian. Tolsa explained that the instructions on how to use the magic square call to mind Indian ritual. As for the Psalmic verses, they appear as a Nestorian Christian adaptation of an equivalent Indian text. The adaptor may have been Sahl Rabbān himself. Accordingly, Tolsa dismissed Vinel's arguments and preferred an Indian origin for magic squares at large.

More recently, in his own investigation of the early history of magic squares in Arabic, Bink Hallum called attention to a recent archeological discovery analyzed by Julia Lougovaya that seems to confirm Vinel's suggestion: a fifth- or sixth-century writing tablet found in Antinoopolis in Egypt contains an arrangement of Greek letters that looks like a proto-magic square.[201] Hallum also emphasized the close connection between magic squares and the "science of letters" (*'ilm al-ḥurūf*). The kinship between the Arabic "science of letters" and practices attested in Greek, Syriac, and Hebrew earlier than the seventh century have been mentioned earlier. This lends further credence to the deduction that magic squares have roots in Late Antiquity. However, this does not condemn us to constantly debating where each element in every intellectual edifice comes from. Neither does it necessarily lead to a recirculation of the centuries-old argument that every intellectual edifice under the sun can be traced to Graeco-Roman, or Chinese, or Egyptian, or Ancient Near Eastern antiquity, whence it was eventually transmitted to the modern "West".

In his discussion of magic squares, Bink Hallum pointed out the following:

> The history of *awfāq* [Arabic for "magic squares"] has most often been discussed solely from the perspective of the history of mathematics, frequently with a focus on locating scientific firsts. But history, be it intellectual, cultural, or social, is not best understood as a series of eureka moments, as if only the first instance of every historical phenomenon is important or that every repetition of an action or idea, with all its contextual variations and permutations, is derivative and therefore of limited interest. Focusing on firsts leads to a general ignorance of the historic and cultural significance of *awfāq* in Islamicate societies.[202]

This agrees with a broader tendency within history of science to pay equal attention to the content of scientific activity and the social context in which this activity occurs. Such a trend leads away from pursuing the transmission of philosophy and science as a linear narrative of ideas passed on from one civilization to the next, becoming more complex in the process, and resulting in progress (a concept effectively indebted to nineteenth-century German idealism). Instead, it implies the need to construe the history of philosophy and science as a series of interconnected episodes, in which some of the same questions were asked but different intellectual, social, and political considerations led to related but effectively different results. Within this new conceptualization

201 Lougovaya 2017; Hallum 2021: 63.
202 Hallum 2021: 59.

of history of science, it becomes possible to find a place for the Indian (and Chinese) engagement with magic squares.

Given the evidence surveyed above, it is possible to argue that magic squares and their related disciplines were imparted to the Islamicate world by both the Greek and the Indian tradition – and likely also other cultures with which the vast geography inhabited by Muslims was in contact with, as suggested by the example of Sahl Rabbān, a Christian inhabiting a geography under Persian cultural influence. After all, the idea that Greek philosophy arrived in Greece from India was expressed already in antiquity and was occasionally revisited by modern scholarship.[203] This suggests that a certain compatibility between the Greek and the Indian thought world was observable to anyone who cared to investigate them. Trade between the Indian Ocean and the Greek-speaking world is attested from the Roman period into the sixth century and creates the possibility of continued intellectual contact.[204] Magic squares (and other ideas) may have been exchanged between the Indian and the Greek cultural spheres during the Alexandrian period, perhaps as part of swapped medical or astrological lore. However, it would be overly simplistic to imagine that ideas traveled in exclusively one direction from India to Greece (following the logic of "Light from the East", which encapsulates an ancient but enduringly popular motif in intellectual history), and that such a transfer happened only once, namely immediately following Alexander's conquests.[205] The reason is that Indian and Greek medicine appear to have influenced one another. In addition, evidence dating from a couple of centuries after Alexander's conquests indicates that Greek astronomy and astrology were indisputably known in India.[206] This was part of an enduring bi-directional exchange: in the 1950s, David Pingree stumbled upon MS *Vaticanus graecus* 1056, a fourteenth-century Byzantine manuscript preserving a twelfth-century astrological collection. On its margins, he noticed references to Indian astrological ideas. He checked the printed edition of a text by Varahamihira, a sixth-century Indian astronomer and mathematician, and found in it transliterations of Greek technical terms in an Indian script used from the eighth century onwards. This suggested a continuous currency of the Greek terms in Indian astrology for centuries after Varahamihira's lifetime. Since the material in MS *Vaticanus graecus* 1056 had clearly

---

203 For a recent reiteration, see Beckwith 2015. Beckwith uses philological evidence to argue that Greek philosophical skepticism is derived from early Buddhism through direct contact between thinkers during the conquests of Alexander the Great.

204 For an account of the primary sources attesting to Graeco-Roman knowledge of India (written for the non-specialist), see Stoneman 2019: 5–35. The latest textual evidence for direct knowledge of India in the Byzantine world is the *Christian Topography* by the so-called "Cosmas Indicopleustes", who traveled several times to India during the reign of Justinian (not discussed by Stoneman). On Indicopleustes, see Kominko 2013: 10–17.

205 On the label "Light from the East" as a historiographical cliché, see Mavroudi 2023: 265, esp. n207.

206 On the influence of Indian on Greek medicine, see Stoneman 2019: 231–232; on the influence of Greek on Indian medicine, astronomy, and astrology, see pp. 400–401.

been translated from the Arabic, it became obvious to Pingree that Greek astrology had traveled to India, from there to Islam, and thence to Byzantium.[207] Understanding how this occurred became his life-long project.

Richard Stoneman's *The Greek Experience of India* discusses the different interpretations of modern scholars regarding the Graeco-Indian encounter, which has repeatedly served as a metaphor for the confrontation of India with Britain as a colonial power.[208] In such a context, "Greek" is identified with the "West" and therefore the British colonizers and Western modernity. Accordingly, Indian historians who viewed British colonialism in purely negative terms insisted on an Indian culture free of Greek influence. This asserts the strength and superiority of "native" Indian culture vis-à-vis the modern colonial establishment. Conversely, Indian historians who were positively inclined towards the values of Western modernity were ready to highlight the absorption of Greek ideas by Indian culture since antiquity. This position does not dwell on the negative aspects of British colonialism but suggests modern India's inherent ability to embrace and emulate Western "progress". For example, before Indian independence, Ananda Kentish Coomaraswamy insisted that there is nothing Greek about the art of Gandhara; after independence, Rafique Ali Jairazbhoy enthusiastically enumerated the Indian disciplines under Greek influence, such as philosophy, literature, astronomy, painting, and sculpture.[209] As an alternative to the cultural polemics implied in such historiographical approaches, Stoneman suggested focusing on the "middle ground" – a concept famously proposed by the historian Richard White in 1991 to describe the native American encounter with the European settlers in North America.[210] This enables the historian to explain how both parties in a cultural or other encounter change and develop as a result of mutual contact.

In order to produce a "middle ground" narrative on the ancient and medieval transmission of magic squares and Jābirian arithmology more generally, one must begin with a chronological outline of the surviving evidence that remains mindful of the gaps in the existing primary sources. This would read as follows: allusions to

207 See Calder III and Heilen 2007: 517: "My interest in the transmission of scientific ideas from one culture to another was awakened in 1955, the year after my graduation from Harvard College where I had concentrated on Classics and Sanskrit. I spent that year in Rome on a Fulbright, studying the paleography of Greek manuscripts in the Vatican Library. In the margins of one such manuscript, *Vaticanus graecus* 1056, I noticed references to Indian astrological ideas. In the library I found a printed edition of Varahamihira's *Brhajjataka*, and found therein Nagari transliterations of Greek technical terms. Since the material in *Vaticanus graecus* 1056 was clearly translated from Arabic, it was clear that Greek astrology had travelled to India, from there to Islam, and thence to Byzantium. As I soon found out, very little research had yet been carried out in this field".

208 On the historiographical approaches to Alexander's campaign to India by British and Indian writers of the late eighteenth and early nineteenth centuries, see Vasunia 2013.

209 Stoneman 2019: 400–401.

210 Stoneman 2019: 401–402.

magic squares occur in Chinese sources from the two centuries BCE, while their earliest known discussion in Chinese is dated to the first century CE. Chronologically next is the earliest allusion to a magic square in Greek, which can be found in the work of Iamblichus from the fourth century CE. A Greek tablet found in Egypt and dating from a century or two later contains a crude type of magic square. The earliest known magic square in Sanskrit is from the mid-sixth century CE. The Chinese and Indian evidence for magic squares is more direct than in the Greek examples but still mathematically very simple. The earliest known magic square in Arabic is a Christianized version of a Hindu example and also mathematically simple. More complex magic squares appear in Arabic during the tenth century. Later Arabic magic squares are mathematically more sophisticated. The earliest mathematical discussion of magic squares in Greek dates from the fourteenth century and is based on the simpler Arabic approaches from the tenth or eleventh century.[211] Given the late antique evidence for a Greek engagement with magic squares, the mathematical simplicity of the fourteenth-century text suggests that its role within the Byzantine intellectual universe was similar to that of other Arabic-to-Greek imports: to confirm a pre-existing Greek tradition while repackaging it in more digestible form.[212] Given how little modern scholarship engaged with Greek mathematics after Late Antiquity, it is reasonable to expect that more evidence on the topic may surface in the future.

These facts suggest that, over many centuries, several different cultural realms that were in direct or indirect contact with one another kept engaging with and reinventing some of the same concepts. Their ultimate origin (if one exists) may not always be traceable. Direct influences can be verified only in fortuitous moments, such as the Indian and Arabic influence on the magic squares of Sahl Rabbān and Moschopulus. When confronted with such instances, one must remain focused on what they tell us about their concrete historical context without generalizing. Consequently, one should always ask whether the recurrence of similar ideas represents direct influences or independent elaborations developed as a matter of common sense.

The next step in tracing the transmission of ideas is imagining the broader intellectual contacts leading to concrete moments of cultural transfer. Regarding magic squares, an important avenue that enabled the exchange of ideas between India and China was Buddhism, which appeared in the Indian subcontinent around the fifth century BCE and was introduced to China around the third century BCE. Beyond magic squares, broader philosophical ideas originating in the BCE period are also shared between the Indian and Greek cultural spheres, as modern scholarship has repeatedly observed. Based on evidence from within the ancient Greek literary sources – the earliest is Diogenes Laertius, an author writing in the third century CE – modern scholars suggested an Indian (perhaps Buddhist) influence on Greek Pythagorean and Platonic

211 These facts are outlined in Hallum 2021. The rest is my interpretation.
212 For a discussion of this Byzantine intellectual phenomenon, see Mavroudi 2023: 266–268.

thought.[213] Pythagoras and Plato have been identified as sources of inspiration for both Greek and Arabic alchemical thought, as well. The most famous instance of medieval (as opposed to ancient) traces of Buddhism on the Greek intellectual universe is the narrative titled *Barlaam and Ioasaph*, a Christianized version of the story of Buddha which develops a number of philosophical themes and was translated from Georgian into Greek in the eleventh century.[214] This underscores that Greek contact with Indian ideas happened over a long stretch of time and through channels (e.g., literary genres, languages) that the old narrative on the medieval transmission of philosophy and science does not account for.

In addition to the Pythagorean and Platonic coloring of Jābirian arithmology, scholars need to also pay attention to its Christian inflexions, which were already acknowledged by Kraus.[215] As Joel Kalvesmaki recently showed, the Platonic and Pythagorean literary tradition provided the impetus for an early Christian theology of arithmetic. Kalvesmaki also argued that second-century Christian debates on the theology of arithmetic influenced the subsequent pagan tradition.[216] Although this second assertion may not find all scholars in agreement,[217] both arguments made by Kalvesmaki invite further reflection on the possible Christian roots of Jābirian arithmology, which also influenced Sufism and the development of a Muslim spirituality. This facet of the Christian-Muslim encounter is as politically charged as the Graeco-Indian example.

The idea that Sufism came from Christianity is as old as Miguel Asín Palacios (1871–1944), the famous Spanish Catholic priest and pioneer of Islamic studies who dedicated much of his scholarship to Christian and Muslim mysticism in Spain, the land of Ibn ʿArabī's birth.[218] Asín Palacios attributed the analogies between Hispano-Arabic Sufism and sixteenth-century Spanish mysticism to the fact that early Sufism was modeled, in terms of its doctrine and way of life, after Eastern Christian monasticism. A further exploration at around the same time and along similar lines was Margaret Smith's *Studies in Early Mysticism in the Near and Middle East*.[219] The topic was revisited in the twenty-first century. Recently, Georg Günther Blum argued that the flowering of Syriac monasticism after the rise of Islam inspired the translation of classics in Egyptian monastic spirituality from Greek to Syriac and the composition of original texts on the same subject in Syriac. According to Blum, these Syriac texts influenced the development of Islamic Sufism shortly thereafter.[220]

---

213 The possible Buddhist origin is explored in Beckwith 2015.

214 *Vita Barlaam et Ioasaph* (ed. Volk 2006–2009).

215 Kraus 1942–1943: 2.262–281, 328, etc.

216 Kalvesmaki 2013: 2.

217 See the review of Kalvesmaki by Slaveva-Griffin 2014.

218 The earliest contribution by Asín Palacios along these lines is his *El Islam cristianizado* (1931). His engagement with this topic is even earlier.

219 Smith 1931.

220 Blum 2009. For some criticism on Blum, see the review by Monferrer-Sala 2010.

Older and more recent scholarship that identifies the roots of Sufism in Christian spirituality stumbles upon an old and broader ideological problem with colonial inflexions that were amply addressed by post-colonial critique: the originality of Islam as a religion and intellectual tradition. An additional complication with Sufism is its label as "popular" religion. This was originally applied to Sufism by colonial Western observers who considered it a form of superstition appropriate for the uneducated masses. In the post-colonial period, the label obtained the positive connotation of "indigenous" and was embraced by educated Muslims.[221] In Western countries, especially from the second half of the twentieth century onwards, Sufism is frequently viewed as a universalist form of "spirituality" in step with the secularizing trends in Western societies and therefore as an alternative to traditional religion, and especially Christianity.[222] In such contexts, highlighting Sufism's Christian roots may be interpreted as undermining its "indigeneity" (as a Muslim phenomenon) and the secularist promise of its universalism (as a modern Western phenomenon). A reply to such labeling would be that insisting on the Christian roots of Sufism is not a form of Christian triumphalism or even an irenic contribution to Christian-Muslim understanding. Rather, it allows a modern researcher to explore how the overlaps between the Christian and the Muslim spiritual experience aided the Islamization of Christian populations after the Arab conquests of the seventh century, or the Turkish conquests of the late medieval period.[223] In this way, it aids a broader understanding of how important ideological and demographic changes occur in any historical context.

In addition to these considerations, the preceding pages argued that much alchemical thought was absorbed into Arabic literary culture through the lens of seventh-to-ninth-century Byzantine literary and intellectual culture. These challenges received wisdom on the transmission of the Greek heritage into Arabic, traditionally understood as a primarily ninth- and tenth-century textual exercise dependent on the Hellenizing curricula of Syriac schools in the early Abbasid period and the recovery of Greek manuscripts preserved in Christian monasteries in Muslim lands. Both sources are imagined as frozen in time and reflecting the intellectual realities prevalent in Byzantine lands when the Muslim conquests took place, untouched by contact with the later Byzantine world. However, it is possible to collect increasing amounts of evidence suggesting that the Byzantine and the early Islamic worlds were, indeed, in close intellectual

221 This brief description applies broadly to different parts of the Islamic world but of course has different inflections in each place. See Ridgeon 2021 for a chronological history of the Sufi tradition, organized into three periods: early, middle, and modern.

222 For a chronicle of this renewal, see Sedgwick 2017: Part 4.

223 On the importance of *zuhd* (ascetic practice) and its overlap with Christianity in the two centuries after Arab conquests (when the term *ṣūfī* was neither frequent nor clear regarding its content), see the overview of the literature in Bin Ramli 2010: 1301. On the role of Sufism in conversions from Christianity to Islam during the Turkish conquests, see Vryonis 1971.

communication.[224] Still, the point of these realizations is not to substitute one linear narrative (ancient Greek-to-Arabic) for another (Byzantine Greek-to-Arabic). Rather, the goal is to invent and build, from the ground up, a completely new narrative.[225] Consistently with such a goal, the picture that this article aimed to convey is one where different linguistic and religious realms are in constant contact with each other. Such contact is easy to follow with alchemy and its related disciplines (Hermetism, theosophy, magic) because, by their very nature, these genres drew authority from quoting "alien" wisdom and for this reason were particularly interested in it. The Byzantine Greek evidence had absorbed such wisdom earlier and siphoned it to Arabic literary culture from the seventh century onwards. At the same time, Arabic literary culture was interacting with Syriac and Coptic that possessed their own share of such lore, itself inspired by contact with Greek lore both before and after the Muslim conquests of the seventh century. These are interconnected literary universes that keep recycling earlier tradition while at the same time intervening to it depending on the needs of their time and the erudition and creativity of the personalities involved. In the present article, the Greek evidence from antiquity into the tenth century received the lion's share of the attention because, compared with the written record in other languages (Arabic, Syriac, Coptic) it is more abundant and continuous. In addition, modern scholarship has studied it for longer and with more intensity. The findings of this earlier scholarship need not be discarded wholesale as the products of colonial mentalities and Eurocentrism. Rather, they can be reframed to serve the new narrative. An example of such reframing was offered earlier, using Sesiano's conclusion on the Arabic provenience of Moschopulus' magic squares.[226]

Bibliography

Sources

Collected Sources

Berthelot and Ruelle 1887–1888: Marcellin Berthelot and Charles-Émile Ruelle (ed. and Fr. transl.), *Collection des anciens alchimistes grecs*, 3 vols. Paris: G. Steinheil, 1887–1888 (reprint: Osnabrück: O. Zeller, 1967).

224 See Mavroudi 2015a; 2015b.

225 On the theoretical considerations that go into constructing such a narrative along the lines attempted for alchemy and its germane disciplines here, see Mavroudi 2023: 281–288.

226 For more details on inventing a new narrative regarding the transmission of philosophy and science, see Mavroudi 2023: 281–288; 2026.

Authors and Works

[Archelaus], *De arte sacra*: Günther Goldschmidt (ed.), *Heliodori carmina quattuor ad fidem codicis Casselani* (Religionsgeschichtliche Versuche und Vorarbeiten 19.2). Giessen: A. Töpelmann, 1923, pp. 50–59.

Bacchius Geron, *Isagoge artis musicae*: Luisa Zanoncelli (ed.), *La manualistica musicale greca* (Ricerche). Milan: Guerini Studio, 1990.

Doctrina Patrum: Franz Diekamp (ed.), *Doctrina Patrum de Incarnatione Verbi. Ein griechisches Florilegium aus der Wende des siebenten und achten Jahrhunderts*. Münster: Aschendorff, 1907.

Eusebius, *Praeparatio evangelica*: Karl Mras, *Eusebius Werke*, vol. 8: *Die Praeparatio evangelica* (Die griechischen christlichen Schriftsteller 43.1–2). Berlin: Akademie-Verlag, 1954–1956.

Iamblichus, *De mysteriis*: Édouard des Places (ed.), *Jamblique. Les mystères d'Égypte*. Paris: Belles Lettres, 1966.

Engl. transl.: Emma C. Clarke, John M. Dillon, and Jackson P. Hershbell, *Iamblichus:* De mysteriis. *Translated with an Introduction and Notes* (Writings from the Greco-Roman World 4). Atlanta, GA: Society of Biblical Literature, 2003.

Ikhwān al-Ṣafā', *De astronomia* (*Ep.* 3): F. Jamil Ragep and Taro Mimura (eds), *Epistles of the Brethren of Purity, On* Astronomia: *An Arabic Critical Edition and English Translation of Epistle 3*. Oxford: Oxford University Press in association with The Institute of Ismaili Studies, 2015.

Ioannes Lydus, *De magistratibus populi Romani*: Anastasius C. Bandy (ed.), *Ioannes Lydus, On Powers or the Magistracies of the Roman State*. Philadelphia, PA: The American Philosophical Society, 1983.

Ioannes Lydus, *De mensibus*: Richard Wünsch (ed.), *Ioannis Lydi Liber de mensibus* (Bibliotheca Scriptorum Graecorum et Romanorum Teubneriana). Leipzig: Teubner, 1898.

Iulianus Imperator, *Epistulae*: Joseph Bidez (ed.), *L'empereur Julien. Œuvres complètes*, vol. 1.2: *Lettres et fragments*. Second edition. Paris: Belles Lettres, 1960.

al-Kindī, Ya'qūb b. Isḥāq, *Fī al-ṣinā'a al-'uẓmā*: Ya'qūb b. Isḥāq al-Kindī, *Fī al-ṣinā'a al-'uẓmā*, ed. 'Azīz Ṭāha al-Sayyid Aḥmad. Nicosia: Dār al-Shabāb li-l-Nashr, 1987.

Mysteria litterarum: Cordula Bandt (ed. and Germ. transl.), *Der Traktat "Vom Mysterium der Buchstaben". Kritischer Text mit Einführung, Übersetzung und Anmerkungen* (Texte und Untersuchungen zur Geschichte der altchristlichen Literatur 162). Berlin and New York: De Gruyter, 2007.

Parastaseis Syntomoi Chronikai: Theodor Preger (ed.), *Scriptores originum Constantinopolitanarum*, vol. 1. Leipzig: Teubner, 1901, pp. 19–73.

Engl. transl.: Averil Cameron and Judith Herrin, *Constantinople in the Early Eighth Century: The* Parastaseis Syntomoi Chronikai. *Introduction, Translation and Commentary* (Columbia Studies in the Classical Tradition 10). Leiden: Brill, 1984.

Photius, *Bibliotheca*: René Henry (ed.), *Photius, Bibliothèque*, 8 vols (Collection byzantine). Paris: Belles Lettres, 1959–1977.

Stephanus Alexandrinus, *De magna et sacra arte*: Maria K. Papathanassiou (ed.), *Stephanos von Alexandreia und sein alchemistisches Werk. Die kritische Edition des griechischen Textes eingeschlossen*. Athens: Cosmosware, 2017.

Vita Barlaam et Ioasaph: Robert Volk (ed.), *Die Schriften des Johannes von Damaskos*, vols 6.1–2: *Historia animae utilis de Barlaam et Ioasaph (spuria)* (Patristische Texte und Studien 60–61). Berlin and New York: De Gruyter, 2006–2009.

Literature

Acevedo 2020: Juan Acevedo, *Alphanumeric Cosmology from Greek into Arabic: The Idea of* Stoicheia *through the Medieval Mediterranean* (Forschungen zum Alten Testament 2. Reihe). Tübingen: Mohr Siebeck, 2020.

Adamson and Pormann 2012: Peter Adamson and Peter E. Pormann, *The Philosophical Works of al-Kindī* (Studies in Islamic Philosophy). Oxford: Oxford University Press, 2012.

Agapitos 1991: Panagiotis A. Agapitos, *Narrative Structure in the Byzantine Vernacular Romances: A Textual and Literary Study of* Kallimachos, Belthandros *and* Libistros (Miscellanea Byzantina Monacensia 34). Munich: Institut für Byzantinistik und Neugriechische Philologie der Universität München, 1991.

Aladdin 2012: Bakri Aladdin, "Oneness of Being (*waḥdat al-wujūd*): The Term and the Doctrine", *Journal of the Muhyiddin Ibn Arabi Society* 51 (2012), pp. 3–26.

Algar 1987: Hamid Algar, "Astarābādī, Fażlallāh", in *Encyclopaedia Iranica*, online edition, 1987, https://iranicaonline.org/articles/astarabadi-fazlallah-sehab-al-din-b.

Anderson 2011: Benjamin Anderson, "Classified Knowledge: The Epistemology of Statuary in the *Parastaseis Syntomoi Chronikai*", *Byzantine and Modern Greek Studies* 35.1 (2011), pp. 1–19.

Asín Palacios 1931: Miguel Asín Palacios, *El Islam cristianizado: estudio del "sufismo" a través de las obras de Abenarabi de Murcia*. Madrid: Editorial Plutarco, 1931.

Atiyeh 1966: George Atiyeh, *Al-Kindī, the Philosopher of the Arabs*. Rawalpindi: Islamic Research Institute, 1966.

Bacchi and Martelli 2009: Eleonora Bacchi and Matteo Martelli, "Il principe Ḫālid bin Yazīd e le origini dell'alchimia araba", in Daniele Cevenini and Svevo d'Onofrio (eds), *Conflitti e dissensi nell'Islam* ('Uyûn al-Akhbâr. Studi sul mondo islamico 3). Bologna: Il Ponte, 2009, pp. 85–120.

Bandt 2007: Cordula Bandt, *Der Traktat "Vom Mysterium der Buchstaben". Kritischer Text mit Einführung, Übersetzung und Anmerkungen* (Texte und Untersuchungen zur Geschichte der altchristlichen Literatur 162). Berlin and New York: De Gruyter, 2007.

Bartelink 1983: Gerard J. M. Bartelink, "Homer in den Werken des Kyrillos von Alexandrien", *Wiener Studien* 96 (1983), pp. 62–68.

Beatrice 2001: Pier Franco Beatrice, *Anonymi Monophysitae Theosophia: An Attempt at a Reconstruction* (Vigiliae Christianae, Supplements 56). Leiden: Brill, 2001.

Beckwith 2015: Christopher I. Beckwith, *Greek Buddha: Pyrrho's Encounter with Early Buddhism in Central Asia*. Princeton, NJ: Princeton University Press, 2015.

Bidez and Cumont 1938: Joseph Bidez and Franz Cumont, *Les Mages hellénisés. Zoroastre, Ostanès et Hystaspe d'après la tradition grecque*. 2 vols. Paris: Belles Lettres, 1938.

Bin Ramli 2010: Harith Bin Ramli, "The Rise of Early Sufism: A Survey of Recent Scholarship on Its Social Dimensions", *History Compass* 8.11 (2010), pp. 1299–1315.

Blum 2009: Georg Günter Blum, *Die Geschichte der Begegnung christlich-orientalischer Mystik mit der Mystik des Islams* (Orientalia Biblica et Christiana 17). Wiesbaden: Harrassowitz, 2009.

Bouché-Leclercq 1899: Auguste Bouché-Leclercq, *L'Astrologie grecque*. Paris: Leroux, 1899.

Boudon-Millot 2016: Véronique Boudon-Millot, "Stephanos d'Athènes. Médecin et philosophe grec commentateur d'Hippocrate", in Richard Goulet (ed.), *Dictionnaire des philosophes antiques*, vol. 6. Paris: CNRS Éditions, 2016, no. S 153, pp. 579–588.

Brock 2004: Sebastian Brock, "Without Mushe of Nisibis, Where Would We Be? Some Reflections on the Transmission of Syriac Literature", *The Journal of Eastern Christian Studies* 56 (2004), pp. 15–24.

Browne 1946: Charles A. Browne, "Rhetorical and Religious Aspects of Greek Alchemy. Including a Commentary and Translation of the Poem of the Philosopher Archelaos upon the Sacred Art, Part I", *Ambix* 2.3–4 (1946), pp. 129–137.

Browne 1948: Charles A. Browne, "Rhetorical and Religious Aspects of Greek Alchemy, Part II", *Ambix* 3.1–2 (1948), pp. 15–25.

Burnett 1992: Charles Burnett, "The Astrologer's Assay of the Alchemist: Early References to Alchemy in Arabic and Latin Texts", *Ambix* 39.3 (1992), pp. 103–109.

Calder III and Heilen 2007: William M. Calder III and Stephan Heilen, "David E. Pingree: An Unpublished Autobiography", *Greek, Roman, and Byzantine Studies* 47.4 (2007), pp. 515–523.

Calzolari 2009: Valentina Calzolari, "David et la tradition arménienne", in Valentina Calzolari and Jonathan Barnes (eds), *L'œuvre de David l'Invincible et la transmission de la pensée grecque dans la tradition arménienne et syriaque*, vol. 1 (Philosophia Antiqua 116). Leiden and Boston, MA: Brill, 2009, pp. 15–36.

Cameron 2019: Averil Cameron, "Byzantinists and Others", in Daniëlle Slootjes and Mariëtte Verhoeven (eds), *Byzantium in Dialogue with the Mediterranean: History and Heritage* (The Medieval Mediterranean 116). Leiden and Boston, MA: Brill, 2019, pp. 6–23.

Carlotta 2022: Vincenzo Carlotta, "Introducing Greek Alchemy to Christianity: Inclusion and Exclusion of Religious Elements in Stephanus's Lessons", *ARYS* 20 (2022), Special Issue: *Chemical Arts and Religion in Antiquity*, pp. 323–348.

Carrara 2019: Laura Carrara, "The *Tübingen Theosophy:* Some Philological Notes on Chronology and Structure", in Lucia M. Tissi, Helmut Seng, and Chiara O. Tommasi (eds), *Stimmen der Götter. Orakel und ihre Rezeption von der Spätantike bis in die frühe Neuzeit* (Bibliotheca Chaldaica 8). Heidelberg: Universitätsverlag Winter, 2019, pp. 193–214.

Chrysos 2002: Evangelos Chrysos, "Illuminating Darkness by Candlelight: Literature in the Dark Ages", in Paolo Odorico and Panagiotis A. Agapitos (eds), *Pour une "nouvelle" histoire de la littérature byzantine* (Dossiers byzantins 1). Paris: Centre d'études byzantines, néo-helléniques et sud-est européennes, École des Hautes Études en Sciences Sociales, 2002, pp. 13–24.

Colinet 2000: Andrée Colinet, "Le *Travail des quatre éléments* ou lorsqu'un alchimiste byzantin s'inspire de Jabir", in Isabelle Draelants, Anne Tihon, and Baudouin Van den Abeele (eds), *Occident et Proche-Orient. Contacts scientifiques au temps des Croisades* (Réminisciences 5). Turnhout: Brepols, 2000, pp. 165–190.

Cottrell 2012: Emily Cottrell, "*Trivium* and *Quadrivium*: East of Baghdad", in Alessandro Musco, Rosanna Gambino, Luciana Pepi, Patrizia Spallino, and Maria Vassallo (eds), *Universalità della Ragione. Pluralità delle Filosofie nel Medioevo*, vol. 3. Palermo: Officina di Studi Medievali, 2012, pp. 11–25.

Cottrell 2015: Emily Cottrell, "L'Hermès arabe" de Kevin van Bladel et la question du rôle de la littérature sassanide dans la présence d'écrits hermétiques et astrologiques en langue arabe", *Bibliotheca Orientalis* 72.3–4 (2015), pp. 336–401.

Cottrell and Ross 2019: Emily J. Cottrell and Micah T. Ross, "Persian Astrology: Dorotheus and Zoroaster, According to the Medieval Arabic Sources (8th–13th Century)", in Pavel B. Lurje (ed.), *Proceedings of the Eighth European Conference of Iranian Studies*, vol. 1: *Studies on Pre-Islamic Iran and on Historical Linguistics*. St Petersburg: The State Hermitage Publishers, 2019, pp. 87–105.

Crowe 1990: Michael J. Crowe, "Duhem and History and Philosophy of Mathematics", *Synthese* 83.3 (1990), pp. 431–447.

Daiber 2014: Hans Daiber, "Possible Echoes of *De mundo* in the Arabic-Islamic World: Christian, Islamic and Jewish Thinkers", in Thom 2014, pp. 169–180.

D'Ancona 2019: Cristina D'Ancona, "Greek Sources in Arabic and Islamic Philosophy", in Edward N. Zalta (ed.), *The Stanford Encyclopedia of Philosophy* (Fall 2019 Edition), https://plato.stanford.edu/archives/fall2019/entries/arabic-islamic-greek/.

Dapsens 2016: Marion Dapsens, "De la *Risālat Maryānus* au *De Compositione alchemiae*. Quelques réflexions sur la tradition d'un traité d'alchimie", *Studia Graeco-Arabica* 6 (2016), pp. 121–140.

Dapsens 2021: Marion Dapsens, "The Alchemical Work of Khālid b. Yazīd b. Muʿāwiya (d. c. 85/704)", *Asiatische Studien – Études Asiatiques* 75.2 (2021), pp. 327–427.

Dapsens and Moureau 2021: Marion Dapsens and Sébastien Moureau, "The Four Signs of the Art: Edition and Translation of an Alchemical Epistle Attributed to Khālid b. Yazīd and Its Latin Translation", *Arabica* 68 (2021), pp. 557–627.

Debié 2017: Muriel Debié, "'La science est commune': sources syriaques et culture grecque en Syrie-Mésopotamie et en Perse par-delà les siècles obscurs byzantins", in Flusin and Cheynet 2017, pp. 87–127.

Dufault 2019: Olivier Dufault, *Early Greek Alchemy, Patronage and Innovation in Late Antiquity* (California Classical Studies 7). Berkeley, CA: Department of Classics, University of California, Berkeley, 2019.

Ebstein 2014: Michael Ebstein, "Ḏū l-Nūn al-Miṣrī and Early Islamic Mysticism", *Arabica* 61 (2014), pp. 559–612.

El-Cheikh 2004: Nadia Maria El-Cheikh, *Byzantium Viewed by the Arabs* (Harvard Middle Eastern Monographs 36). Cambridge, MA: Harvard University Press, 2004.

El Daly 2005: Okasha El Daly, *Egyptology: The Missing Millennium. Ancient Egypt in Medieval Arabic Writings* (UCL Institute of Archaeology Publications 33). London: Routledge, 2005.

Erbse 1941: Hartmut Erbse, *Fragmente griechischer Theosophien*. Hamburg: Hansischer Gildenverlag, 1941.

Erbse 1995: Hartmut Erbse, *Theosophorum Graecorum fragmenta* (Bibliotheca Scriptorum Graecorum et Romanorum Teubneriana). Stuttgart and Leipzig: Teubner, 1995.

Erismann 2017: Christophe Erismann, "Meletius Monachus on Individuality: A Ninth-Century Byzantine Medical Reading of Porphyry's Logic", *Byzantinische Zeitschrift* 110.1 (2017), pp. 37–60.

Evans 2000: James Evans, "The First Astrophysical Synthesis: What Ptolemy Did and Why It Mattered", in *New Perspectives in the History of Ancient Astronomy*. American Physical Society, April Meeting, April 29–May 2, 2000. Long Beach, CA, abstract id. Q22.001.
A summary of the paper can be found in *History of Physics Newsletter. A Forum of the American Physical Society* 8.1 (September 2000), pp. 6–7, https://higherlogicdownload.s3.amazonaws.com/APS/fc77d112-850f-4db6-997c-e434bd4670b5/UploadedImages/Documents/Meeting_Minutes/september00.pdf.

Ferngren 2017: Gary B. Ferngren (ed.), *Science and Religion: A Historical Introduction*. Second edition. Baltimore, MD: Johns Hopkins University Press, 2017.

Fleischer 2015: Cornell Fleischer, "The Mystic Lettrist 'Abd al-Rahman al-Bistami and the Origins of Ottoman Historical Consciousness". Lecture, American School of Classical Studies at Athens, 8 May 2015, https://www.ascsa.edu.gr/News/newsDetails/videocast-the-mystic-lettrist-abd-al-rahman-al-bistami-and-the-origins-of-o.

Fleischer 2018: Cornell H. Fleischer, "A Mediterranean Apocalypse: Prophecies of Empire in the Fifteenth and Sixteenth Centuries", *Journal of the Economic and Social History of the Orient* 61 (2018), pp. 18–90.

Flusin and Cheynet 2017: Bernard Flusin and Jean-Claude Cheynet (eds), *Autour du* Premier humanisme byzantin *et des* Cinq études sur le XIe siècle, *quarante ans après Paul Lemerle* (= *Travaux et mémoires* 21.2). Paris: Association des Amis du Centre d'histoire et civilisation de Byzance, 2017.

Gannagé (Forthcoming): Emma Gannagé, "Al-Kindī's Prologue to *On the Great Art*", in Alan Bowen and Emma Gannagé (eds), *"In Praise of the Divine Beauty": The Philosophy of Ptolemy and Its Greek, Arabic, and Hebrew Reception*. Leiden and Boston, MA: Brill (Forthcoming), pp. 171–208.

Gibbon 1909–1913: Edward Gibbon, *The History of the Decline and Fall of the Roman Empire*, ed. John Bagnell Bury, 12 vols. London: Methuen, 1909–1913.

Goulet 2016: Richard Goulet, "Stéphanos. Commentateur tardif de la *Rhétorique* d'Aristote", in Richard Goulet (ed.), *Dictionnaire des philosophes antiques*, vol. 6. Paris: CNRS Éditions, 2016, no. S 151, p. 563.

Gutas 1998: Dimitri Gutas, *Greek Thought, Arabic Culture: The Graeco-Arabic Translation Movement in Baghdad and Early 'Abbāsid Society (2nd–4th/8th–10th Centuries)*. London: Routledge, 1998.

Gutas 2004: Dimitri Gutas, "Geometry and the Rebirth of Philosophy in Arabic with al-Kindī", in Rüdiger Arnzen and Jörn Thielmann (eds), *Words, Texts and Concepts Cruising the Mediterranean Sea: Studies on the Sources, Contents and Influences of Islamic Civilization and Arabic Philosophy and Science Dedicated to Gerhard Endress on His Sixty-Fifth Birthday* (Orientalia Lovaniensia Analecta 139). Leuven: Peeters, 2004, pp. 195–209. Reprinted in Peter E. Pormann (ed.), *Islamic Medical and Scientific Tradition*, vol. 4 (Critical Concepts in Islamic Studies). London: Routledge, 2010, pp. 3–16.

Gutas et al. 2017: Dimitri Gutas, Anthony Kaldellis, and Brian Long, "Intellectual Exchanges with the Arab World", in Anthony Kaldellis and Niketas Siniossoglou (eds), *The Cambridge Intellectual History of Byzantium*. Cambridge: Cambridge University Press, 2017, pp. 79–98.

Hallum 2021: Bink Hallum, "New Light on Early Arabic Awfāq Literature", in Liana Saif, Francesca Leoni, Matthew Melvin-Koushki, and Farouk Yahya (eds), *Islamicate Occult Sciences in Theory and Practice* (Handbook of Oriental Studies, Section 1: The Near and Middle East 140). Leiden and Boston, MA: Brill, 2021, pp. 57–161.

Harrison 2015: Peter Harrison, "Myth 24. That Religion Has Typically Impeded the Progress of Science", in Numbers and Kampourakis 2015, pp. 195–201.

al-Hassan 2004: Ahmad Y. Al-Hassan, "The Arabic Original of *Liber de compositione alchemiae*: The Epistle of Maryanus, the Hermit and Philosopher, to Prince Khālid ibn Yazīd", *Arabic Sciences and Philosophy* 14.2 (2004), pp. 213–223.

Hatlie 2007: Peter Hatlie, *The Monks and Monasteries of Constantinople, ca. 350–850*. Cambridge: Cambridge University Press, 2007.

Heilen 2010: Stephan Heilen, "Problems in Translating Ancient Greek Astrological Texts", in Annette Imhausen and Tanja Pommerening (eds), *Writings of Early Scholars in the Ancient Near East, Egypt, Rome, and Greece: Translating Ancient Scientific Texts* (Beiträge zur Altertumskunde 286). Berlin and New York: De Gruyter, 2010, pp. 299–332.

Hemmerdinger 1962: Bertrand Hemmerdinger, "Une mission scientifique arabe à l'origine de la renaissance iconoclaste", *Byzantinische Zeitschrift* 55.1 (1962), pp. 66–67.

Hemmerdinger 1964: Bertrand Hemmerdinger, "La culture grecque classique du VIIe au IXe siècle", *Byzantion* 34 (1964), pp. 125–133.

Henry 2010: John Henry, "Religion and the Scientific Revolution", in Peter Harrison (ed.), *The Cambridge Companion to Science and Religion* (Cambridge Companions to Religion). Cambridge: Cambridge University Press, 2010, pp. 39–58.

Hilkens 2013: Andy Hilkens, "Syriac *Iliou persides*: The Fall of Troy in Syriac Historiography", *Le Muséon* 12 (2013), pp. 285–317.

Hunger 1954: Herbert Hunger, "Allegorische Mythendeutung in der Antike und bei Johannes Tzetzes", *Jahrbuch der Österreichischen Byzantinischen Gesellschaft* 3 (1954), pp. 35–54.

Hunger 1955: Herbert Hunger, "Johannes Tzetzes, Allegorien zur Odyssee, Buch 13–24. Kommentierte Textausgabe", *Byzantinische Zeitschrift* 48 (1955), pp. 4–48.

Ierodiakonou 2012: Katerina Ierodiakonou, "Byzantine Philosophy Revisited (A Decade After)", in Börje Bydén and Katerina Ierodiakonou (eds), *The Many Faces of Byzantine Philosophy* (Papers and Monographs from the Norwegian Institute at Athens, Series 4, 1). Athens: The Norwegian Institute at Athens, 2012, pp. 1–21.

Ierodiakonou and Bydén 2018: Katerina Ierodiakonou and Börje Bydén, "Byzantine Philosophy", in Edward N. Zalta (ed.), *The Stanford Encyclopedia of Philosophy* (Fall 2018 Edition), https://plato.stanford.edu/archives/fall2018/entries/byzantine-philosophy/.

Jokisch 2007: Benjamin Jokisch, *Islamic Imperial Law: Harun-Al-Rashid's Codification Project* (Studien zur Geschichte und Kultur des islamischen Orients, NF 19). Berlin and New York: De Gruyter, 2007.

Kalligas 2016: Paul Kalligas, "Platonic Astronomy and the Development of Ancient *Sphairopoiia*", *Rhizomata* 4.2 (2016), pp. 176–200.

Kalvesmaki 2013: Joel Kalvesmaki, *The Theology of Arithmetic: Number Symbolism in Platonism and Early Christianity* (Hellenic Studies Series 59). Washington, DC: Center for Hellenic Studies, 2013.

Kazhdan 1991: Alexander Kazhdan, "Erotapokriseis", in Alexander P. Kazhdan (ed.-in-chief), *The Oxford Dictionary of Byzantium*, 3 vols. New York and Oxford: Oxford University Press, 1991, vol. 1, p. 727.

Kominko 2013: Maja Kominko, *The World of Kosmas: Illustrated Byzantine Codices of the* Christian Topography. Cambridge: Cambridge University Press, 2013.

König 2018: Götz König, "The Pahlavi Literature of the 9th Century and Greek Philosophy", *Iran and the Caucasus* 22.1 (2018), pp. 8–37.

Koutalis et al. 2018: Vangelis Koutalis, Matteo Martelli, and Gerasimos Merianos, "Graeco-Egyptian, Byzantine and Post-Byzantine Alchemy: Introductory Remarks", in Efthymios Nicolaidis (ed.), *Greek Alchemy from Late Antiquity to Early Modernity* (De Diversis Artibus 104). Turnhout: Brepols, 2018, pp. 11–43.

Kraemer 1956: Jörg Kraemer, "Arabische Homerverse", *Zeitschrift der Deutschen Morgenländischen Gesellschaft* 106 (1956), pp. 259–316.

Kraus 1942–1943: Paul Kraus, *Jābir ibn Ḥayyān, Contribution à l'histoire des idées scientifiques dans l'Islam*, 2 vols (Mémoires de l'Institut d'Égypte 44–45). Cairo: Institut français d'archéologie orientale, 1942–1943.

Lauxtermann 2019: Marc D. Lauxtermann, *Byzantine Poetry from Pisides to Geometres. Texts and Contexts*, vol. 2 (Wiener Byzantinistische Studien 24.2). Vienna: Austrian Academy of Sciences Press, 2019.

Lemerle 1971: Paul Lemerle, *Le premier humanisme byzantin. Notes et remarques sur enseignement et culture à Byzance, des origines au Xe siècle* (Bibliothèque byzantine, Études 6). Paris: Presses universitaires de France, 1971.

Letrouit 1995: Jean Letrouit, "Chronologie des alchimistes grecs", in Didier Kahn and Sylvain Matton (eds), *Alchimie: art, histoire et mythes* (Textes et travaux de *Chrysopoeia* 1). Paris: S. É. H. A.; Milan: Archè, 1995, pp. 11–93.

Levey 1962: Martin Levey, "The Aqrābādhīn of al-Kindī and Early Arabic Chemistry", *Chymia* 8 (1962), pp. 11–20.

Lévy 2018: Carlos Lévy, "Philo of Alexandria", in Edward N. Zalta (ed.), *The Stanford Encyclopedia of Philosophy* (Spring 2018 Edition), https://plato.stanford.edu/archives/spr2018/entries/philo/.

Lewy 2011: Hans Lewy, *Chaldean Oracles and Theurgy*. Troisième édition par Michel Tardieu avec un supplément "Les Oracles chaldaïques 1891–2011" (Collection des Études Augustiniennes, Série Antiquité 77). Paris: Institut d'Études Augustiniennes, 2011.

Lloyd 1978: Geoffrey E. R. Lloyd, "Saving the Appearances", *The Classical Quarterly* 28.1 (1978), pp. 202–222.

Long 1992: A. A. Long, "Stoic Readings of Homer", in Robert Lamberton and John J. Keaney (eds), *Homer's Ancient Readers: The Hermeneutics of Greek Epic's Earliest Exegetes*. Princeton, NJ: Princeton University Press, 1992, pp. 41–66.

Lory 1983: Pierre Lory, *Jābir ibn Ḥayyān: Dix Traités d'alchimie. Les dix premiers Traités du Livre des Soixante-dix*. Paris: Sindbad, 1983.

Lory 2003: Pierre Lory, *Alchimie et mystique en terre d'Islam* (Folio essais 425). Paris: Gallimard, 2003 (original edition: Paris: Verdier, 1989).

Lory 2008: Pierre Lory, "Kimiā", in *Encyclopaedia Iranica*, online edition, 2008, https://www.iranicaonline.org/articles/kimia.

Lougovaya 2017: Julia Lougovaya, "A Perfect Pangram: A Reconsideration of the Evidence", *Greek, Roman, and Byzantine Studies* 57.1 (2017), pp. 162–190.

Louth 2002: Andrew Louth, *St. John Damascene: Tradition and Originality in Byzantine Theology* (Oxford Early Christian Studies). Oxford: Oxford University Press, 2002.

Luck 1991: Georg Luck, "The Doctrine of Salvation in the Hermetic Writings", *The Second Century: A Journal of Early Christian Studies* 8.1 (1991), pp. 31–41.

Macé and De Vos 2013: Caroline Macé and Ilse De Vos, "Pseudo-Athanasius, *Quaestio ad Antiochum* 136 and the *Theosophia*", *Studia Patristica* 66 (2013), pp. 319–332.

Magdalino 2006: Paul Magdalino, "Occult Science and Imperial Power in Byzantine History and Historiography (9th–12th Centuries)", in Paul Magdalino and Maria Mavroudi (eds), *The Occult Sciences in Byzantium*. Geneva: La Pomme d'or, 2006, pp. 119–162.

Majercik 1998: Ruth Majercik, "A Reminiscence of the *Chaldean Oracles* at Gregory of Nazianzus, *Or.* 29,2 οἷον κρατήρ τις ὑπερερρύη", *Vigiliae Christianae* 52.3 (1998), pp. 286–292.

Mango 1975: Cyril Mango, "The Availability of Books in the Byzantine Empire, A. D. 750–850", in William C. Loerke et al. (eds), *Byzantine Books and Bookmen: A Dumbarton Oaks Colloquium*. Washington, DC: Dumbarton Oaks, Center for Byzantine Studies, 1975, pp. 29–45.

Martelli 2013: Matteo Martelli, *The* Four Books *of Pseudo-Democritus* (Sources of Alchemy and Chemistry 1; Ambix 60, Supplement 1). Leeds: Maney Publishing, 2013.

Martelli 2014a: Matteo Martelli, "L'assimilazione al dio attraverso le τέχναι. Gli 'scienzati' in età ellenistica e imperiale", in Marco Beretta, Francesco Citti, and Alessandro Iannucci (eds), *Il culto di Epicuro. Testi, iconografia e paesaggio* (Biblioteca di Nuncius, Studi e testi 75). Florence: Leo S. Olschki, 2014, pp. 15–38.

Martelli 2014b: Matteo Martelli, "L'alchimie en syriaque et l'oeuvre de Zosime", in Émilie Villey (ed.), *Les sciences en syriaque* (Études syriaques 11). Paris: Geuthner, 2014, pp. 191–214.

Martelli 2016: Matteo Martelli, "Stéphanos. Alchimiste et astronome/astrologue", in Richard Goulet (ed.), *Dictionnaire des philosophes antiques*, vol. 6. Paris: CNRS Éditions, 2016, no. S 150, pp. 557–563.

Matus 2012: Zachary A. Matus, "Alchemy and Christianity in the Middle Ages", *History Compass* 10.12 (2012), pp. 934–945.

Mavroudi 2006: Maria Mavroudi, "Occult Science and Society in Byzantium: Considerations for Future Research", in Paul Magdalino and Maria Mavroudi (eds), *The Occult Sciences in Byzantium*. Geneva: La Pomme d'or, 2006, pp. 39–95.

Mavroudi 2012: Maria Mavroudi, "The Naples Dioscorides", in Helen Evans, with Brandie Ratliff (eds), *Byzantium and Islam: Age of Transition (7th–9th Centuries). Catalogue of the Exhibition at the Metropolitan Museum of Art.* New Haven, CT: Yale University Press, 2012, pp. 22–26.

Mavroudi 2013: Maria Mavroudi, "Pletho as Subversive and His Reception in the Islamic World", in Dimiter Angelov and Michael Saxby (eds), *Power and Subversion in Byzantium* (Society for the Promotion of Byzantine Studies, Publications 17). Farnham: Ashgate, 2013, pp. 177–203.

Mavroudi 2015a: Maria Mavroudi, "Greek Language and Education under Early Islam", in Behnam Sadeghi, Asad Q. Ahmed, Adam Silverstein, and Robert Hoyland (eds), *Islamic Cultures, Islamic Contexts: Essays in Honor of Professor Patricia Crone* (Islamic History and Civilization: Studies and Texts 114). Leiden and Boston, MA: Brill, 2015, pp. 295–342.

Mavroudi 2015b: Maria Mavroudi, "Translations from Greek into Latin and Arabic during the Middle Ages: Searching for the Classical Tradition", *Speculum* 90.1 (2015), pp. 28–59.

Mavroudi 2020: Maria Mavroudi, "Homer in Greece from the End of Antiquity 1: The Byzantine Reception of Homer and His Export to Other Cultures", in Corinne Ondine Pache (ed.), *The Cambridge Guide to Homer*. Cambridge: Cambridge University Press, 2020, pp. 444–472.

Mavroudi 2021: Maria Mavroudi, "The Modern Historiography of Byzantine and Islamic Philosophy: A Comparison", *Al-Masāq* 33.3 (2021), pp. 282–299.

Mavroudi 2023: Maria Mavroudi, "Byzantine Translations from Arabic into Greek: Old and New Historiography in Confluence and in Conflict", *Journal of Late Antique, Islamic and Byzantine Studies* 2 (2023), pp. 215–288.

Mavroudi 2024a: Maria Mavroudi, "Arabic Terms in Byzantine *Materia Medica*: Oral and Textual Transmission", in Petros Bouras-Vallianatos and Dionysios Stathakopoulos (eds), *Drugs in the Medieval Mediterranean: Transmission and Circulation of Pharmacological Knowledge.* Cambridge: Cambridge University Press, 2024, pp. 130–183.

Mavroudi 2024b: Maria Mavroudi, "The Complexities of Retrieving Byzantine Science out of Medieval Greek Manuscripts and a Comparison with the Situation in Arabic", in Despina-Eirini Tsourka-Papastathi, Myrto Malouta, and Nikolaos Toutos (eds), *Κολοφών. Τόμος πρὸς τιμὴν τοῦ Καθηγητῆ Εὐθυμίου Κ. Λίτσα.* Thessaloniki: Greek Paleography Society, 2024, pp. 565–585.

Mavroudi 2026: Maria Mavroudi, "Moving beyond Post-Colonial Approaches: Grammar as an Item of Cultural Brokerage between Byzantium and the Islamic World (Seventh to Eleventh Centuries)", in Uriel Simonsohn and Luke Yarbrough (eds), *Cultural Brokerage in Premodern Islamic Societies* (Edinburgh Studies in Classical Islamic History and Culture). Edinburgh: Edinburgh University Press, 2026, pp. 149–208.

Mavroudi (Forthcoming a): Maria Mavroudi, "The Byzantine Reception of Ptolemy's *Almagest* between the Seventh and the Ninth Centuries", in Alan Bowen and Emma Gannagé (eds), *"In Praise of the Divine Beauty": The Philosophy of Ptolemy and Its Greek, Arabic, and Hebrew Reception*. Leiden and Boston, MA: Brill (Forthcoming), pp. 90–144.

Mavroudi (Forthcoming b): Maria Mavroudi, "John of Damascus: The Intellectual Context", in Ken Parry (ed.), *The Brill Companion to John of Damascus*. Leiden: Brill (Forthcoming).

McCown 1922: Chester Charlton McCown, *The Testament of Solomon*. Leipzig: J. C. Hinrichs, 1922.

McKitterick and Quinault 1997: Rosamond McKitterick and Roland Quinault (eds), *Edward Gibbon and Empire*. Cambridge: Cambridge University Press, 1997.

Merianos 2017: Gerasimos Merianos, "Alchemy", in Anthony Kaldellis and Niketas Siniossoglou (eds), *The Cambridge Intellectual History of Byzantium*. Cambridge: Cambridge University Press, 2017, pp. 234–251.

Merianos 2022: Gerasimos Merianos, "The Christianity of the Philosopher Christianos. Ethics and Mathematics in Alchemical Methodology", *ARYS* 20 (2022), Special Issue: *Chemical Arts and Religion in Antiquity*, pp. 271–322.

Merianos 2024: Gerasimos Merianos, "Aspects of the Relationship between Alchemy and Christianity in Byzantium", in Kostas Tampakis and Ronald Numbers (eds), *Orthodox Christianity and the Study of Nature: Histories of Interaction* (Science and Orthodox Christianity 4). Turnhout: Brepols, 2024, pp. 21–54.

Merianos and Sakorrafou 2013: Gerasimos Merianos and Sandy Sakorrafou, "Μαρτυρίες περί αλχημείας στο Βυζάντιο σε μη αλχημικά κείμενα", in Eirini Mergoupi-Savaidou, Gerasimos Merianos, Faidra Papanelopoulou, and Christiana Christopoulou (eds), *Επιστήμη και Τεχνολογία. Ιστορικές και Ιστοριογραφικές Μελέτες* (Εταιρεία Μελέτης και Διάδοσης της Ιστορίας των Επιστημών και της Τεχνολογίας 1). Athens: Ekdotike Athenon, 2013, pp. 45–65.

Miller 2017: Timothy S. Miller, "Medical Thought and Practice", in Anthony Kaldellis and Niketas Siniossoglou (eds), *The Cambridge Intellectual History of Byzantium*. Cambridge: Cambridge University Press, 2017, pp. 252–268.

Monferrer-Sala 2010: Juan Pedro Monferrer-Sala, "Review of Blum 2009", *Collectanea Christiana Orientalia* 7 (2010), pp. 326–329.

Moran 2011: Bruce T. Moran, "Introduction", *Isis* 102.2 (2011), pp. 300–304.

Moreschini 2011: Claudio Moreschini, *Hermes Christianus: The Intermingling of Hermetic Piety and Christian Thought*, transl. Patrick Baker (Cursor Mundi 8). Turnhout: Brepols, 2011.

Munkholt Christensen 2021: Maria Munkholt Christensen, "Meditatio mortis. Meditating on Death, Philosophy and Gender in Late Antique Hagiography", *Philosophy and Society* 32.2 (2021), pp. 177–193.

Newman 2011: William R. Newman, "What Have We Learned from the Recent Historiography of Alchemy?", *Isis* 102.2 (2011), pp. 313–321.

Newman and Principe 1998: William R. Newman and Lawrence M. Principe, "Alchemy vs. Chemistry: The Etymological Origins of a Historiographic Mistake", *Early Science and Medicine* 3.1 (1998), pp. 32–65.

Nomanul Haq 1990: Syed Nomanul Haq, *A Critical Study of Jābir ibn Ḥayyān's* Kitāb al-Aḥjār 'alā Ra'y Balīnās. PhD Dissertation. University College London, 1990.

Nomanul Haq 1994: Syed Nomanul Haq, *Names, Natures, Things. The Alchemist Jābir ibn Ḥayyān and His* Kitāb al-Aḥjār *(Book of Stones)* (Boston Studies in the Philosophy and History of Science 158). Dordrecht: Kluwer Academic Publishers, 1994.

Nomanul Haq 2011: Syed Nomanul Haq, "Alchemical Manuscripts: The Specific Case of Corpus Jabirianum", in *The Earth and Its Sciences in Islamic Manuscripts*. London: Al-Furqān Islamic Heritage Foundation, 2011, pp. 221–232.

Numbers and Kampourakis 2015: Ronald L. Numbers and Kostas Kampourakis (eds), *Newton's Apple and Other Myths about Science*. Cambridge, MA: Harvard University Press, 2015.

Nummedal 2013: Tara Nummedal, "Alchemy and Religion in Christian Europe", *Ambix* 60.4 (2013), pp. 311–322.

Odorico 2014: Paolo Odorico, "Du recueil à l'invention du texte: le cas des *Parastaseis Syntomoi Chronikai*", *Byzantinische Zeitschrift* 107.2 (2014), pp. 755–784.

Papadogiannakis 2013: Yiannis Papadogiannakis, "Erotapokriseis", in Roger S. Bagnall, Kai Brodersen, Craige B. Champion, Andrew Erskine, and Sabine R. Huebner (eds), *The Encyclopedia of Ancient History*, First Edition. Oxford: Blackwell, 2013, pp. 2491–2493.

Pérez Martín and Manolova 2020: Inmaculada Pérez Martín and Divna Manolova, "Science Teaching and Learning Methods in Byzantium", in Stavros Lazaris (ed.), *A Companion to Byzantine Science* (Brill's Companions to the Byzantine World 6). Leiden and Boston, MA: Brill, 2020, pp. 53–104.

Pingree 1983: David Pingree, "Abū Sahl Nawbaḵt", *Encyclopaedia Iranica*, online edition, 1983, http://www.iranicaonline.org/articles/abu-sahl-b.

Plass 1982: Paul Plass, "A Greek Alchemical Formula", *Ambix* 29.2 (1982), pp. 69–73.

Pontani 2005: Filippomaria Pontani, *Sguardi su Ulisse. La tradizione esegetica greca all'*Odissea (Pleiadi 13). Rome: Edizioni di Storia e Letteratura, 2005.

Pontani 2014: Filippomaria Pontani, "L'Homère de Pléthon", *Scriptorium* 68 (2014), pp. 25–48.

Pormann and Savage-Smith 2007: Peter E. Pormann and Emilie Savage-Smith, *Medieval Islamic Medicine*. Washington, DC: Georgetown University Press, 2007.

Price 1986: Simon R. F. Price, "The Future of Dreams: From Freud to Artemidorus", *Past and Present* 113 (1986), pp. 3–37.

Principe 2015: Lawrence M. Principe, "Myth 4. That Alchemy and Astrology Were Superstitious Pursuits That Did Not Contribute to Science and Scientific Understanding", in Numbers and Kampourakis 2015, pp. 32–39.

Principe and Newman 2001: Lawrence M. Principe and William R. Newman, "Some Problems with the Historiography of Alchemy", in William R. Newman and Anthony Grafton (eds), *Secrets of Nature: Astrology and Alchemy in Early Modern Europe* (Transformations). Cambridge, MA: MIT Press, 2001, pp. 385–431.

Raggetti 2019: Lucia Raggetti, "Apollonius of Tyana's *Great Book of Talismans*", *Nuncius* 34.1 (2019), pp. 155–182.

Raven 2003: Wim Raven, "*De mundo*. Tradition syriaque et arabe", in Richard Goulet (ed.), *Dictionnaire des philosophes antiques. Supplément*. Paris: CNRS Éditions, 2003, pp. 481–483.

Reitzenstein 1919: Richard Reitzenstein, "Zur Geschichte der Alchimie und des Mystizismus", *Nachrichten von der Königlichen Gesellschaft der Wissenschaften zu Göttingen, Philologisch-Historische Klasse* (1919), pp. 1–37.

Rescher 1963: Nicholas Rescher, "Al-Kindī's Sketch of Aristotle's Organon", *The New Scholasticism* 37.1 (1963), pp. 44–58.

Richter 2015: Tonio Sebastian Richter, "The Master Spoke: 'Take One of *the Sun* and One Unit of *Almulgam*'. Hitherto Unnoticed Coptic Papyrological Evidence for Early Arabic Alchemy", in Alexander T. Schubert and Petra M. Sijpesteijn (eds), *Documents and the History of the Early Islamic World* (Islamic History and Civilization: Studies and Texts 111). Leiden and Boston, MA: Brill, 2015, pp. 158–194.

Ridgeon 2021: Lloyd Ridgeon (ed.), *Routledge Handbook on Sufism* (Routledge Handbooks). London: Routledge, 2021.

Roberts 2019: Alexandre M. Roberts, "Framing a Middle Byzantine Alchemical Codex", *Dumbarton Oaks Papers* 73 (2019), pp. 69–102.

Roberts 2020: Alexandre M. Roberts, "A Greek Alchemical Epigram in Its Middle Byzantine Context", *Journal of the Warburg and Courtauld Institutes* 83 (2020), pp. 1–36.

Roberts 2022: Alexandre M. Roberts, "Hierotechnicians by Name and Their Middle Byzantine Fame", *Journal of Late Antique, Islamic and Byzantine Studies* 1.1–2 (2022), pp. 167–199.

Rochberg 2016: Francesca Rochberg, "The Brown School of the History of Science: Historiography and the Astral Sciences", in John M. Steele (ed.), *The Circulation of Astronomical Knowledge in the Ancient World* (Time, Astronomy, and Calendars 6). Leiden and Boston, MA: Brill, 2016, pp. 5–17.

Rosenthal 1956: Franz Rosenthal, "Al-Kindī and Ptolemy", in *Studi orientalistici in onore di Giorgio Levi della Vida*, vol. 2 (Pubblicazioni dell'Istituto per l'Oriente 52). Rome: Istituto per l'Oriente, 1956, pp. 436–456.

Rosenthal 1963: Franz Rosenthal, "From Arabic Books and Manuscripts", *Journal of the American Oriental Society* 83.4 (1963), pp. 452–457.

Roueché 1974: Mossman Roueché, "Byzantine Philosophical Texts of the Seventh Century", *Jahrbuch der Österreichischen Byzantinistik* 23 (1974), pp. 61–76.

Roueché 1980: Mossman Roueché, "A Middle Byzantine Handbook of Logic Terminology", *Jahrbuch der Österreichischen Byzantinistik* 29 (1980), pp. 71–98.

Ruska 1923: Julius Ruska, "Sal ammoniacus, Nušādir und Salmiak", *Sitzungsberichte der Heidelberger Akademie der Wissenschaften, Philologisch-Historische Klasse 5. Abhandlung*. Heidelberg: C. Winter, 1923.

Ruska 1924: Julius Ruska, *Arabische Alchemisten*, 1: *Chālid ibn Jazīd ibn Mu'āwija*. Heidelberg: C. Winter, 1924.

Saif 2015: Liana Saif, *The Arabic Influences on Early Modern Occult Philosophy* (Palgrave Historical Studies in Witchcraft and Magic). Basingstoke: Palgrave Macmillan, 2015.

Schulze 2006: Christian (Bochum) Schulze, "Ahron", in Hubert Cancik and Helmuth Schneider (eds), *Brill's New Pauly Online*. Leiden: Brill, 2006, https://doi.org/10.1163/1574-9347_bnp_e12219020.

Searby 2016: Dennis Searby, "Stéphanos d'Alexandrie", in Richard Goulet (ed.), *Dictionnaire des philosophes antiques*, vol. 6. Paris: CNRS Éditions, 2016, no. S 152, pp. 563–579.

Sedgwick 2017: Mark Sedgwick, *Western Sufism from the Abbasids to the New Age*. Oxford: Oxford University Press, 2017.

Sesiano 1998: Jacques Sesiano, "Les carrés magiques de Manuel Moschopoulos", *Archive for History of Exact Sciences* 53.5 (1998), pp. 377–397.

Sezgin 1967–1984: Fuat Sezgin, *Geschichte des arabischen Schrifttums*, 9 vols. Leiden: Brill, 1967–1984.

Shank 2015: Michael Shank, "Myth 1. That There Was No Scientific Activity between Greek Antiquity and the Scientific Revolution", in Numbers and Kampourakis 2015, pp. 7–15.

Signes Codoñer 2014: Juan Signes Codoñer, *The Emperor Theophilos and the East, 829–842: Court and Frontier in Byzantium during the Last Phase of Iconoclasm* (Birmingham Byzantine and Ottoman Studies 13). Farnham and Burlington, VT: Ashgate, 2014.

Siorvanes 1996: Lucas Siorvanes, *Proclus: Neo-Platonic Philosophy and Science*. Edinburgh: Edinburgh University Press, 1996.

Slaveva-Griffin 2014: Svetla Slaveva-Griffin, "Review of Kalvesmaki 2013", *The Classical Review* 64.2 (2014), pp. 429–431.

Smith 1931: Margaret Smith, *Studies in Early Mysticism in the Near and Middle East*. London: The Sheldon Press, 1931.

Smith 1978: Robin Smith, "The Mathematical Origins of Aristotle's Syllogistic", *Archive for History of Exact Sciences* 19.3 (1978), pp. 201–209.

Smith 2002: Morton Smith, "Ostanes", in *Encyclopaedia Iranica*, online edition, 2002, https://www.iranicaonline.org/articles/ostanes.

Stapleton et al. 1949: Henry Ernest Stapleton, G. L. Lewis, and Frank Sherwood Taylor, "The Sayings of Hermes Quoted in the *Mā' al-Waraqī* of Ibn Umail", *Ambix* 3.3–4 (1949), pp. 69–90.

Stoneman 2019: Richard Stoneman, *The Greek Experience of India: From Alexander to the Indo-Greeks*. Princeton, NJ: Princeton University Press, 2019.

Stoyas 2013: Yannis Stoyas, "Ο Νικόλαος Ραβδάς και το *βασιλικόν χρυσουργίον*. Προβλήματα αριθμητικά και ζητήματα κοπής νομισμάτων στην υστεροβυζαντινή περίοδο", in Eirini Mergoupi-Savaidou, Gerasimos Merianos, Faidra Papanelopoulou, and Christiana Christopoulou (eds), *Επιστήμη και Τεχνολογία. Ιστορικές και Ιστοριογραφικές Μελέτες* (Εταιρεία Μελέτης και Διάδοσης της Ιστορίας των Επιστημών και της Τεχνολογίας 1). Athens: Ekdotike Athenon, 2013, pp. 67–95.

Strohmaier 1968: Gotthard Strohmaier, "Die griechischen Götter in einer christlich-arabischen Übersetzung. Zum Traumbuch des Artemidor in der Version des Ḥunain ibn Isḥāḳ", in Franz Altheim and Ruth Stiehl (eds), *Die Araber in der alten Welt*, vol. 5.1. Berlin: De Gruyter, 1968, pp. 127–162.
Strohmaier 1991: Gotthard Strohmaier, "'Umāra ibn Ḥamza, Constantine V, and the Invention of the Elixir", *Graeco-Arabica* 4 (1991), pp. 21–24.
Strohmaier 2016: Gotthard Strohmaier, "Elixir, Alchemy and the Metamorphoses of Two Synonyms", *Al-Qanṭara* 37.2 (2016), pp. 423–434.
Stroumsa 2008: Rachel Stroumsa, *People and Identities in Nessana*. PhD Dissertation. Duke University, 2008.
Stroumsa 2015: Rachel Stroumsa, "Greek and Arabic in Nessana", in Alexander T. Schubert and Petra M. Sijpesteijn (eds), *Documents and the History of the Early Islamic World* (Islamic History and Civilization, Studies and Texts 111). Leiden and Boston, MA: Brill, 2015, pp. 143–157.
Sypiański 2011: Jakub Sypiański, "Les mouvements des érudits et des manuscrits entre Byzance et le monde arabe au IXe siècle et l'hypothèse des influences du mouvement de traduction gréco-arabe au califat abbasside sur la première renaissance byzantine". Le mémoire de Master 1. Université Paris 1 Panthéon-Sorbonne, 2011.
Sypiański 2012: Jakub Sypiański, "Arabo-Byzantine Traffic of Manuscripts and the Connections between the Graeco-Arabic Translation Movement and the First Byzantine 'Renaissance' (9th–10th Centuries)", in Michał Janocha, Aleksandra Sulikowska, and Irina Tatarova (eds), *Byzantium and Renaissances: Dialogue of Cultures, Heritage of Antiquity. Tradition and Modernity*. Warsaw: University of Warsaw, 2012, pp. 187–194.
Terzes 2023: Chrestos Terzes, "The Bacchius Fragment: A Critical Edition", *Greek and Roman Musical Studies* 11 (2023), pp. 117–138.
Thom 2014: Johan C. Thom (ed.), *Cosmic Order and Divine Power: Pseudo-Aristotle,* On the Cosmos. *Introduction, Text, Translation and Interpretative Essays* (SAPERE 23). Tübingen: Mohr Siebeck, 2014.
Tolsa 2020–2021: Cristian Tolsa, "The Earliest Arabic Magic Squares", *Suhayl* 18 (2020–2021), pp. 7–24.
Trizio 2007: Michele Trizio, "Byzantine Philosophy as a Contemporary Historiographical Project", *Recherches de Théologie et Philosophie médiévales* 74.1 (2007), pp. 247–294.
Ullmann 1972: Manfred Ullmann, *Die Natur- und Geheimwissenschaften im Islam* (Handbuch der Orientalistik, 1. Abt.: Der Nahe und der Mittlere Osten, Ergänzungsband 6.2). Leiden: Brill, 1972.
Ullmann 1979: Manfred Ullmann, *Islamic Medicine* (Islamic Surveys 11). Edinburgh: Edinburgh University Press, 1979.
van Bladel 2009: Kevin van Bladel, *The Arabic Hermes: From Pagan Sage to Prophet of Science* (Oxford Studies in Late Antiquity). Oxford: Oxford University Press, 2009.
Van Deun 2003: Peter Van Deun, "Review of Beatrice 2001", *Kernos* 16 (2003), pp. 387–389.
Van Rompay 2011: Lucas Van Rompay, "Theophilos of Edessa", in Sebastian P. Brock, Aaron M. Butts, George A. Kiraz, and Lucas Van Rompay (eds), *Gorgias Encyclopedic Dictionary of the Syriac Heritage*. Piscataway, NJ: Gorgias Press, 2011, pp. 409–410.
Vasilikopoulou-Ioannidou 1971: Agne Vasilikopoulou-Ioannidou, *Ἡ ἀναγέννησις τῶν γραμμάτων κατὰ τὸν ΙΒ' αἰῶνα εἰς τὸ Βυζάντιον καὶ ὁ Ὅμηρος* (Βιβλιοθήκη Σοφίας Ν. Σαριπόλου 14). Athens: Ethnikon kai Kapodistriakon Panepistemion Athenon, Philosophike Schole, 1971.
Vasunia 2013: Phiroze Vasunia, *The Classics and Colonial India*. Oxford: Oxford University Press, 2013.
Viano 2018: Cristina Viano, "Graeco-Egyptian Alchemy", in Alexander Jones and Liba Taub (eds), *The Cambridge History of Science*, vol. 1: *Ancient Science*. Cambridge: Cambridge University Press, 2018, pp. 468–482.
Vinel 2005: Nicolas Vinel, "Un Carré Magique Pythagoricien? Jamblique précurseur des témoins Arabo-Byzantins", *Archive for History of Exact Sciences* 59.6 (2005), pp. 545–562.
Vryonis 1971: Speros Vryonis, Jr., *The Decline of Medieval Hellenism in Asia Minor and the Process of Islamization from the Eleventh through the Fifteenth Century*. Berkeley, CA: University of California Press, 1971.

Matteo Martelli

Homeric Alchemy between Byzantium and Baghdad

Homeric poems are rarely mentioned in the Byzantine anthologies of Greek alchemical writings. In the first chapter of his *Authentic Memoirs*, Zosimus of Panopolis implicitly paraphrases a verse from the *Iliad* to explain the allegorical meaning of the letter *Omega*. The letter corresponds to "what is called Ocean, the origin and seed of all the gods",[1] an explanation that resonates with what is said in Book 14 of the *Iliad*: "Ocean, the origin of gods" (*Il.* 14.201 Ὠκεανόν τε, θεῶν γένεσιν [...]; see also v. 246). Towards the end of the same chapter,[2] Zosimus quotes another verse from the *Iliad* (4.320) to illustrate that every art is practiced with different methods and levels of expertise. However, neither quotation conveys a technical or specifically alchemical meaning. Instead, they are introduced to substantiate universal truths about the cosmos and human nature.

In fact, Homer is not included among the esteemed heroes of alchemy who are mentioned in the lists of alchemical authors which open various Byzantine manuscripts.[3] Furthermore, he is never portrayed as a skilled alchemist in the Greek texts transmitted by these sources. This absence may easily go unnoticed when considering this material in isolation. However, when compared with Syriac and Arabic writings, the silence of Byzantine alchemical sources becomes loud and compels us to explore possible lines of continuity and discontinuity in the intellectual exchanges between Byzantium and the Islamicate world. Both Syriac and Arabic alchemical collections, indeed, feature Homer as an authoritative figure in the field, thereby expanding the spectrum of ancient, almost mythical figures who were involved in alchemy.

1 Zosimus Panopolitanus, *Hypomnemata authentica* 1.1.5–6 (*De littera omega*) (ed. Mertens 1995: 1; see also 57n5). Unless otherwise noted, translations are by the author.
2 Zosimus Panopolitanus, *Hypomnemata authentica* 1.17.163–164 (*De littera omega*) (ed. Mertens 1995: 8; along with the correction proposed by Letrouit 2002: 93).
3 See, in particular, the lists of names in MSS *Marcianus graecus* 299 (coll. 584), f. 7 verso (ed. Berthelot 1885: 128) and *Parisinus graecus* 2327, ff. 195 verso and 294 recto (ed. Berthelot and Ruelle 1887–1888: 2.25–26).

https://doi.org/10.1515/9783111589619-009

1 Homer's Poetry in Arabic and Syriac Alchemical Texts

In the list of sixty alchemical authors provided by the poet and alchemist al-Ṭuġrā'ī (eleventh–twelfth century) in his *Lantern of Wisdom* (*Maṣābīḥ al-ḥikma*),[4] we also find Homer, along with many pre-Islamic alchemists (e.g., Hermes, Ostanes, Democritus) to whom al-Ṭuġrā'ī refers in his writings.[5] Homer is also mentioned in the *Tome of Life* (*Muṣḥaf al-ḥayāt*), a dialogue between Āras and the Byzantine King Theodorus, which collects alchemical sayings attributed to Greek *auctoritates*, from Hermes and Democritus to Apollonius of Tyana and Plato.[6] The exact role played by the poet in these two writings as well as the extent of the alchemical teaching attributed to him remain unknown. Both texts are unedited as unedited is a large portion of the alchemical literature transmitted in Arabic manuscripts, which might likely include further references to Homer. In her preliminary study of MS Istanbul, Süleymaniye, Fatih 3435, an important witness of the *Tome of Life*, Marion Dapsens has already detected various passages where the poet is mentioned, along with different deities and heroes of his poems. For instance, Homer is presented as the author of a work entitled *Key of the Treasures* (*Miftāḥ al-kunūz*),[7] while another passage refers to his account of the Olympian gods – Athena, Aphrodite and Zeus are mentioned – who gathered to drink the "poison of life" (*samm al-ḥayāh*). The gods "were seated on a golden floor" (وكانوا جلوسا على ارض من ذهب), while Aphrodite "made them to drink in a golden cup that contained the poison of life" (تسقيهم بكاس ذهبي فيه سم الحياة). Zeus is presented as the leader of this divine council, which at some point seems to be equated to "a council of stars" (*ma'šar al-nuǧūm*).[8] The entire passage may refer to the opening verses of the fourth book of the *Iliad* (4.1–4):

> Οἱ δὲ θεοὶ πὰρ Ζηνὶ καθήμενοι ἠγορόωντο / χρυσέῳ ἐν δαπέδῳ, μετὰ δέ σφισι πότνια Ἥβη / νέκταρ ἐῳνοχόει· τοὶ δὲ χρυσέοις δεπάεσσι / δειδέχατ' ἀλλήλους, Τρώων πόλιν εἰσορόωντες· [...].
>
> Now the gods, seated by the side of Zeus, were holding assembly on the golden floor, and among them the queenly Hebe poured them nectar, and they with golden goblets pledged each other as they looked out on the city of the Trojans.[9]

---

4 On al-Ṭuġrā'ī, see Ullmann 1972: 229–231, where the work *Keys of Mercy and Lantern of Wisdom* (*Mafātīḥ al-raḥma wa maṣābīḥ al-ḥikma*) is also discussed; this title refers to two separate works according to Razook 1963: 240–264.

5 Cf. Razook 1963: 248 (and 261n40); Inoue 2017: 66n255.

6 Forster 2017: 283–284.

7 MS Istanbul, Süleymaniye, Fatih 3435, f. 145 verso.

8 MS Istanbul, Süleymaniye, Fatih 3435, f. 147 verso. I warmly thank Marion Dapsens, who provided me with this information about Homer in the *Muṣḥaf al-ḥayāt* along with her preliminary transcription of the relevant passages.

9 Engl. transl. Murray and Wyatt 1999: 1.165.

These verses also captured the attention of various Byzantine commentators. In the eleventh century, Michael Psellus devoted a short essay to the allegorical interpretation of *Iliad* 4.1–4, where the name of Zeus is explained as stemming from the term "life" (ζωή) and gods are equated to angels.[10] John Tzetzes (twelfth century), on the other hand, offered an astrological reading of the verses, where gods were interpreted as referring to the stars and their position in the Zodiac.[11]

An astronomical reading of Homer also emerges from the pseudo-Platonic alchemical treatise entitled *Book of the Fourths* (*Kitāb al-rawābīʿ*), known as *Liber quartorum* in its Latin translation.[12] A section of Book 4 deals with the formation of what the Latin translator rendered with *imaginatio*.[13] The ancients identified it with the round bodies (Ar. *aǧrām mustadīra* = Lat. *corpora sphaerica*), which are always in motion. Indeed, the author specifies:

> فهو قول مجاز مشهور في اليونانيين: تصوروا في أوهامهم أن الفلك إنما سرعته في الدوران لطلب الجزء منه ما يليه فالكل طالب لا يدرك، وذلك كثير في كلام الشعراء. قال أوميرس الشاعر: «طلبتك طلب الأفق للأفق» – يعنى به أفق السماء.
>
> There is a well-known metaphorical saying among the Greeks: in their delusion, they believe that the celestial sphere (*al-falak*) rotates swiftly, striving to reach the portion that lies ahead, yet each seeking portion does not reach (what lies ahead). This (saying) is frequent in the words of the poets. The poet Homer said, "your desire is to seek the horizon of the horizon", which refers to the sky's horizon.[14]

The exact identification of Homer's quote remains uncertain. However, astronomical interpretations of Homeric verses as allegorical descriptions of the celestial sphere can be traced back to the Alexandrian exegetical tradition. Crates of Mallus' exegesis of Homer's poems was even labelled as *sphairopoiia* ("the construction of the sphere") or *sphairikos logos* ("theory of the sphere") in antiquity.[15]

---

10 Michael Psellus, *Philosophica minora 1, op.* 42 (*Allegoria in Iliadem IV 1–4*) (ed. Duffy 1992: 148–152). See Cesaretti 1981: 65–69. On allegory in Byzantine and Arabic literary culture, see also the contribution of Maria Mavroudi to this volume.

11 See Ioannes Tzetzes, *Allegoriae Iliadis* 4.45–86 (ed. Boissonade 1851: 102–104).

12 See [Plato], *Book of the Fourths* (*Kitāb al-rawābīʿ*) (ed. Badawī 1955: 119–239). No critical edition exists for the Latin translation, [Plato], *Liber quartorum* (ed. Zetzner 1601–1661: 5.114–208). On this text, see Hasse 2002: 31–65; Thillet 2005: 201–232.

13 See [Plato], *Liber quartorum* (ed. Zetzner 1601–1661: 5.185). It is likely that the Latin translator read *al-taḫayyul* ("imagination, phantasy", often used to translate the Greek φαντασία). Instead, the Arabic text has *al-taǧabbul* in Badawī's edition (1955: 210).

14 [Plato], *Book of the Fourths* (*Kitāb al-rawābīʿ*) (ed. Badawī 1955: 210, 17–20). The Latin translator rendered the Homeric verse as follows: "Dicit Homerus: quaestio tua est quaestio extremitatum, scilicet extremitatum coeli"; [Plato], *Liber quartorum* (ed. Zetzner 1601–1661: 5.185, 42–186, 1).

15 See Porter 1992: 85–91; see Crates Mallotes, frag. 74–75 (ed. Broggiato 2001: 76–77 and LI–LVII).

Homer is also associated with a "Platonic" alchemical context within the so-called Ǧābirian *Corpus*, which confirms the poet's alchemical expertise. The *Corpus* even contained an entire treatise titled *Homer's Rectifications* (*Kitāb muṣaḥḥaḥāt Amūras*), presumably focusing on the alchemical teachings attributed to Homer.[16] Unfortunately, the book has been lost. However, another Ǧābirian treatise, *Plato's Rectifications* (*Kitāb muṣaḥḥaḥāt Iflāṭūn*), features Homer as an important interpreter of alchemical procedures. This text situates the epic poet in a complex geography of pseudepigraphic attributions, where a recipe ascribed to Plato is explained – rectified according to Ǧābir – through the lens of different interpretations by Homer and Porphyry.[17]

Furthermore, another reference to Homer's alchemical knowledge occurs in Ibn Umayl's *Book on the Explanation of the Symbols* (*Kitāb ḥall al-rumūz*, tenth century). The poet is mentioned in the framework of a discussion about the names that alchemists gave to the "water of the sages" (*mā' al-ḥukamā'*).[18] Ibn Umayl writes:

وسموا هذا الماء لازم الأوابق ولازم كل صبغ وهو المفتاح الذي به تفتح أبواب الحكمة، وهو الذي قال فيه أمورس: هو غلق لايفتح الأبواب إلابه

> And they named this water the holder of the fugitives and the holder of every dye. It is the key with which the doors of wisdom are opened. It is that about which Homer said: "It is a lock without which doors cannot be opened".[19]

The source of this alleged Homeric quotation is not specified in the text. Ibn Umayl could refer either to a passage of Homer's poems (here interpreted alchemically), or to a pseudepigraphal work attributed to the poet. It is tempting to draw a comparison between this quote and the first lines (vv. 166–169) of the famous episode in Book 14 of the *Iliad*, known since antiquity as "the deception of Zeus" (Διὸς ἀπάτη). Before seducing Zeus on Mount Ida, Hera adorned herself within her bedroom, which had gleaming doors safeguarded by a secret lock (κληίς) devised by Hephaestus, ensuring that no other deity could enter the room.[20] The episode has been subject to numerous allegorical

16 See Kraus 1942–1943: 1.66. The title is reported in the famous catalogue of Ǧābirian treatises provided by al-Nadīm's *Liber Catalogi* (*Kitāb al-fihrist*) 10.12 (ed. Flügel 1871: 1.357, 4). The same title also appears in a similar list of "rectifications" transmitted by the MS Cambridge University Library, T.-S. Ar. 34.104 (Genizah Collection) and recently studied by Ferrario 2021: 531.

17 On this text and the references to Homer in the Ǧābirian *Corpus*, see Bojidar Dimitrov's contribution to this volume. We must note that both Plato and Porphyry are listed among the fathers of alchemy in the list transmitted in the Byzantine MS *Parisinus graecus* 2327 (see above, note 3). An alchemical recipe is attributed to Plato in the Syriac MS Cambridge University Library, Mm. 6.29 (f. 30 verso); see Berthelot and Duval 1893: 221.

18 The quotation was already noticed by Rodríguez Guerrero 2011: 342. The Spanish version of this paper is Rodríguez Guerrero 2002–2007.

19 Muḥammad ibn Umayl, *Liber explicationis signorum* (*Kitāb ḥall al-rumūz*) (ed. and Engl. transl. Abt et al. 2003: 76–77).

20 *Ilias* 14.166–169: βῆ δ' ἴμεν ἐς θάλαμον, τόν οἱ φίλος υἱὸς ἔτευξεν Ἥφαιστος, / πυκινὰς δὲ θύρας σταθμοῖσιν ἐπῆρσε / κληῗδι κρυπτῇ, τὴν δ' οὐ θεὸς ἄλλος ἀνῷγεν· / ἔνθ' ἥ γ' εἰσελθοῦσα θύρας ἐπέθηκε

interpretations, ranging from Heraclitus' *Homeric Questions*[21] to Neoplatonic philosophers. Proclus offered a detailed philosophical interpretation divided into five distinct questions,[22] but he did not provide an explanation for the lock that sealed Hera's bedroom doors.[23] In his *Commentary on Homer's Iliad*, the twelfth-century Byzantine scholar Eustathius of Thessaloniki, on the contrary, recorded a twofold interpretation of the "secret lock": it alludes both to theories in natural philosophy which are difficult to grasp and to the concealed nature of the events occurring within Hera's chamber.[24]

At least a century before Ibn-Umayl's treatise, Homer's reputation as a skilled alchemist already circulated in Eastern Syriac monastic milieux, as emerges from the portrait of the poet offered by the ninth-century chronographer Thomas of Marga: "By means of drugs, fire and a furnace, Homer turned lead into silver and brass into gold, and formed precious stones by using plant products from these common (plants) and other things from other materials".[25] In the same period, alchemy is also mentioned in Syriac chronicles dealing with the West Syriac Church, in particular with reference to Isaac, patriarch of Antioch (754–756 CE).[26] No references, however, are made to Homer or to any specific alchemical author. On the other hand, Thomas of Marga's account of the Greek poet is embedded in a broader argument highlighting the advantages of an ascetic life, even for sages who lived prior to Christianity. Pythagoras, Homer, Plato, and Hippocrates, despite their extensive education, each attained the pinnacle of wisdom after spending many years in the desert.[27] Surprisingly, Homer's expertise lies in the alchemical arts, while Pythagoras stresses the importance of silence as a path to true wisdom, and Plato contemplates the mystery of Trinity after meditating on a biblical quote (Deuteronomy 6:4).[28] Thanks to his ascetic lifestyle, Hippocrates gained a deep understanding of how a child is nourished in its mother's womb.

φαεινάς ("So she set out to go to her chamber that her dear son Hephaestus had fashioned for her, fitting strong doors to the doorposts with a secret bolt that no other god could open. There she entered in, and closed the bright doors"; Engl. transl. Murray and Wyatt 1999: 2.79).

21 Heraclitus Paradoxographus, *Quaestiones Homericae* 39 (ed. Pontani 2005: 122–124).

22 See Lamberton 1986: 208–214.

23 The lock is not mentioned in Proclus' fourth question of his sixth essay, which deals with Hera's adornments (Proclus, *In Platonis Rem publicam commentarii* 6.1.14 [ed. Kroll 1899–1901: 1.137, 2–139, 19 = ed. Lamberton 2012: 186–191]).

24 Eustathius Thessalonicensis, *Commentarii ad Homeri Iliadem* 14.168 (ed. Van der Valk 1971–1987: 3.601, 17–602, 3).

25 Thomas Margensis, *Liber superiorum* 5.13 (ed. Bedjan 1901: 312); Engl. transl. (slightly modified) Budge 1893: 2.530–531 (Syriac text in vol. 1.298).

26 See Martelli 2014: 192–199.

27 See Becker 2006: 197; Bettiolo 2004.

28 Since at least the late sixth century, Syriac sources maintained that Plato endorsed the mystery of Trinity, as emerges, for instance, in Barḥadbšabbā, *Causa fundationis scholarum* (ed. Scher 1908: 363); Engl. transl. Becker 2008: 134. On Plato's and Pythagoras' prophecies, see also Brock 1983: 229–230.

Plato, Homer, and Pythagoras are also mentioned in a long list of Greek philosophers which opens MS London, British Library, Add. 14620 (ninth century),[29] which also includes poets and mythological figures such as Minos and Rhadamanthus, who are credited with a treatise on philosophy (*maktbānūtā d-pīlsūpūtā*). We find no mention of the physician Hippocrates, who is again associated with Homer in a peculiar alchemical text transmitted by the Syriac manuscript, Cambridge, University Library, Mm. 6.29 (fifteenth century).[30] Indeed, this manuscript is the unique witness to a short alchemical treatise with a twofold structure (ff. 133 recto–135 verso). The first part praises the Greek physician Hippocrates as a man blessed by God, who used his alchemical and medical knowledge to take care of human beings (ff. 133 recto–134 verso).[31] By contrast, the second part is a critique against Homer (ff. 134 verso–135 verso), who adulterated the alchemical science and, more importantly, promoted a false and blasphemous doctrine against the true Christian religion. The beginning of the passage reads:

> ܡܢ ܒܬܪ ܠܗܢܐ ܐܒܐ ܕܟܠܗܝܢ ܛܒ̈ܬܐ ܐܝܦܘܩܪܐܛܝܣ ܐܘܡܝܪܘܣ ܥܒܕ ܒܝ̈ܫܬܐ ܓܒܪܐ ܒܝܫܐ. ܠܐ ܡܪܚܡܢܐ: ܠܐ ܫܘܐ ܗܘܐ ܒܬܪ
> ܡܘܬܗ ܠܬܫܒܘܚܬܐ. ܫܪܝ ܕܢܦܫܘܩ ܠܐܘܡܢܘܬܐ ܐܠܗܝܬܐ. ܘܫܚܠܦܗ ܘܫܢܝܗ ܘܚܫܟܗ. ܟܕ ܠܐ ܐܬܒܝܢ ܡܕܡ
> ܡܢ ܗܠܝܢ ܕܒܢܝ̈ܢܫܐ ܘܙܒܢܝ̈ܐ ܐܢܝܢ. ܗܘ ܠܗܠܝܢ: ܗܘ ܡܚܒܠܢܐ:
>
> After Hippocrates, the father of every good thing, Homer, the evildoer, a wicked and merciless man, who does not deserve any praise after his death, began explaining the divine art, but he changed, altered, and darkened (it), without considering any of the human and temporal things.[32]

The attack on Homer's blasphemy brings to mind certain passages from Greek Christian apologetic writings that were translated into Syriac. For example, Pseudo-Justin's *Oratio ad Graecos*, also known as Ambrose's *Hypomnemata* in its Syriac translation, starts with a vehement critique of the poet, whose "foolishness" (*saklūtā*) was unparalleled. The heroes of the *Iliad* personify the epitome of human passions and insanity, and, even more significantly, the tales about the gods narrate their malevolence and violence.[33] Pseudo-Justin presents the reader with a broad range of misbehaviours drawn from Homer and Greek mythology. He even goes as far as to paraphrase a few verses from the *Odyssey* (10.306–309), where Hephaestus exposes Aphrodite's unfaithfulness to the Olympian gods, as she shamelessly commits adultery with Ares.[34]

29 For an edition and English translation, see Anonymus, *Dicta Philosophorum Graecorum* (ed. Arzhanov 2019: 131–132). For a description of the manuscript, see Wright 1870–1872: 2.800–803 (item DCCC).

30 Fr. transl. Berthelot and Duval 1893: 314–318.

31 See Martelli 2017.

32 MS Cambridge, University Library, Mm. 6.29, f. 134 verso, ll. 11–16.

33 [Iustinus], *Oratio ad Graecos* (ed. Pouderon et al. 2009: 302–313).

34 [Iustinus], *Oratio ad Graecos* 3 (ed. Pouderon et al. 2009: 308): ܐܡܪ ܕܝܢ ܗܦܐܣܛܘܣ ܟܕ ܐܬܚܙܝ:. ܬܘ ܚܙܘ ܥܒܕܐ ܓܘܚܟܢܝܐ ܘܠܐ ܡܬܒܥܐ ܗܘܐ: ܐܝܟܢܐ ܕܠܝ ܕܐܝܬܝ ܚܓܝܪܐ ܠܝ ܒܠܚܘܕ ܒܪܬܗ ܕܐܪܐ ܐܦܪܘܕܝܛܐ: ܕܠܝܬܝ ܐܢܐ: ܘܡܒܣܪܐ ܠܐܪܐܘܣ ܕܫܦܝܪ ܗܘ (“When he saw [it], Hephaestus said: ‘come to see a ridiculous and senseless

Alongside Homer's blasphemy, the author of the alchemical text fiercely criticizes those who followed Homer's teachings and praised the value of his poetry for education. This criticism is particularly directed at Homer's role as a misguided teacher of alchemy, evident in his own poems which were seemingly interpreted in an alchemical context. The anonymous author quotes and comments on the opening verse of the *Iliad* to expose Homer's deceitful practices:

> ܐܡܪ ܥܠ ܡܘܣܐ. ܐܡܪ ܠܢ ܚܡܬܗ ܕܒܪ ܦܠܘܤ. ܚܡܬܐ ܕܝܢ: ܠܗ̇ ܠܐܠܗܘܬܐ ܐܡܪ ܥܝܢܐܝܬ. ܘܥܠܗ̇ ܒܗ̇ ܠܗ̇ ܠܐܠܗܘܬܐ ܡܠܠ. ܒܪ ܦܠܘܤ ܕܝܢ ܡܠܬܐ ܟܣܐ ܕܐܣܝܘܬܐ ܐܠܗܝܬܐ. ܐܦ ܠܡܥܒܕܘܬܐ ܕܚܡܫܐ ܐܠܗܐ. ܘܡܥܒܕܢܘܬܐ ܚܡܫܐ ܠܫܢܝܢ ܡܠܠ.

> He speaks of the "muses", he tells us about "the wrath of Peleus' son"; he clearly says "wrath" for the divinity: indeed, he talks of the divinity in person. By the word "Peleus' son", he secretly refers to (literally, "conceals", "covers") the divine remedy, and he also talks of the action of the five gods and the operation of the five languages.[35]

The overall sense of the proposed interpretation is difficult to grasp. Some Homeric terms appear to be interpreted allegorically, suggesting a reference to the "divine medicine", possibly the *pharmakon* used for alchemical transmutations, as well as to the number five. This latter reference may imply the application of arithmology to alchemical practices, which is also evident in the Homeric quotations included in the Ǧābirian *Corpus*.[36] A few lines after the quoted passage, the anonymous author draws a comparison between Homer, who employed only five words in his alchemical practice, and Hermes, who used seven words instead.[37] Indeed, the number seven plays an important role in other Syriac alchemical texts, such as the tale about Ostanes' discovery of Hermes' alchemical teachings, which were written on seven tablets, each associated with a different metal and planet.[38]

Despite the challenges, the passage quoted above shows that Homer's expertise in alchemy was woven into his poems, which were meant to be interpreted allegorically to reveal their hidden alchemical meaning. It is worth noting that an allegorical interpretation of Peleus and his son Achilles also appears in the Syriac *Liber scholiorum* (Book of Scholia) by Theodore bar Konai (eighth century). According to this view, Peleus symbolizes the earth (or clay), which, when mixed with water (Tethys, daughter of the Nereids), gives rise to human bodies (Achilles).[39] Furthermore, in the following paragraph of the book, Homer is explicitly mentioned as a pagan sage (alongside Hesiod and

---

affair, how Aphrodite [here referred to as *Bīltī*], the daughter of the lord of the gods, dishonours me, who I am her [husband]. I am her husband, but she honours Ares, who is a stranger to her'").

35 MS Cambridge, University Library, Mm. 6.29, f. 135 recto, ll. 9–14.

36 See Kraus 1942–1943: 2.117–118; also Dimitrov's paper in this volume.

37 MS Cambridge, University Library, Mm. 6.29, f. 135 verso; Fr. transl. Berthelot and Duval 1893: 317.

38 See Berthelot and Duval 1893: 311. On the Arabic version of this story, see van Bladel 2009: 54–57.

39 Bar Konai, *Liber scholiorum* 11.5 (ed. Scher 1910–1912: 2.288); Fr. transl. Hespel and Draguet 1981–1982: 2.215. This interpretation already appears in Pseudo-Clement's *Homilies*; see De Vos 2020.

Plato): a verse from the *Iliad* (9.99) is cited to exemplify Homer's idea that the cosmos is a combination of a moist and earthly substance.[40]

The allegorical interpretation of Homer's poems and Greek mythology was employed to reveal hidden meanings related to natural philosophy as well as alchemy, as we have observed. To gain a better understanding of how and when Homer's expertise was expanded to include alchemy, we will explore late antique and Byzantine sources in search of illustrative examples of a specific inclination to read ancient myths alchemically. This inclination became more common during the Renaissance, when Byzantine alchemy was first discovered through classical mythology. It is known, in fact, that humanists dealt with the entry of the *Suda* lexicon (Δ 250),[41] which provides an alchemical interpretation of the golden fleece.[42] In Vincenzo Percolla's *Auriloquio* (1560–1570), there exists an extensive collection of classical myths interpreted as allegories for alchemical processes.[43] The book includes 209 *favole* ("tales/myths"), such as the stories of Minos, Narcissus, Jason, and Midas to name but a few. Interestingly, Byzantine authors had already associated Midas' power of turning everything he touched into gold with alchemy. This connection is evident in the technical verb *chrysopoiein* ("to make gold"), which was mainly used in Graeco-Egyptian and Byzantine alchemical literature. The verb is introduced by Maximus Planudes (ca. 1260–1305) in his own translation of Ovid's *Metamorphoses* into Byzantine Greek. Sick of having wheat, meat and even water immediately transformed into gold, Midas begged the god Dionysus to be freed of this power. When he followed Dionysus' orders and plunged in a river in Sardis,

> Ἡ δὲ τοῦ χρυσοποιεῖν δύναμις τὸν ποταμὸν ἔβαψεν ἐκ τοῦ ἀνθρωπείου σώματος πρὸς τὸ ῥεῖθρον αὐτοῦ μεταστᾶσα.
>
> The capacity of making gold (*chrysopoiein*) dyed the river when it was transferred from (Midas') human body to the stream of the river.[44]

2 Alchemical Mythology in Byzantium

As already mentioned, the myth of Jason and the Argonauts gives us the most famous example of an alchemical interpretation of ancient mythology that can be reliably traced back to Byzantine sources. According to the *Suda* lexicon (Δ 250), the Golden

40 Bar Konai, *Liber scholiorum* 11.6 (ed. Scher 1910–1912: 2.289). See already Nöldeke 1898.
41 *Suda lexicon*, s.v. Δέρας (Δ 250) (ed. Adler 1928–1938: 2.24). On the entries of the *Suda* lexicon pertaining to alchemy, including the one under discussion, see Alexandre Roberts' contribution to this volume.
42 On the circulation of the *Suda* entry among humanists, see Matton 1995: 309–314.
43 Vincenzo Percolla, *Auriloquio* (ed. Anzuini 1996).
44 Maximus Planudes, *Publii Ovidii Nasonis Metamorphoseon libri XV Graece versi* 11.142–143 (ed. Papathomopoulos and Tsavare 2002: 399).

Fleece that Jason and Medea brought to Greece from Colchis was, in fact, a parchment book containing instructions on how to use the alchemical art to produce gold. John of Antioch has been tentatively recognized as one of the sources for the *Suda*'s entry, as he might have reported this information in his seventh-century *Chronicle*, which has only survived in fragments.[45] Moreover, the same interpretation of the Golden Fleece is also mentioned in an anonymous collection of mythographic passages, which is nowadays known as either "Vatican excerpts" (*Excerpta Vaticana*) or *On Unbelievable Things* (*Περὶ ἀπίστων*). The third chapter of this collection reads:

> Περὶ τοῦ χρυσοῦ δέρους. Ὅτι τὸ ἐν Κόλχοις φυλασσόμενον οὐκ ἦν δέρας ἀληθῶς χρυσοῦν (τοῦτο γὰρ ποιητικευόμενόν ἐστιν), ἀλλὰ βιβλίον ἐν δέρμασι γεγραμμένον περιέχον ὅπως δεῖ γίνεσθαι διὰ χειμείας χρυσόν. εἰκότως οὖν οἱ τότε χρυσοῦν ὠνόμαζον αὐτὸ διὰ τὴν ἐξ αὐτοῦ ἐνέργειαν.

> On Golden Fleece. The object guarded in Colchis was not a truly golden fleece (this is a poetic fabrication), but it was a book written on leather containing instructions on how to produce gold through alchemy. And so, in all likelihood, the men of that time called it "golden" because of the (alchemical) power it possessed.[46]

The date of the compilation is uncertain, and the work clearly includes material later than the fifth century – since the Neoplatonic philosopher Proclus is explicitly mentioned as the source of Chapter 20.[47] As for the Golden Fleece, Eustathius of Thessaloniki, the famous Byzantine commentator of Homer, attributed a similar exegesis of the myth to the second-century historian Charax of Pergamon.[48] In his commentary on Dionysius Periegetes, Eustathius explains that the Golden fleece would have referred to chrysography according to Charax:

> Ὁ Χάραξ τὸ χρυσοῦν δέρμα μέθοδον εἶναι λέγει χρυσογραφίας μεμβράναις ἐμπεριειλημμένην, δι' ἥν, ὡς λόγου ἀξίαν, τὸν τῆς Ἀργοῦς καταρτισθῆναι στόλον φησί.

> Charax says that the Golden Fleece is a method for writing in gold letters transmitted in leather folia. Because of this, he says, the renowned expedition of Argos was prepared.[49]

Charax of Pergamon, along with two other important mythographers, Heraclitus and Palaephatus, is mentioned once again by Eustathios in his commentary on the *Odyssey*.

45 Ioannes Antiochenus, *Fragmenta*, frag. 26.3 (ed. Roberto 2005: 64–67). The fragment, however, has been considered spurious in the edition by Mariev (2008: 595).

46 Anonymus paradoxographus, *De incredibilibus* (*Excerpta Vaticana*) 3 (ed. Festa 1902: 89); Engl. transl. Hawes 2014: 240 (slightly modified).

47 Anonymus paradoxographus, *De incredibilibus* (*Excerpta Vaticana*) 20 (ed. Festa 1902: 98, 10).

48 Charax is also mentioned in *De incredibilibus* (*Excerpta Vaticana*) 16 (ed. Festa 1902: 95, 15); Engl. transl. Hawes 2014: 244–245, with reference to the myth of Io.

49 Eustathius Thessalonicensis, *Commentarium in Dionysii Periegetae orbis descriptionem* 689 (ed. Müller 1861: 340, 17–19 = Charax Pergamenus (ed. Jacoby 1923–1958: 103 F 37); see already, Charax Pergamenus, frag. 14 (ed. Müller 1841–1873: 3.639).

In his second reference to them in this commentary, Eustathios remarks that these three authors offered rationalizing interpretations of ancient poetry and myths by reading them allegorically, thus "healing mythology into history" (θεραπεύσαντα μύθους πρὸς ἱστορίαν).[50] It is worth noting that the anonymous treatise titled *On Unbelievable Things* (*Περὶ ἀπίστων*) is only preserved in MS *Vaticanus graecus* 305 (thirteenth century; f. 197 verso), which transmits our work along with other ancient exegetical material on mythology and Homer's poems. This compilation includes Porphyry's *Homeric Questions* and *On the Cave of the Nymphs* along with various writings attributed to the aforementioned mythographers. Two treatises with the same title, *On Unbelievable Things* (*Περὶ ἀπίστων*), are respectively attributed to Palaephatus and Heraclitus, who is also credited with a work entitled *Homeric Questions*.[51]

These works do not explicitly mention alchemy, but they do offer rational explanations for ancient myths by treating them as allegorical stories about the invention and use of specific technologies, many of which were practical and beneficial. This interpretation often revolves around the motif of the "first inventor". For example, Palaephatus interpreted the myth of Pandora, who was supposedly made from earth, as a tale about a wealthy Greek woman who discovered how to apply a substantial amount of earth-based cosmetic to her face.[52] Similarly, according to the myth, Medea possessed the ability to restore youth to old men, but in reality, she was the first to discover a red and black plant dye (ἄνθος τὸ πυρρὸν καὶ τὸ μέλαν) that made elderly men appear to have black and rosy hair. Furthermore, Palaephatus adds that Medea also discovered the health benefits of steam baths for men.[53]

Even Homer's verses had the capacity to convey technological knowledge. In the work titled *Homeric Questions*, Heraclitus tackles one of the most debated scenes of the *Odyssey*, where Hephaestus discovers the affair between his wife Aphrodite and Ares (*Od.* 8.266ff.).[54] This scene had been strongly criticized for its impious and disrespectful portrayal of the gods. However, upon a careful reading of the scene, this strong criticism becomes unfounded. After presenting a philosophical interpretation suggesting that the two gods symbolize Empedocles' cosmological forces of love and strife, Heraclitus adds a second, more technical interpretation. According to this reading, Homer allegorically describes the art of metalwork, in particular iron metallurgy. Ares is likened to iron,

---

50 Eustathius Thessalonicensis, *Commentarii ad Homeri Odysseam* 19.163 (ed. Stallbaum 1825–1826: 2.195–196). The first reference to the three authors occurs at Eustathius, *Commentarii ad Odysseam* 4.450 (ed. Stallbaum 1825–1826: 1.176).

51 On Heraclitus the Paradoxographer, see Nilsson 2025: 69–72. On the reception of Palaephatus in Byzantium, see Alganza-Roldán 2019.

52 Palaephatus, *De incredibilibus* 34 (ed. Festa 1902: 52). The application of a substantial amount of earth-based cosmetic (ἐχρίετο πολλῇ τῇ γῇ) might refer to the use of white lead or ceruse (ψιμμύθιον; *cerussa* in Latin); cf. Stern 1996: 66.

53 Palaephatus, *De incredibilibus* 43 (ed. Festa 1902: 64–65); Engl. transl. Stern 1996: 75–76.

54 Heraclitus Paradoxographus, *Quaestiones Homericae* 69 (ed. Pontani 2005: 162).

which undergoes initial melting by the fire personified as Hephaestus. Once the iron is softened, skilled metalworkers expertly shape it, with their mastery exemplified by Aphrodite. Finally, the hot mass of iron is quenched (βαπτίζω) into cold water, represented by Poseidon, who, in Homer's scene, saves Ares from Hephaestus' anger. The same interpretation is summarized in the *Scholia on Odyssey*,[55] and briefly mentioned in Eustathius' commentary, which only refers to iron and fire.[56] On the contrary, the role of Poseidon–water is not explained, even though Eustathius explicitly credits Homer with a certain expertise on how to harden and quench iron. After explaining the three types of medicaments known by Homer – namely, powders, plasters, and drinks – he specifies that various terms stem from *pharmaka*: *pharmakeuein* (φαρμακεύειν), typical of the medical lexicon (ἰατρικὴ λέξις), *pharmassein*, that is, to temper iron in a quenching bath (φαρμάσσειν, τὸ στομοῦν σίδηρον ἐκ βαφῆς) or simply "to dye" (βάπτειν) along with "dye-houses" (βαφεῖα, once called φαρμακῶνες).[57] Homer, for instance, referred to the second meaning in *Odyssey*, Book 9 (vv. 391–393), where he described the blinding of Polyphemus.[58] Moreover, along with iron, Homeric verses refer to a wide range of metals which captured the attention of Eustathius. Interestingly, while commenting on v. 43 of Book 8 of the *Iliad*, he reports that each metal corresponds to a specific planet, thus explicitly associating sun with gold (and with Jupiter/Zeus, which often allegorically refers to the sun), moon with silver, and Saturn with lead.[59] The full set of associations between the seven planets and the seven metals is reported in Tzetzes' short treatise titled *On the Birth of Gods* (*Περὶ γεννήσεως τῶν θεῶν*),[60] or *Collection of the*

55 *Scholia Graeca in Odysseam* EX 8.267f (ed. Pontani 2007–2022: 4.144). The verb στομόω is here used to refer to the hardening of iron (ὁ δὲ Ποσειδῶν ἤτοι τὸ ὕδωρ στομοῖ αὐτόν, δηλονότι τὸν σίδηρον).

56 Eustathius Thessalonicensis, *Commentarii ad Homeri Odysseam* 8.367 (ed. Stallbaum 1825–1826: 1.298–299).

57 Eustathius Thessalonicensis, *Commentarii ad Homeri Iliadem* 4.724 (ed. van der Valk 1971–1987: 1.734).

58 Eustathius Thessalonicensis, *Commentarii ad Homeri Odysseam* 9.393 (ed. Stallbaum 1825–1826: 1.352): Φαρμάσσειν δὲ οὐχ' ἁπλῶς ἐνταῦθα τὸ βάπτειν ὅθεν καὶ φαρμακῶνες τὰ βαφεῖα ὡς καὶ προεδηλώθη, ἀλλὰ τὸ στομοῦν καὶ στεῤῥοποιεῖν. φησὶ γοῦν. ὕδατι ψυχρῷ βάπτει φαρμάσσων. στομοῦται γὰρ σίδηρος τοιαύτῃ βαφῇ ("The verb *pharmassein* does not simply means 'to dye', from which the term 'dye-house' derives as explained above; it rather means 'to quench', 'to harden'. Indeed, he [i.e., Homer] says: 'he dips [iron] in cold water by tempering it' [*Od.* 8.391]. In fact, iron is tempered in a quenching bath").

59 Eustathius Thessalonicensis, *Commentarii ad Homeri Iliadem* 8.43 (ed. van der Valk 1971–1987: 2.523): Ἰστέον δὲ ὅτι οἱ ἐκ μαθημάτων ἀνήκειν φασὶν Ἡλίῳ τὸν χρυσόν, ὡς τῇ Σελήνῃ ἄργυρον, Κρόνῳ δὲ μόλιβδον, καὶ ἄλλῳ τῶν πλανωμένων ἕτερόν τι τῶν μετάλλων. Διὸ χαίρειν χρυσῷ λέγεται καὶ ὁ Ζεύς, καθότι καὶ εἰς ἥλιον ἀλληγορεῖταί ποτε κατὰ τὸν εἰπόντα "Ὁ μὲν δὴ μέγας ἐν οὐρανῷ Ζεὺς πτηνὸν ἅρμα ἐλαύνων". See also Eust. *ad Iliad.* 1.15 (ed. van der Valk 1971–1987: 1.40), where only gold–sun and silver–moon are mentioned.

60 Two editions of Tzetzes' *De generatione deorum* have been published: (1) ed. Cramer 1839–1841: 3.101–113, based on MS *Parisinus graecus* 2556 (thirteenth century; ff. 30 recto–34 recto); (2) ed. Matranga 1850: 2.364–371, based on MS *Angelicus graecus* 122 (thirteenth century; ff. 7 verso – 8 verso). For a description of these witnesses, see Sciarra 2005: 20–26 and 39–44. These two witnesses are part of a

Allegories by John Tzetzes the Grammarian (*Συλλογὴ ἀλληγοριῶν Ἰωάννου γραμματικοῦ τοῦ Τζέτζου*) in MS *Vaticanus graecus* 1316 (thirteenth century; ff. 31 recto–36 verso):[61]

> Περὶ τῶν ἀνακειμένων ὕλων καὶ τοῖς θεοῖς ἑπτὰ ἄστρασι· καὶ τοῦτο ἐστὶ τὸ τῆς ἀλληγορίας τοῦ τετάρτου Ἰλιάδος θέματος. Τούτων τῶν ἀστέρων ἑνὶ ἑκάστῳ ἔκειντο καὶ ὕλαι· τῷ Κρόνῳ, ὁ χαλκός· τῷ Διί, ὁ χρυσός· τῷ Ἄρεϊ, ὁ σίδηρος· τῷ Ἡλίῳ, ὁ ἤλεκτρος· τῇ Ἀφροδίτῃ, ὁ κασσίτερος· τῷ Ἑρμῇ, ὁ μόλιβδος· τῇ Σελήνῃ, ὁ ἄργυρος.
>
> On the substances (*hylai*)[62] that are also ascribed to the Gods, the seven stars. This is the kind of allegory of the fourth *thema* of the *Iliad*. Substances (*hylai*) also depend[63] on each one of these stars: copper on Saturn, gold on Jupiter, iron on Mars, electrum on the Sun, tin on Venus, lead on Mercury, silver on the Moon.[64]

Tzetzes briefly referred to these associations also in his *Allegories of the Iliad*, when commenting on Achilles' weapons that were borrowed by Patroclus for the deadly duel against Hector. The weapons were given by gods, which represent the stars of the sky that rule the human fate. They "were made of gold, silver, and other substances" (ἅπερ ἦσαν ἐκ χρυσοῦ, ἀργύρου καὶ τῶν ἄλλων), "for – Tzetzes argues – to each of the stars a certain substance (*hylē*) is ascribed" (ἑκάστῳ τῶν ἀστέρων γὰρ ἀνάκειταί τις ὕλη).[65] Similar associations also appear in the late-antique Neoplatonic commentators on Plato and Aristotle, in the astrological tradition, and in Byzantine alchemical anthologies.[66]

---

group of manuscripts copied in the Otranto area, in which Tzetzes' short treatise introduces a group of scholia on the *Iliad* belonging to the so-called **h**-family in Erbse's critical edition (*Scholia vetera in Iliadem* [ed. Erbse 1969–1988]), where this family is largely neglected; see, e.g., Montanari et al. 2017).

61 On this manuscript, which hands down a more complete version of Tzetzes' work (with a different arrangement of the sections), see Sciarra 2005: 64–72 and 131–133.

62 Tzetzes uses the term *hylē* (as well as the plural *hylai*) in reference to metals in his exegetical works on Homer. For instance, in the *Allegoriae Iliadis* 15.65–67 (ed. Boissonade 1851: 182), he describes Achilles' weapons and specifies that Hephaestus melted "copper, gold, tin, and the other substances" (χαλκόν, χρυσόν, κασσίτερον, καὶ τὰς λοιπὰς τὰς ὕλας). Achilles' weapons, indeed, represent the physical world. Tzetzes identifies its components with the elements, namely earth, the sea (i.e., water), air, and fire, and interestingly associates each element with a metal: earth–tin, sea–silver, air–gold, fire–copper (*Allegoriae Iliadis*, 18.674–676 [ed. Boissonade 1851: 251]).

63 The verb is omitted in Cramer's edition (1839–1841: 3.113).

64 Ioannes Tzetzes, *De generatione deorum* (ed. Matranga 1850: 2.371; also Cramer 1839–1841: 3.113). The text has been collated with MSS *Vaticanus graecus* 1316 (f. 32 verso) and *Bodmerianus* 85 (f. 4 recto). For a list of witnesses, see Sciarra 2005: 127n230.

65 Ioannes Tzetzes, *Allegoriae Iliadis* 18.127–129 (ed. Boissonade 1851: 226). On astrology in Tzetzes' *Allegoriae*, see Komorowska 2019: 123–142 (p. 139 on metals). On the other hand, in his brief commentary notes, Boissonade explained the passage by referring to magic. He probably had in mind a passage, where Tzetzes explains that the vessels for lecanomancy were made of "any precious substance (*hylē*), gold, silver, electrum, crushed pearls, (gem)stones, copper, and every substance of this kind" (*Exegesis in Homeri Iliadem* 1.63 [ed. Papathomopoulos 2007: 168, 14–16]).

66 See Halleux 1974: 149–156.

To sum up, ancient myths, as well as Homeric poems, were read and interpreted by referring to various areas of ancient technology, ranging from cosmetics to astrology and from metallurgy to ink-making and chrysopoeia. Mythology, moreover, was not only a repository of technical knowledge. It also served to highlight the transformative power of alchemy, which was compared with the metamorphoses described in various mythological accounts. A catalogue of mythological figures known for their abilities to transform reality is provided by the fourth-century philosopher and rhetorician Themistius, who offers an important reference to alchemy, counted among the early mentions of this discipline in non-alchemical literature. In a speech dating to the winter between 366 and 367 CE,[67] Themistius praises the clemency and forgiveness of Emperor Valens (r. 364–378 CE). Valens was lauded for showing mercy to the supporters of Procopius, who had taken control of Constantinople in 365 and even declared himself emperor. Through this act, Valens transformed enemies into friends, thereby demonstrating a power (*dynamis*) that can be observed in other mythical royal figures:

> Καὶ ταῦτα ἐγὼ μεταμόρφωσιν βασιλικωτέραν νενόμικα ἧς μετεμόρφου Ξέρξης τὴν ἤπειρον καὶ τὴν θάλασσαν, καὶ ἧς ἡ Κίρκη τοὺς ἀνθρώπους καὶ ἧς ἡ Μήδεια τὸ γῆρας καὶ ἧς Αὐτόλυκος τὰ φώρια. οὐ γάρ ἐστι πρὸς τὴν φύσιν φιλονεικία, ἀλλὰ θεόσδοτος ἀκριβῶς ἡ δύναμις αὕτη καὶ Ἑρμοῦ δῶρον, ψυχὰς ἐξ ἀγρίων ἡμέρους κατασκευάζειν καὶ ἐξ ἀλλοτρίων ἐπιτηδείους. νῦν δὲ τοῦ χαλκὸν μὲν εἰς ἀργύριον μεταβαλεῖν καὶ τοῦ τὸ ἀργύριον εἰς χρυσίον ἀσμένως ἄν τινα ἐξεύροιμεν τέχνην, [...].
>
> I consider these (achievements) as a transformation more deserving of a king than the transformation of sea and land carried out by Xerxes, the transformation of men by Circe, the transformation of old age by Medea, the transformation of stolen goods by Autolycus. Indeed, the power to make ferocious souls gentle and enemy souls friendly is not a rivalry against nature, but rather a clear power given by gods, a gift of Hermes. Today, it might even be possible to readily discover an art (*technē*) that tuns copper into silver and silver into gold, etc.[68]

Xerxes turned the sea into land by building his famous pontoon bridges, which was made by tying boats together. He also turned the land into sea by excavating a canal across the Mount Athos peninsula. Circe possessed the power to transform men into pigs, while Medea could rejuvenate herself by using her *pharmaka*.[69] Autolycus, a skilled thief and the grandfather of Odysseus, had the knack for altering the shape or colour of the animals he stole.[70] In this collection of royal and mythical figures, they all possessed transformative powers, although distinct from Valens' capacity of turning enemies into friends. Valens' power, in fact, did not involve a competition with nature. The role of alchemy remains somewhat uncertain within this context. Themistius was undoubtedly

---

67 Themistius, *Oratio* 7 (ed. Downey et al. 1965–1974: 1.127–151). See Vanderspoel 1995: 162–167.

68 Themistius, *Oratio* 7, 96d–97a Harduin (ed. Downey et al. 1965–1974: 1.145–146).

69 As seen above, according to the mythographer Palaephatus, Medea's ability to rejuvenate old men was due to the use of plant-dyes and steam baths.

70 Themistius, *Oratio* 29, 344a–b Harduin (ed. Downey et al. 1965–1974: 2.173).

aware of various techniques employed for counterfeiting gold, purple, and precious stones, which are encompassed by the Graeco-Egyptian alchemical texts.[71] However, in the quoted passage, he appears to refer to an actual transformation of metals. The mention of Hermes, often associated with alchemy, might serve as a link between the mythical exempla and the alchemical art, which can perfect metals through its transformative *dynamis*, similar to how Valens – according to Themistius' suggestion – can transform human souls.

Almost five centuries later, in his *Letter on the Making of Gold*, Michael Psellus utilizes again similar rhetorical devices to extol the value of the chrysopoetic instructions he provided to Patriarch Michael I Cerularius (1043–1058).

> Ἀλλ' ὁρᾷς ὅπερ ἐγὼ σοι πεποίηκα; χρυσοῦ γὰρ πηγὰς ἀναδούς, οὔτε τὸν Ἄθω διέσεισα, οὔτε τὸ Πάγγαιον μετεκίνησα, οὐδὲ φλέβας τινὰς ὑπογείους χρυσίτιδας ἀνεστόμωσα, ἀλλὰ λίθους ἀλλήλοις ἐντρίψας καὶ πόας δή τινας μίξας, ἀφελῶς σοι καὶ εὐώνως τὸν πολύτιμον ἐξειργασάμην χρυσόν.

> But do you consider what I have done for you? Indeed, I provide you with sources of gold without shaking Mount Athos, relocating Mount Pangaion, or excavating underground gold veins. By grinding stones together and combining herbs, I can effortlessly and affordably produce highly valuable gold for you.[72]

As pointed out by Albini, Psellus probably referred to Xerxes' expedition, with specific reference to the same episode already highlighted in Themistius' passage, namely the excavation of a canal across the Mount Athos peninsula.[73] The Pangaion Hills, on the other hand, were known for their rich mines of gold and silver, largely exploited in antiquity. The art of alchemy save humankind from the need of acting against nature and violating the earth and mountains, being, at the same time, a technology that deals with natural alterations of matter, as Psellus already emphasized at the beginning of his letter.[74] Here, in fact, he explicitly links alchemy with "the science of nature" (ἡ περὶ

---

71 Cf. Themistius, *Oratio* 21, 247b Harduin (ed. Downey et al. 1965–1974: 2.23): ἢ πρὸς Διός, εἰ μέν τις χρυσίον ὑπόχαλκον ἢ πορφύραν ἐψευσμένην τὸ ἄνθος ἢ λίθον ἐπιβουλεύσας εἰς τὴν ἀγορὰν ὑμῖν εἰσαγάγοι, ὀργιεῖσθέ τε καὶ οὐκ ἀνέξεσθε, ἀλλ' ὡς κακοῦργόν τινα καὶ παλίμβολον μαστιγοῦν παραδώσετε τῷ τεταγμένῳ· καὶ διὰ τοῦτο ἐξεύρηται ὑμῖν πολλὰ μὲν τοῦ χρυσοῦ, πολλὰ δὲ πορφύρας, πολλὰ δὲ ὑακίνθων βασανιστήρια, καὶ βασανισταί τινες ἐπὶ τούτῳ καθήμενοι, οὕς, ὅταν ὠνῆσθε, παρακαλεῖτε ξυμπαρεῖναι ὑμῖν καὶ συνδοκιμάζειν τὰ συναλλάγματα ("By Zeus, if someone should intentionally bring into your marketplace debased gold or garment colored with imitation purple dye or a counterfeit stone, you will get angry and not put up with it. You will deliver such a person to your official flogger to be punished as a criminal and a deceiver. To guarantee the punishment of such deceivers, you have devised many tests for purity of gold, or purple dye and of aquamarines. You have a committee of examiners to do the testing. Whenever you make a purchase, you summon them to be at your side and to get their approval of the transactions"; Engl. transl. Penella 2000: 65).

72 Michael Psellus, *Epistula de chrysopoeia* 15 (ed. Bidez 1928: 40, 20–24).

73 Albini 1988: 49n58.

74 Michael Psellus, *Epistula de chrysopoeia* 1–3 (ed. Bidez 1928: 26–28).

τὴν φύσιν ἐπιστήμη) and introduces the four elements as the basic principles of any alchemical transformation of base metals into gold and silver.

In doing so, Psellus, who was also a fine interpreter of the Homeric poems, does not keep himself from embellishing his technical account with learned references to classical mythology. As a new Eurystheus, the Patriarch Cerularius asked Psellus to bring him the golden apples, by turning lead, tin, and any other natural substance into gold.[75] Psellus, moreover, explicitly refers to the myth of the judgment of Paris at the end of a technical section describing a chrysopoetic procedure to turn silver into gold:

> Ἡ κιννάβαρις καὶ ὁ χρυσίζων ἰὸς τοῦ χαλκοῦ, ὥσπερ τινὰ φυσικὰ εἴδη σεληναίᾳ ὕλῃ ἐπιβληθέντα, σῶμα ποιοῦσι χρυσοῦν. Ἂν οὖν ἄργυρον τήξας, τὰ φάρμακα μίξας ἀνακεράσῃς, ἥλιός σοι ἡ σελήνη γενήσεται, καὶ τεμὼν διὰ βάθους εὑρήσεις τὸ χρῶμα καὶ πρὸς ἅπασαν εὐχρηστίαν χρυσοφυῆ ἀποχρυσωθέντα τὸν ἄργυρον. Καὶ σελήνη μὲν ἥλιον, ἥλιος δὲ σελήνην οὐ δρᾷ· μόνη γὰρ αὕτη ἐκεῖθεν ἔχει τὸ φῶς. Οὐ σελήνη δὲ μόνη, ἀλλὰ καὶ Ἀφροδίτη διακονεῖται τῷ μεγάλῳ φωτί. Τὸ τοίνυν τοῦ χαλκοῦ σῶμα ἐλάσας ὅσον εἰκός, οἷά τινα γλῶτταν ἐπιμήκη ἀπέργασαι, εἶτα δὴ ἐπ' ἀνθράκων θέμενος, ἐπ' αὐτὴν τὸν Ἥφαιστον ὑπερέθιζε, ἐπιρραίνων νῦν μὲν ὀρυκτοῦ ἅλατός <τι> νῦν δὲ Ἀττικῆς ὤχρας ἐφεξῆς ἢ ἐναλλὰξ νῶτα καὶ στέρνα τῆς Παφίας κοσμῶν· ἡ δὲ σοι καλλίων ἀθρόον γενήσεται, καὶ τὸ γλαυκὸν ἀποβαλοῦσα, φανεῖταί σοι χρυσίζουσα. Τοιαύτην ἄρα ἴσως ὁ Πάρις τὴν Ἀφροδίτην ἰδών, τῆς Ἥρας καὶ τῆς Ἀθηνᾶς προτετίμηκεν.
>
> Cinnabar and golden rust of copper, like some other natural substances when casted upon lunar matter, make a (metallic) body gold. If you melt silver, mix the drugs and blend together, the moon will become sun and, if you cut deep into it, you will find that the colour of silver has turned into gold that is suitable for all the uses of natural gold. The moon can produce the sun, but the sun cannot produce the moon: only the moon, indeed, receives the light from it. Not only the moon, but also Aphrodite is supplied with a big light. Now after hammering the body of copper as necessary to make it like a long strip and then putting it on (hot) coals, provoke Hephaestus against it[76] and sprinkle now a bit of rock salt now a bit of Attic ochre – in a row or alternatively – thus embellishing the back and the breast of the goddess of Paphos [i.e., Aphrodite]. Suddenly, she will become more beautiful and, after losing her *glaukos* appearance, she will look like gold. As such Aphrodite was seen by Paris, when he preferred her to Hera and Athena.[77]

The passage is rich of elements that resonate with classical mythology, mentioning traditional gods like Aphrodite and Hephaestus. Psellus first describes the process of transforming silver into gold, emphasizing the connection of these two metals with the moon and the sun, respectively. Olympiodorus had previously noted the correspondence between metals and planets, stating that silver is associated with the moon because,

75 Michael Psellus, *Epistula de chrysopoeia* 1 (ed. Bidez 1928: 26, 12–14).
76 The feminine could refer both to γλῶτταν ἐπιμήκη (literally, "long tongue") or to Aphrodite, mentioned a line above.
77 Michael Psellus, *Epistula de chrysopoeia* 11 (ed. Bidez 1928: 36, 23–38, 12).

like silver shines brighter when placed near gold, the moon is illuminated by the sun.[78] Similarly, Psellus explains that only the moon receives the light of the sun.[79]

Then, Psellus shifts his focus to the transmutation of copper, referred to as the goddess Aphrodite. This connection is further emphasized by mentioning the goddess' birthplace, the city of Paphos in Cyprus, known in antiquity for its copper mines. Furthermore, Aphrodite's radiance could be attributed to her connection with the planet Venus, which, as described by Olympiodorus in his commentary on Aristotle's *Meteorology*, shines almost as brightly as the sun, much like copper glows with a brilliance almost akin to gold.[80] Notably, Psellus seems to play with the traditional epithet of the goddess, who was already referred to as *chrysē* ("golden") in Homer's poems.[81] Aphrodite takes on a golden hue through the alchemical transformation she undergoes after being exposed to a specific chrysopoetic process. Interestingly, Psellus describes Aphrodite's colour before her alchemical transformation as *glaukos* ("gleaming"), a term rarely associated with the goddess.[82] This adjective rather recalls the Homeric epithet for Athena, usually known as *glaukōpis* ("with gleaming eyes") in the poems. Perhaps, Psellus is subtly comparing Aphrodite with Athena: they both had a "gleaming" aspect before the alchemical transformation that turned the first goddess into a golden colour, causing Paris to prefer her over Athena and Hera.

3 Concluding Remarks

The passage from Psellus quoted above illustrates how classical mythology and Homeric references could find their way into an alchemical text, even in a technical recipe. This interweaving of technology and mythology should not come as a surprise when we consider how ancient myths were rationalized as accounts on the discovery of various crafts and skills. Homer, often regarded as the father of ancient mythology, emerged as a skilled expert in many of these domains.

78 See Olympiodorus, *In Aristotelis Meteora commentaria* 3.6 (ed. Stüve 1900: 267, 7–9): τῇ δὲ σελήνῃ τὸν ἄργυρον, ἐπειδὴ καὶ ὁ ἄργυρος παρὰ χρυσῷ τιθέμενος δοκεῖ καταλάμπεσθαι ὑπὸ τοῦ χρυσοῦ καὶ φωτεινότερος γίνεσθαι, ὥσπερ ἡ σελήνη ὑπὸ τοῦ ἡλίου καταλάμπεται.

79 See also Michael Psellus, *De omnifaria doctrina* 135.2 (ed. Westerink 1948: 70): ἡ σελήνη μόνη παρὰ τοῦ ἡλίου φωτίζεται ("only the moon is enlightened by the sun").

80 See Olympiodorus, *In Aristotelis Meteora commentaria* 3.6 (ed. Stüve 1900: 267, 3–5).

81 See *Ilias* 3.64, 5.427, 10.389, 19.282, 22.470, 24.699; *Odyssea* 4.14, 9.337 and 342, 17.37, 19.54. The *scholia* explain the epithet as referring to the beauty of Aphrodite's garment (e.g., *Scholia vetera in Iliadem* 3.64c [ed. Erbse 1969–1988: 1.372]). A Christian polemical reading takes the epithet as referring to the money (χρυσόν) that prostitutes gain with their activity (see, e.g., *Scholia Graeca in Odysseam* 4.14b [ed. Pontani 2007–2022: 2.184]).

82 The adjective γλαυκὸς is also rare in the *Corpus alchemicum*, where it never refers to copper.

Pseudo-Plutarch's treatise *On Homer*, for instance, praises the poet's proficiency in natural philosophy, rhetoric, medicine, and the art of painting.[83] The work is still read and summarized in the *Introduction to Homer* by the Byzantine scholar Isaac Porphyrogenitus,[84] while Neoplatonic interpreters make it clear that the poems, when interpreted allegorically, hold philosophical teachings about the soul and the natural world. For instance, Porphyry's *On the Cave of the Nymphs* offers a comprehensive allegorical reading of *Odyssey*, Book 13 (vv. 96–112), drawing upon earlier Neopythagorean interpretations. Additionally, Homer's verses are cited and commented on in various scientific treatises, ranging from astronomical to medical writings.[85] This attests to Homer's mastery across a wide range of disciplines, a quality that continues to be praised by Byzantine exegetes. According to John Tzetzes' *Exegesis of the Iliad*, the poet travelled as far as Egypt to complete his education and learn "medicine, natural philosophy, music, geometry, astronomy, divination, and the highly praised magic by Orpheus – to put it short, all those (branches of knowledge) in which his poetry shows him to be an expert".[86] Among the arts, Tzetzes argues a bit further in his *Exegesis*, Homer was only concerned with those that were useful in life, namely, grammar, poetry, rhetoric, metalwork, mechanics, and the others that are dealt with in the *Iliad*.[87]

Nearly three centuries before Tzetzes, in Thomas of Marga's *Chronicle*, Homer was not merely recognized as a skilled metalworker; he was portrayed as a full-fledged alchemist with the ability to transform lead into gold, and his verses were taken as evidence for this remarkable skill. This did not necessarily depend on the circulation of Eastern translations of Homer's poems. Indeed, they were never translated into Arabic during the Abbasid period. Moreover, Theophilus of Edessa's Syriac translation of two books by Homer on the conquest of Troy,[88] if never existed, has not survived in manuscripts. Nevertheless, Homer's reputation as a master of various crafts spread to the Eastern Christian and Islamic worlds through translations of Greek treatises

83 [Plutarchus], *De Homero* 2.93–102 (philosophy), 161–174 (rhetoric), 200–211 (medicine), 216–217 (painting) (ed. Kindstrand 1990: 45–49, 87–95, 106–112, 115–116 respectively); see also Hillgruber 1994–1999: 2.213–437.

84 See Pontani 2005: 161–162.

85 See, for instance, Bacigalupo and Savio 2022. As for astronomy, Homer is often quoted in Cleomedes' *Caelestia* (ed. Todd 1990) as well as in Achilles' *Astronomica et in Aratum opuscula* (ed. Di Maria 2012); as for medicine, see, for instance, Moraux 1987: 25–37.

86 Ioannes Tzetzes, *Exegesis in Homeri Iliadem*, Prolog. (ed. Papathomopoulos 2007: 25, 15–18): ἰατρικήν, φυσικήν, μουσικήν, γεωμετρίαν, ἀστρονομίαν, οἰωνοσκοπίαν καὶ τὴν παρὰ τῷ Ὀρφεῖ μεγαλεγκωμίαστον μαγικὴν καὶ ἁπλῶς εἰπεῖν ὅσα ἡ τούτου ποίησις αὐτὸν ἡμῖν εἰδέναι ὑποδεικνύει.

87 Ioannes Tzetzes, *Exegesis in Homeri Iliadem* 1.394 (ed. Papathomopoulos 2007: 343, 12–15): Τῶν ἄλλων δὲ τεχνῶν τῶν βιωφελῶν μόνον φροντίζει, γραμματικῆς, ποιητικῆς τε καὶ ῥητορικῆς, μεταλλικῆς, μηχανικῆς, μαγικῆς καὶ τῶν λοιπῶν, περὶ ὧν ἐν τῇ βίβλῳ διαλαμβάνει.

88 The information on Theophilus of Edessa is reported in Bar Hebraeus' *Chronography* (ed. Bedjan 1890: 126, 26–127, 4). See already Kraemer 1956: 261. Hilkens has recently argued that Theophilus did not translated Homer's poems, but the Cyclic poem *Ilioupersis*: see Hilkens 2013.

on science and philosophy, which contributed to his legendary status.[89] For example, verses from Homer's epics are quoted in Sergius of Rēš'aynā's Syriac translations of Pseudo-Aristotle's *On the Cosmos*[90] as well in a Neoplatonic untitled treatise, possibly the translation of Porphyry's work *On Principles and Matter*.[91] Homeric references also found their way into original Syriac treatises. In the second chapter of his *Treatise on the Constellations*, Severus Sebokht may have referred to the *Iliad* (18.486–487) to explain the position of the Pleiades, the Hyades and the chariot.[92] However, he subsequently criticised the ancient poets, in particular Aratus, for the stories they told about the stars.[93]

Homeric poems were undoubtedly incorporated into the curriculum of Eastern Christian schools. Antony of Tagrit, in his handbook on rhetoric, emphasized the significance of mastering the art of reading poetry as an essential aspect of a well-rounded education. In support of his point, he cited numerous verses from Homer, some of which were also subject to allegorical interpretations.[94] Syriac scholars were well aware of using allegory as a means to interpret classical poetry and mythology. We can find euhemeristic interpretations of ancient myths in Pseudo-Nonnus' *Scholia*,[95] and the name of Palaephatus is specifically mentioned in Michael the Syrian's *Chronicle*.[96]

Homer's *polymatheia* continued to be emphasized in the works of the Eastern Christian translators of Greek writings into Arabic. In his letter addressed to Ibn al-Munağğim, the Melkite physician Qusṭā ibn Lūqā made it explicit that Homer's verses encapsulated teachings on many arts:[97] on medicine, as emerges from "Galen's Book *On the Medicine of Homer*" (*Kitāb Ǧālīnūs fī ṭibb Ūmīrus*); on the art of argument (*ṣināʿat al-ǧadal*), as explained in a *Book of Homer's Argument* (*Kitāb ǧadal Ūmīrus*) attributed to a certain Pilatus (Bīlātus);[98] on many other disciplines, such as grammar, poetics, and rhetoric. Galen's book on the medicine of Homer was also known to the Nestorian translator Ḥunayn ibn Isḥāq, who was famous for having learn the *Iliad* and *Odyssey*

89 See Mavroudi 2020: 459.

90 [Aristoteles], *De mundo* 6, 397b26; Syriac translation edited in Lagarde 1858: 150, 4–5. On the Syriac and Arabic translations, see Takahashi 2014.

91 Porphyrius, *De principiis et materia* (ed. Arzhanov 2021). Homer (*Ilias* 14.246) is quoted in § 6 (p. 86) with reference to the idea that water was the first principle.

92 French translation of Severus Sebokht's treatise *De constellationibus* (On the Constellations), with the edition of a few Syriac sections, in Nau 1929–1930: 327–410 (see p. 349n2 on the possible reference to Homer); see also Villey 2017: 214–221.

93 In particular, Aratus is often mentioned in Chapter 4 of Severus Sebokht's *De constellationibus* (ed. and Fr. transl. Nau 1929–1930: 355–367).

94 See, e.g., Hilkens 2013: 286–287 and Debié 2005. On Antony of Tagrit, see Nicosia 2024.

95 [Nonnus], *Scholia mythologica* (ed. Brock 1971).

96 Michael Syriacus, *Chronicon* 4.7 (ed. and Fr. transl. Chabot 1899–1910: 4.33–34 [Syriac text]; 1.57 [transl.]).

97 Qusṭā ibn Lūqā, *Responsio (jawāb) ad Ibn al-Munağğim* 208–210 (ed. Samir and Nwyia 1981: 666–667).

98 On this name, see Samir and Nwyia 1981: 667n83–86.

by heart. He lists the work *On Medicine According to the Views of Homer* (*fī al-ṭibb ʿalā raʾy Ūmīrūs*), in two books, among Galen's writings whose authenticity was disputable.[99] References to and quotations from Homer also occur in Ḥunayn's translations of Galen's medical treatises,[100] where Ḥunayn also shows awareness on allegorical interpretations applied to Greek mythology, such as Asclepius' and Dionysus' deification.[101] Allegory was also well known to al-Bīrūnī, who, in his *Indica*, often refers to Homer. For instance, he quotes and comments on *Iliad* 19.347 within a discussion on the meaning of the incipit of Aratus' *Phaenomena*.[102]

Although none of these sources explicitly mention alchemy, both Byzantine and Syro-Arabic sources strongly concur in highlighting Homer's vast knowledge, which spanned nearly every field of ancient science and technology. Within this intricate cultural context, the Syriac and Arabic traditions appear to go a step further, imbuing alchemical writings with allusions to Homer and ancient mythology. After all, portraying Homer as an alchemist aligns seamlessly with the image of the poet depicted by both Byzantine and Syro-Arabic scholars. Their accounts portray him as a true master of every art and field of knowledge, leaving little reason to exclude alchemy from the tapestry of his expertise.

Bibliography

Sources

Collected Sources

Berthelot and Duval 1893: Marcellin Berthelot and Rubens Duval (ed. and Fr. transl.), *La chimie au Moyen-Âge*, vol. 2: *L'alchimie syriaque*. Paris: Imprimerie Nationale, 1893.

Berthelot and Ruelle 1887–1888: Marcellin Berthelot and Charles-Émile Ruelle (ed. and Fr. transl.), *Collection des anciens alchimistes grecs*, 3 vols. Paris: G. Steinheil, 1887–1888 (reprint: Osnabrück: O. Zeller, 1967).

Jacoby 1923–1958: Felix Jacoby (ed.), *Die Fragmente der griechischen Historiker (FGrHist)*, 3 parts in 15 vols. Berlin: Weidmann; Leiden: Brill, 1923–1958.

Müller 1841–1873: Karl Müller (ed.), *Fragmenta Historicorum Graecorum*, 5 vols. Paris: A. Firmin-Didot, 1841–1873.

99 Ḥunayn ibn Isḥāq, *Epistula de libris Galeni a se conversis* 164 (ed. and Germ. transl. Bergsträsser 1932: 90 [Arabic text] and 97 [transl.]); see also Meyerhof 1928: 10.

100 See Strohmaier 1980: 196–200.

101 See Strohmaier 2012: 172–173.

102 Al-Bīrūnī, *Indica* 8 (ed. Sachau 1887: 47–48; revised version: 1958: 73–74).

Authors and Works

Achilles, *Astronomica et in Aratum opuscula*: Giorgio Di Maria (ed.), *Achillis quae feruntur Astronomica et in Aratum opuscula: De universo, De Arati vita, De phaenomenorum interpretatione*. Puurs: Peleman Industries NV, 2012.

Anonymus, *Dicta Philosophorum Graecorum* (Syriac): Yury Arzhanov (ed. and Engl. transl.), *Syriac Sayings of Greek Philosophers. A Study on Syriac* Gnomologia *with Edition and Translation* (Corpus Scriptorum Christianorum Orientalium 669, Subsidia 138). Louvain: Peeters, 2019.

Anonymus paradoxographus, *De incredibilibus* (*Excerpta Vaticana*): Nicola Festa (ed.), *Palaephati Περὶ ἀπίστων. Heracliti qui fertur libellus Περὶ ἀπίστων. Excerpta Vaticana (vulgo anonymus De incredibilibus)* (Mythographi Graeci 3.2). Leipzig: Teubner, 1902, pp. 88–99.
Engl. transl.: Greta Hawes, *Rationalizing Myth in Antiquity*. Oxford: Oxford University Press, 2014, pp. 239–248.

[Aristoteles], *De mundo* (Syriac): Paul de Lagarde (ed.), *Analecta Syriaca*. Leipzig: Teubner, 1858, pp. 134–158.

Bar Konai, Theodorus, *Liber scholiorum* (Syriac): Addaï Scher (ed.), *Theodorus bar Kōnī, Liber scholiorum*, 2 vols (Corpus Scriptorum Christianorum Orientalium, Scriptores Syri, Series secunda 65–66). Paris: Imprimerie nationale, 1910–1912.
Fr. transl.: Robert Hespel and René Draguet, *Théodore bar Koni, Livres des scholies (recension de Séert)*, 2 vols (Corpus Scriptorum Christianorum Orientalium, Scriptores Syri 187–188). Louvain: Peeters, 1981–1982.

Barḥadbšabbā, *Causa fundationis scholarum* (Syriac): Addaï Scher (ed. and Fr. transl.), *Mar Barḥadbšabba ʿArbaya, Cause de la fondation des écoles* (Patrologia Orientalis 4.4). Paris: Firmin-Didot, 1908, pp. 319–404.
Engl. transl.: Adam H. Becker, *Sources for the Study of the School of Nisibis* (Translated Texts for Historians 50). Liverpool: Liverpool University Press, 2008, pp. 86–160.

Barhebraeus, *Chronicon Syriacum* (Syriac): Paul Bedjan (ed.), *Gregorii Barhebraei Chronicon Syriacum*. Paris: Maisonneuve, 1890.

al-Bīrūnī, *Indica* (Arabic): Edward C. Sachau (ed. and Engl. transl.), *Alberuni's India: An Account of the Religion, Philosophy, Literature, Chronology, Astronomy, Customs, Laws and Astrology of India about A. D. 1030*. London: K. Paul, Trench, Trübner & Co., 1887. Revised version: *Taḥqīq mā li-l-Hind*. Hyderabad: Dā'irat al-Maʿārif il-Osmania, 1958.

Cleomedes, *Caelestia*: Robert B. Todd (ed.), *Cleomedis Caelestia (Μετέωρα)* (Bibliotheca Scriptorum Graecorum et Romanorum Teubneriana). Leipzig: Teubner, 1990.

Crates Mallotes, *Fragmenta*: Maria Broggiato (ed. and It. transl.), *Cratete di Mallo, I frammenti, edizione, traduzione e note* (Pleiadi 2). La Spezia: Agorà, 2001.

Eustathius Thessalonicensis, *Commentarii ad Homeri Iliadem*: Marchinus van der Valk (ed.), *Eustathii archiepiscopi Thessalonicensis commentarii ad Homeri Iliadem pertinentes ad fidem codicis Laurentiani editi*, 4 vols. Leiden: Brill, 1971–1987.

Eustathius Thessalonicensis, *Commentarii ad Homeri Odysseam*: Gottfried Stallbaum (ed.), *Eustathii archiepiscopi Thessalonicensis commentarii ad Homeri Odysseam ad fidem exempli Romani editi*, 2 vols. Leipzig: Weigel, 1825–1826.

Eustathius Thessalonicensis, *Commentarium in Dionysii Periegetae orbis descriptionem*: Karl Müller (ed.), *Geographi Graeci Minores*, vol. 2, Paris: Firmin-Didot, 1861, pp. 201–407.

Heraclitus Paradoxographus, *Quaestiones Homericae*: Filippomaria Pontani (ed. and It. transl.), *Eraclito, Questioni omeriche. Sulle allegorie di Omero in merito agli dèi* (Il Mito. Testi e saggi 1). Pisa: ETS, 2005.

Homerus, *Ilias*: Augustus T. Murray (Engl. transl.), William F. Wyatt (revision), *Homer, Iliad*, 2 vols (Loeb Classical Library 170–171). Second edition. Cambridge, MA: Harvard University Press, 1999.

Ḥunayn ibn Isḥāq, *Epistula de libris Galeni a se conversis* (*Risāla* [...] *fī ḏikr mā turǧima min kutub Ǧālīnūs*): Gotthelf Bergsträsser (ed. and Germ. transl.), *Neue Materialien zu Ḥunain ibn Isḥāq's Galen-Bibliographie* (Abhandlungen für die Kunde des Morgenlandes 19.2). Leipzig: Deutsche Morgenländische Gesellschaft, 1932.

Ibn al-Nadīm, *Liber Catalogi* (*Kitāb al-fihrist*): Gustav Flügel (ed.), *Kitāb al-Fihrist*, 2 vols. Leipzig: F. C. W. Vogel, 1871–1872.

Ioannes Antiochenus, *Fragmenta*: Umberto Roberto (ed. and It. transl.), *Ioannis Antiocheni fragmenta ex Historia chronica* (Texte und Untersuchungen zur Geschichte der altchristlichen Literatur 154). Berlin and New York: De Gruyter, 2005.
Sergei Mariev (ed. and Engl. transl.), *Ioannis Antiocheni fragmenta quae supersunt omnia* (Corpus Fontium Historiae Byzantinae 47). Berlin and New York: De Gruyter, 2008.

[Iustinus], *Oratio ad Graecos* (Syriac): Bernard Pouderon, in collaboration with Cécile Bost-Pouderon, Marie-Joseph Pierre, and Pierre Pilard (ed. and Fr. transl.), *Pseudo-Justin, Ouvrages apologétiques* (Sources chrétiennes 528). Paris: Les Éditions du Cerf, 2009, pp. 302–313.

Michael Syriacus, *Chronicon* (Syriac): Jean-Baptiste Chabot (ed. and Fr. transl.), *Chronique de Michel le Syrien, Patriarche Jacobite d'Antioche*, 4 vols. Paris: E. Leroux, 1899–1910.

Muḥammad ibn Umail, *Liber explicationis signorum* (*Kitāb ḥall al-rumūz*): Theodor Abt, Wilferd Madelung, and Thomas Hofmeier (eds), Salwa Fuad and Theodor Abt (Engl. transl.), *Muḥammad ibn Umail, Book of the Explanation of the Symbols* (Corpus Alchemicum Arabicum 1). Zurich: Living Human Heritage Publications, 2003.

[Nonnus], *Scholia mythologica* (Syriac): Sebastian Brock (ed. and Engl. transl.), *The Syriac Version of the Pseudo-Nonnos Mythological Scholia* (University of Cambridge Oriental Publications 20). Cambridge: Cambridge University Press, 1971.

Olympiodorus, *In Aristotelis Meteora commentaria*: Wilhelm Stüve (ed.), *Olympiodori in Aristotelis Meteora commentaria* (Commentaria in Aristotelem Graeca 12.2). Berlin: Reimer, 1900.

Palaephatus, *De incredibilibus*: Nicola Festa (ed.), *Palaephati Περὶ ἀπίστων, Heracliti qui fertur libellus Περὶ ἀπίστων, Excerpta Vaticana (vulgo anonymus De incredibilibus)* (Mythographi Graeci 3.2). Leipzig: Teubner, 1902, pp. 1–72.
Engl. transl.: Jacob Stern, *Palaephatus, On Unbelievable Tales. Translation, Introduction and Commentary*. Wauconda: Bolchazy-Carducci Publishers, 1996.

Percolla, Vincenzo, *Auriloquio*: Carlo A. Anzuini (ed.), *Vincenzo Percolla, Auriloquio. Nel quale si tratta dello ascoso secreto dell'Alchimia. Trattato manoscritto del '500 d'interpretazione dei miti greci et romani* (Textes et travaux de *Chrysopoeia* 2). Paris: S. É. H. A.; Milan: Archè, 1996.

Planudes, Maximus, *Publii Ovidii Nasonis Metamorphoseon libri XV Graece versi*: Manoles Papathomopoulos and Isabella Tsavare (eds), *Ὀβιδίου Περὶ Μεταμορφώσεων, ὃ μετήνεγκεν ἐκ τῆς Λατίνων φωνῆς εἰς τὴν Ἑλλάδα Μάξιμος μοναχὸς ὁ Πλανούδης* (Ἑλληνικὴ Βιβλιοθήκη). Athens: Academy of Athens, 2002.

[Plato], *Book of the Fourths* (*Kitāb al-rawābī'*): 'Abd al-Raḥmān Badawī (ed.), *Neoplatonici apud Arabes* (*Al-Afflāṭūnīya al-muḥdaṯa 'inda al-'arab*). Cairo: Maktabat an-nahḍa al-miṣrīya, 1955, pp. 119–239.

[Plato], *Liber quartorum*: Lazarus Zetzner (ed.), *Theatrum chemicum*, 6 vols. Strasburg: 1601–1661, vol. 5, pp. 114–208.

[Plutarchus], *De Homero*: Jan Fredrik Kindstrand (ed.), *[Plutarchi] De Homero* (Bibliotheca Scriptorum Graecorum et Romanorum Teubneriana). Leipzig: Teubner, 1990.

Porphyrius, *De principiis et materia* (Syriac): Yury Arzhanov (ed.), *Porphyry, On Principles and Matter. A Syriac Version of a Lost Greek Text with an English Translation, Introduction and Glossaries* (Scientia Graeco-Arabica 34). Berlin and Boston, MA: De Gruyter, 2021.

Proclus, *In Platonis Rem publicam commentarii*: Wilhelm Kroll (ed.), *Procli Diadochi in Platonis Rem publicam commentarii*, 2 vols. Leipzig: Teubner, 1899–1901.

Robert Lamberton (ed. and Engl. transl.), *Proclus the Successor on Poetics and the Homeric Poems: Essays 5 and 6 of His* Commentary on the Republic of Plato. Text, Translation, Notes and Introduction (Writings from the Greco-Roman World 34). Atlanta, GA: Society of Biblical Literature, 2012.

Psellus, Michael, *De omnifaria doctrina*: Leendert G. Westerink (ed.), *Michael Psellus, De omnifaria doctrina*. Nijmegen: Centrale Drukkerij N. V., 1948.

Psellus, Michael, *Epistula de chrysopoeia*: Joseph Bidez (ed.), *Catalogue des manuscrits alchimiques grecs*, vol. 6: *Michel Psellus, Épître sur la Chrysopée, Opuscules et extraits sur l'alchimie, la météorologie et la démonologie*. Brussels: M. Lamertin, 1928, pp. 1–43.

It. transl.: Francesca Albini, *Michele Psello, La crisopoea ovvero come fabbricare l'oro. Traduzione, introduzione e commento* (Nuova Atlantide, I classici del pensiero magico, esoterico e simbolico). Genova: ECIG, 1988.

Psellus, Michael, *Philosophica minora 1*: John M. Duffy (ed.), *Michaelis Pselli Philosophica minora*, vol. 1: *Opuscula logica, physica, allegorica, alia* (Bibliotheca Scriptorum Graecorum et Romanorum Teubneriana). Leipzig: Teubner, 1992.

Qusṭā ibn Lūqā, *Responsio (jawāb) ad Ibn al-Munağğim* (Arabic): Khalil Samir and Paul Nwyia (ed. and Fr. transl.), *Une correspondance islamo-chrétienne entre Ibn al-Munağğim, Ḥunayn ibn Isḥāq et Qusṭā ibn Lūqā* (Patrologia Orientalis 40.4). Turnhout: Brepols, 1981.

Scholia Graeca in Odysseam: Filippomaria Pontani (ed.), *Scholia Graeca in Odysseam*, 5 vols (Pleiadi 6.4). Rome: Edizioni di Storia e Letteratura, 2007–2022.

Scholia vetera in Iliadem: Hartmut Erbse (ed.), *Scholia Graeca in Homeri Iliadem (scholia vetera)*, 7 vols. Berlin: De Gruyter, 1969–1988.

Severus Sebokht, *De constellationibus* (Syriac): François Nau (partial ed. and Fr. transl.), "Le traité sur les 'constellations' écrit, en 611, par Sévère Sébokt, évêque de Qennesrin", *Revue de l'Orient chrétien* 27 (1929–1930), pp. 327–410.

Suda lexicon: Ada Adler (ed.), *Suidae Lexicon*, 5 vols (Lexicographi Graeci 1.1–5). Leipzig: Teubner, 1928–1938.

Themistius, *Orationes*: Glanville Downey, Albert F. Norman, and Heinrich Schenkl (eds), *Themistii orationes quae supersunt*, 3 vols (Bibliotheca Scriptorum Graecorum et Romanorum Teubneriana). Leipzig: Teubner, 1965–1974.

Engl. transl.: Robert J. Penella, *The Private Orations of Themistius* (The Transformation of the Classical Heritage 29). Berkeley, CA: University of California Press, 2000.

Thomas Margensis, *Liber superiorum* (Syriac): Paul Bedjan (ed.), *Liber superiorum, seu Historia monastica, auctore Thoma, episcopo Margensi; Liber fundatorum monasteriorum in regno Persarum et Arabum; Homiliae Mar-Narsetis in Joseph; Documenta patrum de quibusquam verae fidei dogmatibus*. Paris and Leipzig: Harrassowitz, 1901.

Engl. transl.: Ernest A. Wallis Budge, *The Book of Governors: The Historia Monastica of Thomas, Bishop of Margâ, A. D. 840*, 2 vols. London: Kegan Paul, Trench, Trübner & Co, 1893.

Tzetzes, Ioannes, *Allegoriae Iliadis*: Jean F. Boissonade (ed.), *Tzetzae Allegoriae Iliadis: accedunt Pselli Allegoriae quarum una inedita*. Paris: Dumont, 1851.

Tzetzes, Ioannes, *De generatione deorum*: John A. Cramer (ed.), *Anecdota graeca e codd. manuscriptiis Bibliothecae regiae Parisiensis*, 4 vols. Oxford: Oxford University Press, 1839–1841, vol. 3, pp. 101–113.

Pietro Matranga (ed.), *Anecdota graeca e mss. Bibliothecis Vaticana, Angelica, Barberiniana, Vallicelliana, Medicea, Vindobonensi deprompta*, 2 vols. Rome: C. A. Bertinelli, 1850, vol. 2, pp. 364–371.

Tzetzes, Ioannes, *Exegesis in Homeri Iliadem*: Manoles Papathomopoulos (ed.), *Ἐξήγησις Ἰωάννου Γραμματικοῦ τοῦ Τζέτζου εἰς τὴν Ὁμήρου Ἰλιάδα* (Ἑλληνικὴ Βιβλιοθήκη). Athens: Academy of Athens, 2007.

Zosimus Panopolitanus, *Hypomnemata authentica* (Greek): Michèle Mertens (ed. and Fr. transl.), *Zosime de Panopolis, Mémoires authentiques* (Les Alchimistes grecs 4.1). Paris: Belles Lettres, 1995.

Literature

Alganza-Roldán 2019: Minerva Alganza-Roldán, "On the Tradition and Reception of Palaephatus in Byzantium", *Byzantinoslavica* 77.1–2 (2019), pp. 5–25.

Bacigalupo and Savio 2022: Valeria Bacigalupo and Martina Savio, "Omero e le scienze: andata e ritorno", *Aevum* 96.1 (2022), pp. 3–30.

Becker 2006: Adam H. Becker, *Fear of God and the Beginning of Wisdom: The School of Nisibis and Christian Scholastic Culture in Late Antique Mesopotamia*. Philadelphia, PA: University of Pennsylvania Press, 2006.

Berthelot 1885: Marcellin Berthelot, *Les origines de l'alchimie*. Paris: G. Steinheil, 1885.

Bettiolo 2004: Paolo Bettiolo, "Gnomologia 'siriaci': un censimento", in Maria Serena Funghi (ed.), *Aspetti di letteratura gnomica nel mondo antico*, vol. 2 (Accademia toscana di Scienze e Lettere "La Colombaria", Studi 225). Florence: Leo S. Olschki, 2004, pp. 289–304.

Brock 1983: Sebastian Brock, "A Syriac Collection of Prophecies of the Pagan Philosophers", *Orientalia Lovaniensia Periodica* 14 (1983), pp. 203–246.

Cesaretti 1981: Paolo Cesaretti, *Allegoristi di Omero a Bisanzio. Ricerche ermeneutiche (XI–XII secolo)* (Ricerche). Milan: Guerini Studio, 1981.

Debié 2005: Muriel Debié, "Homère chronographe: l'héritage grec antique dans l'historiographie syriaque", in *Les Syriaques transmetteurs de civilisations: l'expérience du Bilâd el-Shâm à l'époque omeyyade*. Antélias and Paris: L'Harmattan, 2005, pp. 69–93.

De Vos 2020: Benjamin De Vos, "Achilles as an Allegorical Anti-Adam in the *Pseudo-Clementine Homilies*", *Greek, Roman, and Byzantine Studies* 60 (2020), pp. 160–182.

Ferrario 2021: Gabriele Ferrario, "Alchemy in the Cairo Genizah: The *Nachlass* of an Untidy Jewish Alchemist", *Asiatische Studien* 75.2 (2021), pp. 513–544.

Forster 2017: Regula Forster, *Wissensvermittlung im Gespräch. Eine Studie zu klassisch-arabischen Dialogen* (Islamic History and Civilization 149). Leiden and Boston, MA: Brill, 2017.

Halleux 1974: Robert Halleux, *Le problème des métaux dans la science antique* (Bibliothèque de la Faculté de philosophie et lettres de l'Université de Liège 209). Paris: Belles Lettres, 1974.

Hasse 2002: Dag Nikolaus Hasse, "Plato arabico-latinus: Philosophy – Wisdom Literature – Occult Sciences", in Stephen Gersh and Maarten J. F. M. Hoenen (eds), *The Platonic Tradition in the Middle Ages: A Doxographical Approach*. Berlin and New York: De Gruyter, 2002, pp. 31–65.

Hilkens 2013: Andy Hilkens, "Syriac Ilioupersides: The Fall of Troy in Syriac Historiography", *Le Muséon* 126.3–4 (2013), pp. 285–317.

Hillgruber 1994–1999: Michael Hillgruber, *Die pseudoplutarchische Schrift De Homero. Einleitung und Kommentar*, 2 vols (Beiträge zur Altertumskunde 57–58). Stuttgart and Leipzig: Teubner, 1994–1999.

Inoue 2017: Takatomo Inoue, "Avicennian Natural Philosophy and the Alchemical Theory of al-Ṭughrāʾī in *Ḥaqāʾiq al-istishhād*". Master's thesis. McGill University, 2017.

Komorowska 2019: Joanna Komorowska, "Genethlialogy, Katarchoscopy, and Astrological Authorities in John Tzetzes's *Allegories of the Iliad*", *International Journal of Divination and Prognostication* 1 (2019), pp. 123–142.

Kraemer 1956: Jörg Kraemer, "Arabische Homerverse", *Zeitschrift der Deutschen Morgenländischen Gesellschaft* 106.2 (1956), pp. 259–316.

Kraus 1942–1943: Paul Kraus, *Jābir ibn Ḥayyān, Contribution à l'histoire des idées scientifiques dans l'Islam*, 2 vols (Mémoires de l'Institut d'Égypte 44–45). Cairo: Institut français d'archéologie orientale, 1942–1943.

Lamberton 1986: Robert Lamberton, *Homer the Theologian: Neoplatonist Allegorical Reading and the Growth of the Epic Tradition* (The Transformation of the Classical Heritage 9). Berkeley, CA: University of California Press, 1986.

Letrouit 2002: Jean Letrouit, "Hermétisme et alchimie: contribution à l'étude du Marcianus graecus 299 (= M)", in Carlos Gilly and Cis Van Heertum (eds), *Magia, alchimia, scienza dal '400 al '700. L'influsso di Ermete Trismegisto*, 2 vols. Florence: Centro Di, 2002, vol. 1, pp. 85–109.

Martelli 2014: Matteo Martelli, "L'alchimie en syriaque et l'œuvre de Zosime", in Émilie Villey (ed.), *Les sciences en syriaque* (Études syriaques 11). Paris: Geuthner, 2014, pp. 191–214.

Martelli 2017: Matteo Martelli, "Hippocrates in Two Syriac Alchemical Collections", *Aramaic Studies* 15.2 (2017), pp. 230–251.

Matton 1995: Sylvain Matton, "L'influence de l'humanisme sur la tradition alchimique", *Micrologus* 3 (1995), pp. 279–345.

Mavroudi 2020: Maria Mavroudi, "Homer in Greece from the End of Antiquity 1: The Byzantine Reception of Homer and His Export to Other Cultures", in Corinne Ondine Pache (ed.), *The Cambridge Guide to Homer*. Cambridge: Cambridge University Press, 2020, pp. 444–472.

Meyerhof 1928: Max Meyerhof, *Über echte und unechte Schriften Galens nach arabischen Quellen*. Berlin: De Gruyter, 1928.

Montanari et al. 2017: Franco Montanari, Fausto Montana, Davide Muratore, and Lara Pagani, "Towards a New Critical Edition of the *Scholia* to the *Iliad*: A Specimen", *Trends in Classics* 9.1 (2017), pp. 1–21.

Moraux 1987: P. Moraux, "Homère chez Galien", in Jean Servais, Tony Hackens, and Brigitte Servais-Soyet (eds), *Stemmata. Mélanges de philologie, d'histoire et d'archéologie grecques offerts à Jules Labarbe*. Liège and Louvain-la-Neuve: L'Antiquité classique, 1987, pp. 25–37.

Nicosia 2024: Mara Nicosia, "Homer, Heliodorus and Pseudo-Callisthenes in Antony of Tagrit's *Rhetoric*", *Ancient Narratives* 20 (2024), pp. 1–23.

Nilsson 2025: Ingela Nilsson, "To Render Unbelievable Tales Believable: The Storyworlds of Paradoxography", in Stavroula Constantinou and Andria Andreou (eds), *Storyworlds in Short Narratives: Approaches to Late Antique and Early Byzantine Tales* (Brill's Series on the Early Middle Ages 31). Leiden and Boston, MA: Brill, 2025, pp. 59–80.

Nöldeke 1898: Theodor Nöldeke, "Bar Chōnī über Homer, Hesiod und Orpheus", *Zeitschrift der Deutschen Morgenländischen Gesellschaft* 52.3 (1898), pp. 501–507.

Pontani 2005: Filippomaria Pontani, *Sguardi su Ulisse. La tradizione esegetica greca all'*Odissea (Pleiadi 13). Rome: Edizioni di Storia e Letteratura, 2005.

Porter 1992: James O. Porter, "Hermeneutic Lines and Circles: Aristarchus and Crates on the Exegesis of Homer", in Robert Lamberton and John J. Keaney (eds), *Homer's Ancient Readers: The Hermeneutics of Greek Epic's Earliest Exegetes*. Princeton, NJ: Princeton University Press, 1992, pp. 67–114.

Razook 1963: Razook Faraj Razook, *Studies on the Works of Al-Ṭughrāʾī*. PhD Dissertation. SOAS University of London, 1963.

Rodríguez Guerrero 2002–2007: José Rodríguez Guerrero, "Algunos datos Desconocidos sobre las Relaciones entre Alquimia y Mitología", *Azogue* 5 (2002–2007), pp. 9–29.

Rodríguez Guerrero 2011: José Rodríguez Guerrero, "Einige unbekannte Fakten zum Verhältnis von Alchimie und Mythologie", *Grenzgebiete der Wissenschaft* 60.4 (2011), pp. 331–354.

Sciarra 2005: Elisabetta Sciarra, *La tradizione degli* scholia *iliadici in Terra d'Otranto* (Bollettino dei Classici, Supplemento 23). Rome: Accademia Nazionale dei Lincei, 2005.

Strohmaier 1980: Gotthard Strohmaier, "Homer in Baghdad", *Byzantinoslavica* 41 (1980), pp. 196–200.

Strohmaier 2012: Gotthard Strohmaier, "Galen the Pagan and the Christian: Specific Transformations in the Commentaries on *Airs, Waters, Places* and *Epidemics*", in Peter E. Pormann (ed.), *Epidemics in Context: Greek Commentaries on Hippocrates in the Arabic Tradition* (Scientia Graeco-Arabica 8). Berlin: De Gruyter, 2012, pp. 171–184.

Takahashi 2014: Hidemi Takahashi, "Syriac and Arabic Transmission of *On the Cosmos*", in Johan C. Thom (ed.), *Cosmic Order and Divine Power: Pseudo-Aristotle, On the Cosmos* (SAPERE 23). Tübingen: Mohr Siebeck, 2014, pp. 153–167.

Thillet 2005: Pierre Thillet, "Remarques sur le *Liber quartorum* du Pseudo-Platon (Kitāb al-rawābīʿ li-Aflāṭūn)", in Cristina Viano (ed.), *L'Alchimie et ses racines philosophiques. La tradition grecque et la tradition arabe* (Histoire des doctrines de l'Antiquité classique 32). Paris: J. Vrin, 2005, pp. 201–232.

Ullmann 1972: Manfred Ullmann, *Die Natur- und Geheimwissenschaften im Islam* (Handbuch der Orientalistik, 1. Abt.: Der Nahe und der Mittlere Osten, Ergänzungsband 6.2). Leiden: Brill, 1972.

van Bladel 2009: Kevin van Bladel, *The Arabic Hermes: From Pagan Sage to Prophet of Science* (Oxford Studies in Late Antiquity). Oxford: Oxford University Press, 2009.

Vanderspoel 1995: John Vanderspoel, *Themistius and the Imperial Court: Oratory, Civic Duty, and Paideia from Constantius to Theodosius*. Ann Arbor, MI: University of Michigan Press, 1995.

Villey 2017: Émilie Villey, "Syriac Astronomical Texts (500–700 CE): Christian Voices Defending Ptolemaic Astronomy", in Hagit Amirav and Francesco Celia (eds), *New Themes, New Styles in the Eastern Mediterranean: Christian, Jewish, and Islamic Encounters, 5th–8th Centuries* (Late Antique History and Religion 16). Leuven: Peeters, 2017, pp. 205–232.

Wright 1870–1872: William Wright, *Catalogue of Syriac Manuscripts in the British Museum Acquired since the Year 1838*, 3 vols. London: British Museum, 1870–1872.

Bojidar Dimitrov

Pseudepigraphical References to Homer in Some Ǧābirian Texts

Pseudepigraphical attribution is among the definitive characteristics of the alchemical tradition[1] and poses considerable challenges to the investigation of textual history. The frequent appearance of Greek authorities and terminology in Islamic alchemical literature illustrates the latter's strong dependency on Graeco-Egyptian and Byzantine sources, yet the patterns of cross-cultural transmission remain to a large extent obscure due to the complex and fragmented nature of the available evidence.[2] Matteo Martelli's contribution to this volume outlines important parameters for the evolution of Homer's role in alchemical discourses: while Homeric works and their mythological content are not explicitly associated with the "art" in Byzantine alchemical collections, the poet was perceived as a polymath versed in various disciplines early on, and the hermeneutical exegesis of poetry was known to late antique alchemists. It is hence possible that alchemical literature may have interacted with Neoplatonic, allegorical interpretations of Homer. His *mise-en-scène* as an alchemist in treatises ascribed to Ǧābir ibn Ḥayyān[3] bears witness to the transformation of an epic bard into an omniscient sage – a process closely associated with the history of Greek philosophy, which encompassed various cultural and intellectual developments that unfolded over many centuries.[4] The present survey intends to accentuate certain aspects of said process with the goal of speculating on Homer's relevance for the Arabo-Islamic alchemical tradition.

1 Halleux 1979: 97–100.

2 For an early study of pseudepigraphical examples from Arabic and Jewish sources, see Steinschneider 1862 (also quoted in Halleux 1979: 97n83; Kraus 1942–1943: 46n7). Overviews of the topic can be found in Ullmann 1972: 146–165 and Sezgin 1971: 15–23.

3 The very historicity of Ǧābir and the authorship of his oeuvre have been the subject of a rigorous scholarly dispute. For an introduction to the issue, see Forster 2018; also Delva 2017. I would like to express my gratitude to Gabriele Ferrario for drawing my attention to Delva's article.

4 Despite his harrowing critique of Homer, Plato nevertheless considered the former to be the "teacher of the Greeks" (*Phaedrus* 95A) – a widespread *topos* exemplified in Plutarch's perception of the *Iliad* and the *Odyssey* as "encyclopaedic" repositories of universal knowledge; see Bagordo 2012; also Ludwich 1917–1918.

https://doi.org/10.1515/9783111589619-010

1 The Ǧābirian Cycle of *Rectifications*

Ǧābir ibn Ḥayyān's lifespan is generally placed in the eighth century CE and the alchemist is supposed to have died in the early ninth century.[5] A vast, heterogeneous body of texts attributed to him, the so-called *Corpus Gabirianum*, comprises thousands of treatises on diverse branches of knowledge such as alchemy, philosophy, medicine, magic, and astrology. Ǧābir's innovative paradigm synthesised elements from older scientific and intellectual traditions and exhibits substantial Neoplatonic and Neopythagorean influences.

The *Rectifications* (*Muṣaḥḥaḥāt*) represent a specific Ǧābirian sub-genre consisting of ten works.[6] In Julius Ruska's opinion, the core premise of the cycle is that each of the books was intended to *rectify* the theories and practices of an established alchemical authority.[7] Eight of the *Rectifications* deal with Greek "masters", i.e., Pythagoras, Socrates, Plato, Aristotle, Archigenes (two separate books!),[8] Homer, and Democritus. Ǧābir devotes the ninth treatise to his mentor Ḥarbī, and the final one to himself (*muṣaḥḥaḥātinā naḥnu*, "our Rectifications").

Plato's Rectifications is the sole extant work of the cycle, and we can only speculate whether the other books were structured in a similar fashion. The treatise contains ninety recipes with mercury grouped in four categories, according to the amount of added substances – procedures involving mercury only, followed by those combining mercury with one, two, or three other ingredients. In general, each recipe begins with theoretical statements ascribed to Plato. In some cases, the Greek philosopher addresses a disciple named Timaeus (without any actual connection to the eponymous dialogue). Ǧābir then proceeds to elucidate the intended meaning of Plato's statement with practical advice concerning the alchemical procedure.

---

5 Forster 2010.

6 Kraus 1942–1943: 1.64–67; 2.45–51.

7 Ruska 1923: 57. Kraus disagrees with Ruska by suggesting that the *Rectifications* appear to be Ǧābirian commentaries on respective pseudepigraphical works; Kraus 1942–1943: 1.64n1.

8 This particular matter will be discussed in greater detail in the section "The Alexandrian Factor: Medicine and Alchemy" below. Archigenes was a Graeco-Syrian physician who flourished in Rome during Trajan's reign (r. 98–117 CE); Wellmann 1895: 19–22.

2 Homer and the Pythagorean τετρακτὺς in Plato's *Fourth Rectification*

Devoting an entire book of *Rectifications*[9] to the alchemy of Homer may indicate the extent of pseudepigraphic authority accorded to the poet in Ǧābir's scientific system. We find a curious case in the fourth chapter of *Plato's Rectifications* which contains a recipe for the dissolution of mercury without added substances. The text[10] begins with Plato's ambiguous statement that the best work is accomplished when mercury is dissolved alone and one fourth of it is preserved. In Ǧābir's opinion, Plato's statement can be interpreted in two ways: (1) according to the school of Homer, who said "I increase"; or (2) according to Porphyry, who said "I decrease". The followers of Homer then explain that four equals ten because the sum of ten can be found in four. According to Ǧābir, their statement implies adding one and two (i.e., three), then adding three and three (i.e., six), and finally six and four, hence a final sum of ten. In other words, this computation represents the increase proposed by Homer's school. Porphyry's adherents interpret the number "four" as a reference to the four natures and see the increase (i.e., ten) as an implied fifth, superfluous nature. They maintain that there is nothing else except the natures, hence the need to decrease. Ǧābir concludes the passage by remarking that Plato followed Homer in this particular procedure and in his overall principles. The chapter ends with the practical application of the arithmological theories outlined above: the mercury is heated in a cucurbit, and one must collect a quarter of the quantity once in every seven days, until the substance is dissolved. In Homer's opinion, the process takes ten days. The substance is thus either to be heated for ten days and cooled on the eleventh, or heated for nine days and cooled on the tenth. Ǧābir recommends the first method.

The summary provided above exemplifies the fundamental role of mathematical thought in Ǧābir's paradigm.[11] This particular constellation of Greek authorities, the views ascribed to each of the figures, and the relationships existing between them, are equally deserving of attention. The successful outcome of the alchemical procedure is subordinated to the application of Pythagorean number symbolism attributed to Homer. The focus on the number "four"[12] in this text amounts to a concise exposition of

---

9 Since the work is lost, we should keep in mind that many of Ǧābir's "books" actually consist of a few pages.

10 Summary according to MS Bibliothèque nationale de France, Arabe 6915, ff. 41 recto, l. 12–41 verso, l. 18. See Appendix for the full text.

11 Kraus 1942–1943: 2.187–223.

12 The fact that such material is found in the fourth recipe is hardly coincidental.

the *tetrad* (τετρακτύς),[13] a core Pythagorean concept: the number "ten" was revered as an embodiment of the perfection of the universal order. Adding the first four numbers (as exemplified in the recipe = 1 + 2 + 3 + 4) endowed the number "four" (i.e., the *tetrad*) with respective sanctity, since it generated the perfection symbolised by the number "ten". In practical terms, Homer suggests that the completion of the alchemical operation in the recipe should take ten days, i.e., the power of the number is projected onto the dissolution of mercury. The wisdom of Homer (and Plato) is hence deemed superior in Ǧābir's pseudepigraphical *mise-en-scène*, and the alchemist recommends conducting the experiment in accordance with Homer's doctrine. Porphyry's allegory, on the other hand, interprets the number as a representation of the four natures – another key notion in the Ǧābirian paradigm.[14]

Finally, we ought to mention that the *Book of Properties* (*Kitāb al-ḫawāṣṣ*) ascribes to Homer a further allegorical statement concerning the number "four": فقد ذكر ذلك أوميرس الشاعر أنّ الأربعيّات [...] من أمهات العلم ("Homer the poet said this: the things related to the number four [...] are the mothers of science").[15] The Greek bard is thus invoked to introduce the Pythagorean τετρακτὺς into alchemical theory and practice in one instance, and in another he associates the number "four" with the mathematical foundations of science,[16] namely, root (*ǧaḏr*), division (*qisma*), multiplication (*ḍarb*), and algebra (*ǧabr*). The statement is then concluded with a typical didactic tone:[17] it is "through geometry" (*bi-l-handasa*) that you attain what you desire from these sciences (*min hāḏihi al-ʿulūm*). The collective reference at the end appears to denote the various branches of mathematical knowledge crowned by geometry. However, it should be added that the expression "what you desire" is employed topically on a regular basis in Ǧābir's recipes to describe the successful outcome of alchemical procedures. In this particular context, Homer can be taken to imply that mathematics (and especially geometry) is essential for alchemy or for the various relevant sciences in the wider Ǧābirian paradigm.

13 Kraus 1942–1943: 2.117n10. See also the chapter on the τετρακτὺς in Delatte 1915: 249–268 (a work which Kraus uses extensively); see further Izdebska 2020; Kalvesmaki 2013: 325–332. For wider treatments of Pythagorean arithmology with a Platonic outlook, see Horky 2013; Bonazzi et al. 2007; Zhmud 2017.

14 Kraus 1942–1943: 2.4–10. On allegory in Byzantine and Arabic literary culture, see the contribution of Maria Mavroudi to this volume.

15 Kraus 1942–1943: 2.118n10: "Déjà le poète Homère a dit que les choses au nombre de quatre [...] représentent les mères de la science". Key excerpts of the work have been edited in Kraus 1935 (see p. 315, 8 for the passage in question). The scholar interprets the term "mothers" as a metaphor for "elements".

16 Ǧābir ibn Ḥayyān, *Kitāb al-ḫawāṣṣ* (ed. Kraus 1935: 315, 8–12); see Kraus 1942–1943: 2.118n10.

17 Ǧābir ibn Ḥayyān, *Kitāb al-ḫawāṣṣ* (ed. Kraus 1935: 315, 12–316, 2); see Kraus 1942–1943: 2.118n10.

3 Homer and Ǧābir's Concept of Artificial Generation

In a treatise titled *Kitāb iḫrāǧ mā fī'l-quwwa ilā'l-fiʿl*, Ǧābir outlines a kind of alchemical curriculum comprising seven arts.[18] The most advanced discipline in the programme is the *science of generation* (*ʿilm al-takwīn*), also known as *science of the forms* (*ʿilm al-ṣuwar*). The other six arts are arranged as follows:[19] (1) *the science of balance* (*ʿilm al-mīzān*);[20] (2) theurgy (*ʿilm al-ṭilasmāt*); (3) *the subjugation of spiritual beings* (*istiḫdām al-rūḥāniyyāt/ʿulwiyyāt*); (4) *the science of properties* (*ʿilm al-ḫawāṣṣ*);[21] (5) medicine (*al-ṭibb*); (6) alchemy (*al-ṣanʿa*). Paul Kraus traces the origins of Ǧābir's *takwīn* to the time-honoured philosophical notions that the purpose of art (τέχνη) is to imitate nature (φύσις), and that philosophy enables man to imitate God.[22] Manifold Platonic elaborations on these ideas (e.g., *Timaeus* 42e;[23] *Theaetetus* 176a–b;[24] *Republic* 10.601c–d[25]) would codify the alchemically relevant analogy between the work of the craftsman as a demiurge and the role of the divine Creator.[26] In the Neoplatonic cosmology of the Ǧābirian paradigm, all being is generated by the interaction of the world soul,

18 Ǧābir ibn Ḥayyān, *Kitāb iḫrāǧ mā fī'l-quwwa ilā'l-fiʿl* (ed. Kraus 1935: 48); cf. Kraus 1942–1943: 1, no. 331; 2.97n1. Kraus' French rendition of the title is *Le livre du passage de la puissance à l'acte.*

19 Ǧābir ibn Ḥayyān, *Kitāb iḫrāǧ* (ed. Kraus 1935: 93); also quoted in Kraus 1942–1943: 2.97n2.

20 Since the *science of balance* represents the innovative application of mathematical knowledge in Ǧābirian alchemy, the position of this discipline at the beginning of the curriculum corroborates Homer's statement in the *Book of Properties* about the branches of mathematics being the "mothers of science".

21 On this topic, see Raggetti 2015.

22 Kraus 1942–1943: 2.99–102.

23 Plato, *Timaeus* 42e (ed. Burnet 1902): καὶ λαβόντες ἀθάνατον ἀρχὴν θνητοῦ ζῴου, μιμούμενοι τὸν σφέτερον δημιουργόν, πυρὸς καὶ γῆς ὕδατός τε καὶ ἀέρος ἀπὸ τοῦ κόσμου δανειζόμενοι; Engl. transl. Bury 1929: 93–95: "They took the immortal principle of the mortal living creature, and imitating their own Maker, they borrowed from the Cosmos portions of fire and earth and water and air".

24 Kraus (1942–1943: 2.99n3) cites only 176b, yet the end of the previous segment is important for the context of the passage in question; see Plato, *Theaetetus* 176a–b (ed. Burnet 1900): διὸ καὶ πειρᾶσθαι χρὴ ἐνθένδε ἐκεῖσε φεύγειν ὅτι τάχιστα. φυγὴ δὲ ὁμοίωσις θεῷ κατὰ τὸ δυνατόν; Engl. transl. Fowler 1921: 127–129: "Therefore we ought to try to escape from earth to the dwelling of the gods as quickly as we can; and to escape is to become like God, so far as this is possible".

25 Kraus (1942–1943: 2.100n2 and n3) uses this passage to support the statement that "l'art plastique imite l'oeuvre de l'Artisan divin", but the text contains no actual reference to a comparable metaphysical figure. See Belfiore 1984: 136–137, for a discussion of the same passage. For "mimesis", see at large Reid and DeLong 2018. Much closer to Kraus' sense is *Respublica* 10.596c (ed. Slings 2003): ὁ αὐτὸς γὰρ οὗτος χειροτέχνης [...] τὰ ἐκ τῆς γῆς φυόμενα ἅπαντα ποιεῖ καὶ ζῷα πάντα ἐργάζεται, τά τε ἄλλα καὶ ἑαυτόν, καὶ πρὸς τούτοις γῆν καὶ οὐρανὸν καὶ θεοὺς καὶ τὰ ἐν οὐρανῷ καὶ τὰ ἐν Ἅιδου ὑπὸ γῆς ἅπαντα ἐργάζεται; Engl. transl. Shorey 1930–1935: 2.423: "this same handicraftsman [...] produces all plants and animals, including himself, and thereto earth and heaven and the gods and all things in heaven and in Hades under the earth". However, Shorey, the *Republic*'s translator, observes that "it is a tempting error to refer this to God" (1930–1935: 2.423 note d).

26 Porphyry's occupation with related concepts may well have been the reason for his pseudepigraphical reception as an adept of artificial generation in Ǧābirian writings (Kraus 1942–1943: 2.122–134).

substance, and the natures (i.e., the four qualities: hot, cold, dry, and humid). The main difference between the divine and the human act of creation (i.e., artificial generation) is that the alchemist has to complete the process step by step, by joining the natures and substances in a series of procedures. Below are two examples in which Homer is associated with the *ʿilm al-takwīn*.

The Little Book of Balance[27]

This treatise features a reference to a group of Greek authorities consisting of Pythagoras, Homer, Archigenes, and Socrates.[28] The pseudepigraphical figures argue with a rival school whose members are not specified. The dispute concerns the artificial generation of "bodies" and the order of the operations involved in the process. The four Greeks maintain that it is first necessary to apply humidity (*ruṭūba*) to substance (*ǧawhar*), since humidity is a nature which can refine substances and bind them together as an adhesive. The other school holds the opinion that it is possible to begin the process with any one of (the four) natures (*ʾayy al-ṭabāʾiʿ*). Homer is thus portrayed as a master of *takwīn* – the supreme art in Ǧābir's alchemical curriculum. The group of authorities to which the poet belongs unites philosophical, mathematical, and medical knowledge in the imitation of the divine Creator.

A comparable reference to the ontological primacy of humidity regarding the origin of matter in a genuine Homeric context suggests that the poet's presence in this Ǧābirian passage may not be entirely pseudepigraphical. The recently published Syriac rendition of a treatise on Platonic theories of matter written by Porphyry[29] begins with a doxographical introduction that positions Thales as the foremost thinker to conceive of water as "the first principle, from which everything came into existence".[30] Porphyry corroborates the statement by citing "some poets" who considered Oceanus and Tethys (i.e., humidity, cf. *raṭībūṯā*) to be "the parents of everything that existed", and therefore "water should be honored *first*" (emphasis added) among the deities.[31] Yury Arzhanov, the editor of the text, has identified the metaphor about the parents in West's edition of the *Iliad* (14.246).[32]

The possibility of a connection between the pseudepigraphical episode from Ǧābir's *Little Book of Balance* presented above and Porphyry's (provisionally titled) *On Principles and Matter* contains significant implications. The Neoplatonic thinker initiated the

27 Ǧābir ibn Ḥayyān, *Kitāb al-mīzān al-ṣaġīr* (ed. Kraus 1935: 425–459); see Kraus 1942–1943: 1, no. 369.
28 Ǧābir ibn Ḥayyān, *Kitāb al-mīzān al-ṣaġīr* (ed. Kraus 1935: 454, 9–10); see Kraus 1942–1943: 2.102n2.
29 Porphyrius, *De principiis et materia* (ed. Arzhanov 2021).
30 Porphyrius, *De principiis et materia* 5 (ed. Arzhanov 2021: 84, 24–86, 1 [text], 85–87 [Engl. transl.]).
31 Porphyrius, *De principiis et materia* 6 (ed. Arzhanov 2021: 86, 5 [text]; 87 [Engl. transl.]).
32 *Ilias* 14.246 (ed. West 1998–2000: vol. 2); see Arzhanov 2021: 87n2.

reconciliation of Plato and Aristotle which came to dominate the Alexandrian discourse in Late Antiquity and shaped the subsequent Arabo-Islamic reception of Greek philosophy.[33] Moreover, Porphyry's extensive scholarship on Homer contributed decisively to the Neopythagorean/Neoplatonic exegesis of the latter,[34] an intellectual development that culminated in Proclus' reconciliation of Plato and Homer as proponents of a single metaphysical truth, articulated by means of an innovative "scientific theology".[35] The doxographical aspect of *On Principles and Matter* speaks for the same interest in the history of thought which informed Porphyry's influential work *Φιλόσοφος ἱστορία* (*Philosophos historia*). Greek fragments of this text have survived, and it is known to have been transmitted (at least partially) into Syriac and Arabic; it thus resonated in Near Eastern wisdom literatures, particularly as a source of Pythagorean material.[36] Kraus' observations on the wide scope of pseudepigraphical authority accorded to Porphyry in Ǧābirian treatises[37] are substantiated by the pivotal role of the historical figure: Porphyry's legacy unites many of the ingredients found in Ǧābir's paradigm (Homeric hermeneutics being one example) and allows for plausible channels of cross-cultural dissemination (i.e., Neoplatonic Aristotelianism, doxography and gnomology). The fact that authors such as Ibn Rušd and al-Šahrastānī summarise the essence of Porphyry's views on prime matter in the Syriac *On Principles and Matter* bears witness to the text's overall impact.[38]

Porphyry's allegorical invocation of Homeric verse in a Presocratic doxographical context illustrates the intellectual mechanics of Homer's transition to a philosophical and scientific authority – a process which would inevitably prepare the ground for manifold pseudepigraphical reception. The omnipresence of myth and epic in ancient Greek-speaking civilisation (and its spheres of influence) meant that Porhyry's erudite readers were expected to recognise the reference to Oceanus and Tethys in the *Iliad*; in other words, Homer's popularity serves to endorse a particular theory of matter. More significantly, the association with the philosophy of Thales also achieves the reciprocal effect of transferring scientific expertise onto Homer. The gradual development of such discourses over many centuries, together with epic poetry's continuous relevance in education, would lead to the subsequent "Byzantine notion of reading Homer as a foundation for understanding natural science, primarily through explicating him allegorically, following an intellectual tradition already established in antiquity".[39]

33 D'Ancona 2020: 1576. See further Karamanolis 2006: 243–330 (also quoted in Arzhanov 2021: 72); Chiaradonna 2016: 329–332.

34 E.g., Porphyrius, *Quaestiones Homericae ad Iliadem* (ed. MacPhail 2011).

35 De Piano 2014: 157, 165, 171. See further Lamberton 1986: 1–43, 162–232.

36 D'Ancona 2020; Arzhanov 2021: 68.

37 See note 26 above.

38 Arzhanov 2021: 77.

39 Mavroudi 2020: 455.

The Book of Concentration[40]

The treatise is among the main writings dealing with the subject of *takwīn/tawlīd.*[41] Porphyry is presented as the eminent Greek sage, and Ǧābir even states explicitly that his own doctrine of artificial generation relies on a (necessarily apocryphal) work composed by the Neoplatonic philosopher: وقد قال فرفوريوس في الكتاب الذي أخذنا هذه الأشياء منه المسمّى بالتوليد ("Porphyry said in the book, from which we took these things, titled *The Book of Generation*").[42] A passage from the *Book of Concentration* joins Homer with Galen,[43] another medical authority, in an allegorical disputation concerning an opinion attributed to Porphyry. According to the latter, endowing artificially generated beings with intelligence ought to be a process based on catarchic astrology and the so-called *threefold wisdom.*[44] Ǧābir proceeds to interpret the term *threefold wisdom* as Porphyry's allegory of the brain[45] and contrasts it with the *fivefold* (*mutaḫammis*) concept ascribed to Homer and Galen, which stands for the five senses.[46] Kraus hypothesises that the term *mutaḫammis* is a reference to an actual passage in the *Iliad* (regarding the fivefold battle formation of the Trojans),[47] whose use in this context may have its origins in earlier Pythagorean or Stoic allegories of the five senses.[48]

4 The Alexandrian Factor: Medicine and Alchemy

The conspicuous inclusion of a presumably "non-occult" discipline such as medicine, just below artificial generation and alchemy in the Ǧābirian classification of the seven arts, and the pseudepigraphical association of Homer with ancient physicians such as Archigenes and Galen[49] command attention. The proximity of medicine to *takwīn* – as

---

40 Ǧābir ibn Ḥayyān, *Kitāb al-taǧmiʿ* (ed. Kraus 1935: 341–391; the work is not preserved in its entirety); see Kraus 1942–1943: 1, no. 398.

41 The second term denotes the artificial generation of animate beings, especially humans.

42 Ǧābir ibn Ḥayyān, *Kitāb al-taǧmiʿ* (ed. Kraus 1935: 364, 3–4); Fr. transl. Kraus 1942–1943: 2.122–123: "Porphyre a dit dans le livre auquel nous avons emprunté ces choses et qui est intitulé *Livre de la Génération*".

43 On the significance of Galen for the alchemical tradition and references to him in the *Corpus Gabirianum*, see Martelli 2019: 586–590.

44 *Muṯallaṯ al-ḥikma*, rendered by Kraus (1942–1943: 2.117n8) as "Triple-Sage" since the term is the "appellation courante de Hermès Trismégiste".

45 Cf. Plato's tripartite division of the soul; see Lorenz 2024.

46 Lorenz 2024.

47 *Ilias* 12.87 (ed. West 1998–2000: vol. 1): πένταχα κοσμηθέντες ἅμ' ἡγεμόνεσσιν ἕποντο ("and marshalled in five companies they followed after the leaders").

48 Kraus 1942–1943: 2.118n10.

49 Alexander of Tralles and Rufus of Ephesus mention that Galen applied Homeric incantations in his medical practice. The belief in the "occult" (esp. healing) properties of Homeric passages relied on

discussed in the preceding subsections on the *Little Book of Balance* and the *Book of Concentration* – has little to do with the art of healing. Rather, the function of medical knowledge in Ǧābir's paradigm is imbedded in a philosophical and cosmological framework which, in a way, resembles the outlook of Galen and Alexandrian Galenism. Galen was interested in intellectual history and doxographical content,[50] argued that physicians require philosophical training in Stoic logic, physics, and ethics, and valued the expertise of natural science because the human body is also constituted by the four elements, hence subjected to cosmological principles.[51] This fusion of medicine and philosophy was elaborated further at the Neoplatonic school of Alexandria prior to the advent of Islam (fifth–seventh centuries) – a development which had a lasting impact on late antique and mediaeval scientific traditions in Byzantium and the Near East.[52] However, the relevance of medical knowledge, particularly pharmacology and metrology,[53] for practitioners of alchemy is documented as early as the first century CE in the writings of Pseudo-Democritus.[54] At a later stage, Byzantine contemporaries of Alexandrian Galenism would contribute further towards the rapprochement of medicine and alchemy. A noteworthy example is presented in a collection of writings on chrysopoeia by the Philosopher Christianus (sixth–eighth centuries), where the author cites Galen's *De locis affectis* in order to compare the medical treatment of the human body with the alchemical work on the metallic "body".[55] Christianus' interpretation brings us back to the Greek pseudepigraphical authorities in the Ǧābirian *Little Book of Balance*, where Pythagoras, Homer, Archigenes, and Socrates discuss the artificial creation of "bodies" (see the earlier discussion on the *Little Book of Balance*). Moreover, the fact that the historical Galen was interested in Stoic philosophy might have been the reason why Ǧābir's *Book of Concentration* would attribute to him the Stoic/Neopythagorean exegesis of Homeric verse (see the above discussion of the *Book of Concentration*). The particular use of *De locis affectis* in a Byzantine alchemical discourse could be significant for the eventual reception of Archigenes in the *Corpus Gabirianum*, since Galen's treatise is a major source for the reconstruction of Archigenes' medical and philosophical

analogies and was commonly thought to originate in Pythagorean traditions; see Collins 2008: 211–214. Eunapius relates another relevant example: Ionicus of Sardis (d. ca. 399 CE) was a physician with a possibly Neoplatonic background known as a practitioner of theurgy; see Overwien 2018: 279.

50 Garofalo 2019: 67.

51 Chandelier 2020: 1138.

52 Chandelier 2020: 1139.

53 The mathematical aspect of mixing substances (in accordance with their properties) and performing the operations laid down in recipes represents a focal point of alchemy, medicine, and natural philosophy.

54 Martelli 2019: 577–578.

55 Martelli 2019: 582–583. In connection with the medical analogy, we have to point out that *Plato's Rectifications* (and the tradition, in general) use the term *remedy* (*dawāʾ*) to designate ingredients employed for the treatment of metals (e.g., mercury) in the course of alchemical procedures (Kraus 1942–1943: 1.64).

outlook.[56] Although Galen's criticism of certain theories advanced by Archigenes is often brought to scholarly attention,[57] it must be noted that Archigenes' works on pharmacology and diagnosing ailing bodily parts and organs (i.e., the very subject matter of *De locis affectis*) influenced Galen's respective expertise to a considerable extent.[58] With further regard to pharmacology and the properties of substances – two related fields of pronounced interest for the alchemical tradition – Archigenes was also known to differ from Galen in having a strong reputation for the preference of natural remedies.[59]

As far as the peculiar position of Archigenes in the cycle of *Rectifications* is concerned, Kraus considered the attribution of two books to the physician (with each title containing a different Arabic transliteration of the name) to indicate Ǧābir's unawareness that *Aršāǧānas* and *Arkāġānīs* are, in fact, one and the same author.[60] However, Archigenes' name is spelt *Aršīǧānis* in the passage from the *Little Book of Balance* discussed above, and other sources contain further variations, e.g., *Aristiǧānis* in Ibn al-Qifṭī's biographical dictionary of philosophers.[61] Fuat Sezgin's take on the matter is that the key factor at hand would be the pseudepigraphical material on which the respective *Rectifications* books were based, i.e., whether Ǧābir's commentary drew on a number of works or built upon a single one in each case.[62] Another possibility would be that each of Archigenes' *Rectifications* was composed by a different contributor, perhaps at a different stage in the gradual formation of the *Corpus Gabirianum*. Finally, Moritz Steinschneider proposes to identify Archigenes with an *Arsicanus*[63] mentioned in connection with Mary the Jewess in a Latin work attributed to Morienus.[64] This finding demonstrates that the pseudepigraphical reception of the Graeco-Syrian physician may have resonated on a wider scale, across diverse alchemical traditions.

---

56 TOPOI, Research Project D-2–2, https://blog.topoi.org/project/d-2-2/. A critical edition of the Arabic *De locis affectis* is currently in preparation; see Ǧumʿa et al. 2020. On Archigenes in the Islamic sphere, see further Ullmann 1970: 69. Another repository for Archigenes' thought was the so-called *Tetrabiblion* by Aetius of Amida (a sixth-century Byzantine court physician educated at Alexandria), which also circulated in Arabic (see Neuburger 1911: 104ff), as well as Ibn Abī Uṣaybiʿah, *ʿUyūn al-anbāʾ fī ṭabaqāt al-aṭibbāʾ* [4.1.10.3] (Engl. transl. Savage-Smith et al. 2020: n182, https://scholarlyeditions.brill.com/reader/urn:cts:arabicLit:0668IbnAbiUsaibia.Tabaqatalatibba.lhom-tr-eng1:4.1-4.5/).

57 Bouras-Vallianatos and Zipser 2019b: 3

58 Wellmann 1895: 20–21.

59 Bouras-Vallianatos 2019: 49.

60 Kraus 1942–1943: 2.46n7.

61 Ibn al-Qifṭī, *Taʾrīḫ al-ḥukamāʾ* (ed. Lippert 1903: 73).

62 Sezgin 1971: 247n1.

63 < *Aršīǧānis*?

64 Steinschneider 1862: 63n19 (quoted also in Kraus 1942–1943: 2.46n7); Steinschneider 1904: 305–306.

A Proclean Shift?

The passages taken into account thus far portray Ǧābir's Homer as a polymath, with emphasised competence in the mathematical sciences and artificial generation (*takwīn*) – a demiurgic discipline dominated by Porphyry,[65] which appears to be epistemologically related to medicine, since Homer shares this expertise with the physicians Archigenes and Galen. While Galen features prominently in various Ǧābirian writings, the pseudepigraphical profile of Archigenes is more narrow, yet distinguished by the fact that the cycle of *Rectifications* comprised two treatises dedicated to him. Both Porphyry and Galen were historically instrumental in the formation of the interdisciplinary background which eventually came to define the school of Alexandria – by establishing the "harmony" of Plato and Aristotle and the archetype of the ἰατροσοφιστής – but the influence of Proclus[66] concerned also particular structural and didactic innovations that inspire a measured extent of speculation with regard to Ǧābir and the *Rectifications.*

Ammonius Hermiae (ca. 435–517/526)[67] took charge of the philosophical school in Alexandria at the cusp of the sixth century and introduced Proclus' Neoplatonic approach to Aristotle.[68] This development concerned not only the interpretation of philosophical teachings, but also the format of commentaries and instruction itself. The changes brought about by Ammonius were respectively adopted by the medical school as well, and that led to the emergence of structurally parallel curricula. The key elements of this "Proclean" reform were in no way revolutionary *per se* – commentaries building on excerpts (λήμματα) from original works certainly existed before Proclus' time. Yet the introduction of (1) a canon of texts processed in such commentaries, together with related study materials designed to break down the complex content; (2) more or less standardised eight-point questionnaires (κεφάλαια) for the assessment

---

65 A prolific exegete of Homer who first designated the latter as θεολόγος; De Piano 2014: 171.

66 His "harmony" of Homer and Plato was embedded in the Athenian school's greater effort to harmonise Neoplatonism and spiritual traditions in a coherent system (De Piano 2014: 157n12; on Proclus' reception of Homer, see also Lamberton 1986: 162–232). The philosopher's lost works thus include titles such as *On Orphic Theology*; *On the Harmony of Orpheus, Pythagoras, and Plato with the Oracles*; *Commentary on the Whole of Homer*; *On the Gods of Homer*; and *On Mythical Symbols*. The spurious *Commentary on Pseudo-Pythagoras' Golden Verses* is preserved in Arabic and *On Sacrifice and Magic* (*On Theurgy*) survives fragmentarily; see Helmig and Steel 2021.

67 Blank 2017. The philosopher studied under Proclus in Athens in the second half of the fifth century. Ammonius' strong background in the mathematical sciences is important for our investigation. According to Damascius' account: "In geometry and astronomy he was superior not only to his contemporaries, but also to [...] Proclus' companions [...] and I would say, to the men of all times"; quoted in Hadot 2015: 15.

68 Overwien 2018: 277–278.

of the content at the beginning of the commentaries;[69] and (3) interactive didactic sessions (πράξεις), which consisted of in-depth (λέξεις) and more general (θεωρίαι) discussions, had a far reaching impact.[70]

The consolidation of the iatrosophical paradigm created an interdisciplinary intellectual environment – to which the mathematical sciences, theological and literary studies indispensably belonged – founded on Neoplatonic Aristotelianism that shaped the respective production and transmission of texts. The rich Arabo-Islamic traditions of bio-bibliographical and wisdom literature have their origins in late antique compilations and attest the enduring *Nachwirkung* of Alexandrian learning in the Near East. To provide an example, the Hippocratic material in works such as al-Mubaššir ibn Fātik's *Choicest Maxims and Best Sayings* is, in fact, related to recensions of the authentic *Hippocratic Aphorisms* used in the medical curriculum.[71] The spread of scientific and philosophical pseudepigraphy was a natural side effect of the popularity enjoyed by bio-bibliography and wisdom literature.[72]

"Iatrosophical" Features of Ǧābir's *Rectifications*

Specific aspects and genres of the Alexandrian curriculum can be compared with key characteristics of the *Rectifications*.

1) Since both commentaries and lectures (πράξεις) were essentially structured around quotations (λήμματα) from canonical works, such passages can often be identified in later translations by formulaic introductions like "Galen said: [...]", e.g., قال جالينوس: إنّ قصد الطبّ التماس الصحّة وغايته إحرازها ("Galen said: 'The purpose of medicine consists in the quest for health, its goal in attaining the latter'"). The *incipit* of the Galenic *De sectis* was transmitted thus in a treatise by the ninth century author al-Ruhāwī.[73] The statements attributed to Plato and other pseudepigraphical authorities in the *Rectifications* are also usually introduced in this manner, cf. قال فلاطن إذا حللت الروح بغير داخل وحفظت الرباعيه كان أفضل الأعمال ("Plato said: 'If you dissolve the spirit without [adding another] ingredient and save a quarter of it, the works are better'". See Appendix). Regardless of whether the Ǧābirian treatise cites actual excerpts from older material or was deliberately

69 The influence of this format amongst Muslim scientific authors is, for instance, evident in the structure of Ibn Sīnā's *Canon of Medicine*; see Pormann 2018: 347; also Overwien 2019: 18.

70 E.g., Ammonius' reform was the reason for similar innovations at other late antique centres of learning, notably Nisibis in Sassanian Persia. This case is rather striking because the institution in question was not secular, but a Nestorian theological school, whose decision to adopt the "Proclean" programme in a Christian setting may have been reached under the immediate influence of Alexandria (Overwien 2018: 282).

71 Pormann 2018: 342–345.

72 Garofalo 2019: 74.

73 Overwien 2016: 203–204.

formatted in this way, its content appears to bear a certain structural semblance to the "Proclean" commentaries introduced by Ammonius.[74] Plato's statements are followed by an exposition comprising two parts: (1) an analysis of the passage whose function would correspond to that of the θεωρίαι, i.e., the positions of Homer's and Porphyry's rival schools; and (2) the alchemical recipe, which Ǧābir (in the capacity of a lecturer) introduces as the practical application of Plato's statement, and can be understood to imitate the more detailed λέξεις.

2) The ἀπορίαι ("questions") was a further late antique genre related to the format of the commentaries that invites a comparison with the *Rectifications*. Such didactic works were designed to tackle difficulties encountered in scientific and philosophical treatises with the goal of providing solutions (λύσεις), and similarly built on respective excerpts.[75] Ḥunayn ibn Isḥāq, who is said to have studied medicine at Alexandria,[76] reportedly composed ἀπορίαι on Galenic material; the Arabic version of the genre is termed *doubts* (*šukūk*).[77] The *Rectifications* appear to resemble this particular genre conceptually and structurally: each recipe serves as a practical solution that applies theories attributed to Plato and other pseudepigraphical authorities in accordance with Ǧābir's expertise.

3) Rival pseudepigraphical schools occur in the *Rectifications* and other writings of the *Corpus*. We may tentatively juxtapose the topical use of such authorities and the concept of Galen's *De sectis*. He approached the development of medical theory and practice from a rather critical perspective and often dismissed as fallacies opinions that disagreed with his own (or Hippocrates').[78] *De sectis* originated as an account of the teachings in ancient medical schools (i.e., the Empirics, the Dogmatics, and the Methodics) with which the students had to begin their education.

The points outlined above are merely intended to highlight some details that deserve attention. The summary of Homer's pseudepigraphical profile (see the previous subsection) acquires further significance against the background of Ammonius' "Proclean" reform and the scope of the combined Alexandrian curricula. The fact that Ǧābir devised a curriculum of his own (see the earlier section on Homer and Ǧābir's concept of artificial generation), whose seven disciplines create the impression of an alchemical answer to the *trivium* and the *quadrivium*, may serve as a metaphor for his indebtedness to older traditions. The traits of Ǧābir's Homer point in the direction of late Athenian Neoplatonism's scientific theology and the profound mathematical knowledge associated with figures such as Ammonius. Finally, in order to substantiate our speculations, we can address the probability that one of Ammonius' illustrious

74 Festugière 1963.
75 Overwien 2018: 270–272.
76 Pormann 2018: 345–348.
77 Overwien 2018: 271–272.
78 Garofalo 2019: 67.

students was an alchemical practitioner. Sergius of Rēšʿaynā (d. 536) "initiated the first systematic program of scientific translations from Greek into an oriental language".[79] The Hellenised Syrian was a polymath in his own right who had at his disposal the interdisciplinary knowledge that would allow for a hypothetical connection of alchemy with medicine and pharmacology, in particular.[80] Sergius is considered to be the author who collected and adapted excerpts from Galen's *De simplicium medicamentorum facultatibus* from a strictly alchemical perspective. This redaction omitted the medical content, focused on Galen's natural science (i.e., the physical as opposed to the healing properties of substances), and appears to confirm the association of Sergius with alchemy in the Ibn al-Nadīm's *Fihrist*.[81] Such a case would represent a testimony to the implementation of Alexandrian iatrosophical epistemology in alchemical discourse and concretise the relevance of the school's curriculum (and intellectual environment) for the reception of Graeco-Egyptian alchemy in Bagdad.

5 Pseudepigraphy and Wisdom Literature

It has been observed that late antique textual production was among the factors shaping the dissemination of scientific knowledge in the Arabo-Islamic sphere. Wisdom literature had a vital function as a channel of cross-cultural transmission and certain passages in early compilations acquire significance with regard to the Ǧābirian representations of Homer.

Qusṭā ibn Lūqā's Translation of the *Placita philosophorum*

The exposition of the τετρακτὺς in the *Fourth Rectification* resembles a passage in Pseudo-Plutarch's late antique doxographical work known as the *Placita philosophorum* (Opinions of the Philosophers).[82] It appears that Ǧābir has ascribed to Homer Pythagorean material which may have been taken from this or a similar source. The relevant Greek text, its Arabic version, and Hans Daiber's German translation are presented synoptically below:

---

79 Bhayro 2019: 163. On Sergius' studies with Ammonius, see Blank 2017.

80 See further Martelli 2020.

81 Martelli 2019: 584–586.

82 Benevich 2022. The nineteenth-century editor of the *Placita* postulated that Pseudo-Plutarch represents an abridged version of a longer work by Aetius. The Arabic translation of the *Placita* was hence titled *Aetius Arabus* by its editor, Hans Daiber (1980). The selected passage is extended with the aim of broadening the context.

εἶναι δὲ τὴν φύσιν τοῦ ἀριθμοῦ δεκάδα· Ἕλληνες, πάντες βάρβαροι ἀριθμοῦσιν, ἐφ’ ἃ ἐλθόντες πάλιν ἀναποδοῦσιν ἐπὶ τὴν μονάδα. καὶ τῶν δέκα πάλιν, φησίν, ἡ δύναμίς ἐστιν ἐν τοῖς τέσσαρσι καὶ τῇ τετράδι. τὸ δὲ αἴτιον· εἴ τις ἀπὸ τῆς μονάδος ἀναποδῶν κατὰ πρόσθεσιν τιθείη τοὺς ἀριθμούς, ἄχρι τῶν τεσσάρων προελθὼν ἐκπληρώσει τὸν δέκα ἀριθμόν· ἐὰν δὲ ὑπερβάληται τὸν τῆς τετράδος, καὶ τῶν δέκα ὑπερεκπεσεῖται. οἷον εἴ τις θείη ἓν καὶ δύο προσθείη καὶ τρία καὶ τούτοις τέσσαρα, τὸν τῶν δέκα πληρώσει ἀριθμόν· […].[83]

وان طبيعة العدد تنتهى الى العشرة، وذلك ان كل اليونانيين وكل الاعاجم ينتهى عددهم الى العشرة، فاذا صاروا اليها رجعوا الى الراحد، ويقول ايضا ان العشرة بالقوة هي ايضا في الاربعة وفي الروابع، والعلة في ذلك انه اذا اجتمعت الاعداد من الواحد الى الاربعة استكمل عدد العشرة، فانه ان جمع الواحد والاثنان ثم زيد على ذلك ثلثة وعلى المجتمع من ذلك اربعة استكملت العشرة

Die Natur der Zahl laufe auf die Zehn hinaus; denn bei allen Griechen und bei allen Barbaren laufe die Zahl auf die Zehn hinaus; und dann, wenn sie zu ihr (der Zehn) gelangen, kehren sie zu der Eins zurück. Ferner behauptet er, dass die potentielle Zehn auch in der Vier und in den Viererzahlen ist. Die Ursache davon sei: wenn die Zahlen von eins bis vier zusammenkommen, wird die Zahl Zehn vollständig. Denn es wird, wenn eins und zwei zusammengezählt werden und wenn dann zu diesem drei und zu der daraus entstandenen Summe vier hinzugefügt wird, die Zehn vollständig.[84]

Ğābir’s possible familiarity with the Arabic *Placita philosophorum* could have significant, albeit not conclusive, implications for the chronological aspect of his problematic historicity. The translator of the doxographical compilation was Qusṭā ibn Lūqā (820–912), a trilingual Syrian Christian of Greek origin.[85] Qusṭā maintained personal acquaintances with other contemporary intellectuals involved in the reception of Greek philosophy and science, such as Ḥunayn ibn Isḥāq, Yaḥyā ibn al-Biṭrīq, and al-Kindī. Situating the translation of the *Placita* in the second half of the ninth century endorses Kraus’ hypothesis that, if Ğābir was in fact a single individual at all (as opposed to a collective of authors), his lifespan cannot be placed in the eighth century.[86] Furthermore, extensive quotations from the *Placita* are to be found in another Ğābirian treatise, the *Kitāb al-ḥāṣil* (Book of the Result),[87] a work that deals primarily with the alchemical connection between the physical composition of the bodies and the letters composing the words describing the bodies.

83 [Plutarchus], *Placita philosophorum*, 876F–877A (Diels 1879: 281–282).

84 Aetius Arabus, *Placita philosophorum* 1.3.8 (ed. Daiber 1980: 100, 19–25 [text], 101 [Germ. transl.]); also quoted in Benevich 2022: 372–373.

85 Daiber 1980: 3–16.

86 Kraus 1942–1943: 1.xxxvi–xlviii; 2.vii–viii.

87 Kraus 1942–1943: 1.82–83; 2.331–339. For a discussion of the Arabic *Homerica* in Qusṭā’s translation, see Kraemer 1956: 269–270.

Ḥunayn ibn Isḥāq's *Maxims of the Philosophers*

Ḥunayn also composed an influential doxographical compilation in the same period by drawing on Greek, Syriac, and Persian material. This text has come down to us in a later recension titled *Ādāb al-falāsifa* (Maxims of the Philosophers).[88] Despite concerns expressed about the work's authenticity,[89] the *Maxims* offer valuable information on the formative stage of the genre and its interaction with related traditions and cultural processes.

According to Oliver Overwien, the Greek canon of wisdom literature reached its completion by the time of Aristotle.[90] The absence of Homer in this canon prompts the scholar to conclude that Ḥunayn's Homer was derived from alternative sources,[91] whose circulation acquires vital importance.[92] For comparison, Martelli's contribution to the present volume demonstrates that Homer was featured also as an alchemist in the Syriac tradition. As far as wisdom literature in Syriac is concerned, however, it developed predominantly in monastic circles and tended to process pagan Greek influences through a Christian perspective.[93] A distinctive feature of this current is that, with minor exceptions, Syriac compilations usually cannot be traced back directly to earlier Greek predecessors.[94] Certain peripheral traditions of popular philosophy were also disseminated among the monastic communities of Egypt during the early Abbasid period.[95] Moreover, the age of the large-scale translation movement in the Near East was paralleled by a cultural revival in Byzantium which brought about heightened literary activity and the re-assessment of late antique doxographical anthologies.[96]

The peculiar structure of Ḥunayn's *Ādāb al-falāsifa* does not follow a chronological or an alphabetical order.[97] Homer is present but does not occupy central stage in the capacity of a universal reference point. The compilation places an emphasis on the seven Greek philosophical schools and groups (i.e., "harmonises") Plato and Aristotle together as Peripatetic thinkers.[98] Overwien argues that the arrangement reflects Ḥunayn's religious upbringing and education, thereby offering a synthesis of Near Eastern Christian

88 Ḥunayn ibn Isḥāq, *Ādāb al-falāsifa* (ed. Badawī 1985). See Overwien 2003: 97.
89 Pietruschka 2014.
90 Overwien 2003: 95–96.
91 Overwien 2003: 113.
92 In this context, we ought to take into account the fact that Ḥunayn, very much like Sergius of Rēšʿaynā before him, had first-hand Alexandrian experience which similarly amounted to a personal involvement in another "systematic program of scientific translations from Greek".
93 Pietruschka 2014.
94 See Arzhanov 2014.
95 Pietruschka 2015: 91.
96 Pietruschka 2014.
97 Overwien 2003: 104.
98 Overwien 2003: 97–98; 104.

traditions and the Alexandrian iatrosophical curriculum.[99] Such a synthesis illustrates a position comfortable with Greek learning which, however, shunned the more radical (e.g., theurgic) aspects of Neoplatonism for understandable reasons. The access to allegorical and pseudepigraphical Homeric traditions comparable to the reception found in Ğābir's paradigm would, otherwise, have been a lesser issue for Ḥunayn. On the other hand, the apocryphal equipping of Socrates with the jar famously inhabited by Diogenes is a curious touch that appears to exemplify the Christian use of a classical *topos*, i.e., an ascetic portrayal of Socrates designed to uphold his moral stature.[100] This ideological collusion will be contrasted with an altogether different misrepresentation in the next entry, which features Homer as well.

Al-Mubaššir ibn Fātik's *Choicest Maxims and Best Sayings*

We have already pointed out that works such as al-Mubaššir's *Muḫtār al-ḥikam wa-maḥāsin al-kalim*[101] (ca. 1048–1049) can incorporate valuable material descended from Alexandrian (and other late antique) sources. To provide an example, Diogenes has been colluded with Homer, yet the anecdotal framework in which the misrepresentation unfolds happens to be derived from the oldest, only fragmentarily preserved Greek material pertaining to Diogenes. The story known as *The Sale of Diogenes* dates back to the third century BCE and is modelled on Heracles' enslavement in the Euripidean play *Syleus*.[102] Al-Mubaššir may have replaced Diogenes with Homer, who "was captured, and the divider took him away to sell him" (وأُسِر فأتى به المقسم ليباع)[103] but it is fundamentally significant that the compiler of the *Choicest Maxims* had access to fragments from *The Sale of Diogenes*. However, al-Mubaššir also attributed to Homer some statements containing vocabulary that can be relevant for the alchemical contexts in the *Corpus Gabirianum*:

الحكمة هي أن تدرك صورة العلم بالعمل	Wisdom is that you accomplish the image of knowledge through the work.[104]
الحيل فوائد الفكر	Stratagems are the benefits of thought.[105]
طول الجدة يميت الحيل	The length of the process kills the stratagems.[106]

99 Overwien 2003: 108.
100 Overwien 2003: 109.
101 Al-Mubaššir ibn Fātik, *Muḫtār al-ḥikam wa-maḥāsin al-kalim* (ed. Badawī 1980).
102 Overwien 2011: 94.
103 Al-Mubaššir ibn Fātik, *Muḫtār al-ḥikam* (ed. Badawī 1980: 30); Engl. transl. Muftić 2018: 27.
104 Al-Mubaššir ibn Fātik, *Muḫtār al-ḥikam* (ed. Badawī 1980: 30).
105 Al-Mubaššir ibn Fātik, *Muḫtār al-ḥikam* (ed. Badawī 1980: 31).
106 Al-Mubaššir ibn Fātik, *Muḫtār al-ḥikam* (ed. Badawī 1980: 31).

We already came across the plural form of the term *ṣūra* ("image/form") in the initial reference to Ǧābir's alchemical curriculum (see the preceding section on Homer and Ǧābir's concept of artificial generation): the science of artificial generation (*ʿilm al-takwīn*) is also known as *ʿilm al-ṣuwar, the science of forms*. Particularly the first sentence in the table above could be interpreted in accordance with the Neoplatonic framework of the Ǧābirian demiurgic/theurgic discourse because it also contains the term *ʿamal* ("work"). The use of a definite article strengthens the potentially alchemical significance of the statement. In other words, artificial generation is accomplished through *the work*. Kraus discusses a poignant passage in *The Book of Concentration*, the treatise on *takwīn* attributed to Porphyry, where a term derived from the Arabic root *ṣwr* designates the craftsmen who imitate the work of the Creator: هم الذين سمّوا أنفسهم مصوّرين يريدون بذلك التشبّه بالعلّة التى إبتدعت هذه الأشياء ("[…] the ones who name themselves makers of images want thus an imitation of the Cause which created these things").[107] The literal meaning of *muṣawwir* designates an individual who produces representative or figurative works, and has been conceived as a pendant of the Greek term εἰδωλοποιός ("sculptor"), the corresponding pendant of εἶδος being *ṣūra*.[108] The animation of statues was, indeed, a component of Neoplatonic theurgy rooted in ancient imitation concepts, and Kraus would have us believe that Porphyry was the major theorist of this doctrine: "A l'époque néoplatonicienne, elle [i.e., 'croyance'] a été élaborée en une théorie philosophique, dont le représentant le plus en vue n'est autre que Porphyre".[109] The subsequent points rely on some of the philosopher's works and are by all means insightful and correct, yet their intention to validate the pseudepigraphical connection with Ǧābir's content can be partially misleading. In fact, while certainly being familiar with theurgy, Porphyry was less interested in it than other Neoplatonic thinkers such as Iamblichus and especially Proclus, its most avid proponent in the fifth century.[110] The discussion of the Alexandrian curriculum has hence accentuated Overwien's conclusions about the likelihood of a Proclean stimulus behind Ammonius' influential reform.

Whether al-Mubaššir was concerned with the possible alchemical dimensions of the statement attributed to Homer is a question beyond the scope of this essay. The other two passages are of interest since they contain the term *ḥiyal* ("stratagems/ruses"),

---

107 Ǧābir ibn Ḥayyān, *Kitāb al-taǧmiʿ* (Kraus 1935: 350, 12–13); Fr. transl. Kraus 1942–1943: 2.123–124: "[…] ceux qui se sont donnés le nom de faiseurs d'images [*muṣawwirūn*]. Par ce nom ils veulent indiquer qu'ils imitent la Cause qui produit ces choses".

108 Kraus 1942–1943: 2.126.

109 Kraus 1942–1943: 2.127.

110 See Johnston 2008 and Sheppard 1982 (also cited in Helmig and Steel 2021). The indices of Kraus' two-volume *opus* contain no references to Proclus or Iamblichus; that is, the two may not have been pseudepigraphically viable – or even known – to Ǧābir. On the other hand, Porphyry's prominence in Ǧābirian writings can reflect the philosopher's leading position in Alexandrian education as the author of the *Isagoge*, the standard Neoplatonic introduction to Aristotelian studies.

singular *ḥīla*.[111] The context in which the term occurs, however, is rather general and the statements do not seem to contain other terms that would point at relevant interpretations. The occurrences are nevertheless noteworthy because the use of *ḥīla* is not all too frequent even in the *Corpus Gabirianum*.

6 Conclusion

Some of Ǧābir's treatises contain references to Homer where the Greek epic poet is pseudepigraphically stylised as a master of the mathematical sciences and artificial generation. Homer appears in various contexts, and in the variable company of other Greek authorities such as Pythagoras, Plato, Porphyry, and Galen. Ǧābir exhibits the topical tendency to group such authorities in rival schools and have them dispute alchemical questions. The intellectual background of the discourse reflects an overall Neoplatonic, interdisciplinary outlook which combines elements of Pythagorean number symbolism, theurgy, natural philosophy, and so forth. The survey has attempted to explore some perspectives of Homer's Ǧābirian reception: the evolution of Homer as a universal philosophical, scientific, and spiritual authority, particularly in Neoplatonic thought; the relationship of such cultural and intellectual processes with pseudepigraphical reception and the latter's transfer from late antique centres of Greek learning to the Arabo-Islamic sphere.

The circumstances surrounding the school of Alexandria in the last pre-Islamic centuries nurtured a favourable environment for the gestation and dissemination of the content with which Homer is associated in the *Corpus Gabirianum*. At the beginning of the sixth century, the Neoplatonic elaboration of Aristotelian philosophy and science (necessarily including a Pythagorean take on the mathematical disciplines) was mated to medicine and supplemented by the syncretistic theology of fifth century Athens under the new leadership of Ammonius. This environment produced the likely alchemist, polymath, and translator Sergius of Rēšʿaynā and standardised the format of commentaries and instruction – factors which would influence the subsequent reception of Greek learning under Islam. Certain characteristics of the Ǧābirian cycle of *Rectifications* exhibit structural similarities with Alexandrian commentaries and related genres.

Traditions of wisdom literature in Arabic (and other languages) facilitated the reception of both authentic and pseudepigraphical Greek material. Some references to Homer in such sources can be relevant for Ǧābir's representation of the poet.

111 Kraus 1942–1943: 2.385 for the index.

Appendix

Fourth Rectification

Plato said: "If you dissolve the spirit without [adding another] ingredient and save a quarter of it, the works are better".
This statement shows the first meaning according to one of two ways: either the school of Homer the Poet – and it said "I increase!" – or Porphyry – and he said "I decrease!" Ǧābir, may God be pleased with him, said: "And to know what we say you need to search for this purpose of the search. We found out through trial, as we compared it, and we will show you the sense in it, if God wills, may He be praised".
As for Homer's school, it said: "The tetrad is ten because we would find the amount of ten in four". But Porphyry's school said: "We say 'through the natures only', and the natures are four, there is no fifth among them. So how would there be a tenth among them?" If this must be so, he who said all [this] is right. So Homer said "ten" and "I increase!", but Porphyry said "four" and "I decrease". And the one that Plato followed in the procedure of this chapter of his and his foundations is [based] only on gradation and increase, not on decrease, and is the one who said "Homer is better in this section".
It is necessary to know from the comments of this statement that the procedure of the first chapter must be [conducted] before the dissolution and in the first work. According to what Plato thinks, it is right when the dissolution is completed *seven in seven*.
Ǧābir, may God be pleased with him, said: "And this is that the meaning of this long statement is that if the tetrad is every seven days of the days for which the procedure that this mercury is dissolved in continues for you, it must be either in ten days, according to Homer's opinion, or in four days, according to Porphyry's opinion. The meaning of ten in four is indeed: one plus two equals three, and three plus three equals six, and six plus four equals ten. You therefore need the fire [on] the cucurbit for every [period of] ten days and you let it cool on the eleventh, or [for] nine, and you let it cool on the tenth, and the first [method] is better. It is best in the work in which you run the procedure for seven [days] and let it cool on the eighth. You let cool on the fifth [day] what you process in [a period of] four.
Know this, your chapter aspires to it and a wonder will occur for the one who considers it – it belongs to the sciences of the soul in the first procedure. Peace.

التصحيح الرابع

قال فلاطن إذا حللت الروح بغير داخل وحفظت الرباعيه كان أفضل الأعمال.
فدلّ هذا القول على المعنى الأوّل على أحد الطريقين: إمّا طائفة أمورس الشاعر فقالت أزيد وإمّا فرفيريوس فقال أنقص. قال جابر رضى الله عنه وتحتاج أن تبحث عن ذلك غاية البحث لتعلم ما نقول. وقد وجدنا في المحنة كما قايسنا عليه ونحن نريك الوجه فيه إن شاء الله تعالى. أمّا طائفة أمورس فقالت الرباعية عشرة لأنّا قد نجد مبلغ العشرة في الأربعة. وأمّا طائفة فرفيريوس فقالت لا نقول إلّا بالطبائع والطبائع أربعة وليس فيها خامس فكيف يكون فيها عاشر. وإذا وجب أن يكون ذلك كذلك فالذي قال الجميع صواب. أمّا أمورس فقال عشرة وقال أزيد وأمّا فرفيريوس فقال اربعة وقال أنقص. والذي سلك فيه فلاطن في تدبير بابه هذا وأركانه. انّما كان على تدريج وزيادة لا على نقصان والذي قال أمورس الأجود في هذا الفصل. وقد يجب أن يعلم من حواشي هذا الكلام أنّ تدبير الباب الأوّل يجب أن يكون قبل الحلّ وفي أوّل العمل. إذا إنتهى الحلّ سبعا في سبع على ما يرأه فلاطن وهو الصواب. قال جابر رضى الله عنه وذلك أنّ معنى هذا الكلام الطويل أنّ الرباعية إن تكون كلّ سبعة أيّام من الأيّام التي مضى لك فيها التدبير حتّى إنحلّ هذا الزيبق فيه. يجب أن يكون في إمّا عشرة أيّام على رأي أمورس وإمّا في أربعة أيّام على رأي فرفيريوس. ومعنى عشرة في أربعة إنّما هو واحد وإثنين فذلك ثلثة وثلثة فذلك ستّة وأربعة فذلك عشرة.
وإذا ألزمت النار القرعة في كلّ عشرة أيّام وبرّدت الحادي عشر أو تسعة وبرّدت اليوم العاشر والأوّل هو الأجود. كان أفضل في العمل في الذي دبّرت في سبعة وبرّد الثامن. ومن الذي تدبّر في أربعة وتبرّد في الخامس. فأعلم ذلك وقسّ عليه بابك فإنّه يكون عجبا لمن فكّر فيه وهو من العلوم النفيسة هذا في تدبير الأوّل والسلام

Bibliography

Sources

Collected Sources

Kraus 1935: Paul Kraus (ed.), *Jābir ibn Ḥayyān. Essai sur l'histoire des idées scientifiques dans l'Islam*, vol. 1: *Textes choisis*. Paris: G. P. Maisonneuve; Cairo: al-Khanji, 1935.

Authors and Works

Aetius Arabus, *Placita philosophorum*: Hans Daiber (ed.), *Aetius Arabus. Die Vorsokratiker in arabischer Überlieferung* (Akademie der Wissenschaften und der Literatur, Veröffentlichungen der Orientalischen Kommission 33). Wiesbaden: F. Steiner, 1980.

Ǧābir ibn Ḥayyān, *Kitāb al-ḫawāṣṣ*: ed. Kraus 1935: 224–332.

Ǧābir ibn Ḥayyān, *Kitāb al-mīzān al-ṣaġīr*: ed. Kraus 1935: 426–459.

Ǧābir ibn Ḥayyān, *Kitāb al-taǧmiʿ*: ed. Kraus 1935: 341–391.

Ǧābir ibn Ḥayyān, *Kitāb iḫrāǧ mā fī'l-quwwa ilā'l-fiʿl*: ed. Kraus 1935: 1–95.

Homerus, *Ilias*: Martin L. West (ed.), *Homeri Ilias*, 2 vols (Bibliotheca Scriptorum Graecorum et Romanorum Teubneriana). Stuttgart: Teubner (vol. 1); Munich: Saur (vol. 2), 1998–2000.

Ḥunayn ibn Isḥāq, *Ādāb al-falāsifa*: ʿAbd al-Raḥmān Badawī (ed.), *Ḥunayn ibn Isḥāq, Ādāb al-falāsifa. Iḫtaṣarahu Muḥammad ibn ʿAlī ibn Ibrāhīm ibn Aḥmad ibn Muḥammad al-Anṣārī*. Kuwait: Publications de l'Institut des manuscrits arabes – ALECSO, 1985.

Ibn Abī Uṣaybiʿah, *ʿUyūn al-anbāʾ fī ṭabaqāt al-aṭibbāʾ*: Emilie Savage-Smith, Simon Swain, and Geert Jan van Gelder (ed. and Engl. transl.), *A Literary History of Medicine*. Leiden and Boston, MA: Brill, 2020, https://doi.org/10.1163/37704_0668IbnAbiUsaibia.Tabaqatalatibba.lhom-tr-eng1.

Ibn al-Qifṭī, *Taʾrīḫ al-ḥukamāʾ*: Julius Lippert (ed.), *Ibn al-Qifṭī's Taʾrīḫ al-Ḥukamāʾ*. Leipzig: Dieterich, 1903.

al-Mubaššir ibn Fātik, *Muḫtār al-ḥikam wa-maḥāsin al-kalim*: ʿAbd al-Raḥmān Badawī (ed.), *Al-Mubaššir ibn Fātik, Muḫtār al-ḥikam wa-maḥāsin al-kalim*. Beirut: al- Muʾassasah al-ʿArabīyah lil-Dirāsāt wa-al-Našr, 1980.

Plato, *Respublica*: Simon R. Slings (ed.), *Platonis Rempublicam* (Oxford Classical Texts). Oxford: Oxford University Press, 2003.
Engl. transl.: Paul Shorey, *Plato, The Republic*, 2 vols (Loeb Classical Library 237, 276). Cambridge, MA: Harvard University Press; London: W. Heinemann, 1930–1935.

Plato, *Theaetetus*: John Burnet (ed.), *Platonis opera*, vol. 1. Oxford: Clarendon Press, 1900.
Engl. transl.: Harold N. Fowler, *Plato, Theaetetus; Sophist* (Loeb Classical Library 123). Cambridge, MA: Harvard University Press; London: W. Heinemann, 1921, pp. 1–257.

Plato, *Timaeus*: John Burnet (ed.), *Platonis opera*, vol. 4. Oxford: Clarendon Press, 1902.
Engl. transl.: Robert G. Bury, *Plato, Timaeus; Critias; Cleitophon; Menexenus; Epistles* (Loeb Classical Library 234). Cambridge, MA: Harvard University Press; London: W. Heinemann, 1929, pp. 1–253.

[Plutarchus], *Placita philosophorum*: Hermann Diels (ed.), *Doxographi Graeci*. Berlin: G. Reimer, 1879, pp. 267–444.

Porphyrius, *De principiis et materia* (Syriac): Yury Arzhanov (ed. and Engl. transl.), *Porphyry*, On Principles and Matter. *A Syriac Version of a Lost Greek Text with an English Translation, Introduction, and Glossaries* (Scientia Graeco-Arabica 34). Berlin and Boston, MA: De Gruyter, 2021.

Porphyrius, *Quaestiones Homericae ad Iliadem*: John A. MacPhail, Jr. (ed. and Engl. transl.), *Porphyry's* Homeric Questions *on the* Iliad. *Text, Translation, Commentary* (Texte und Kommentare, Eine altertumswissenschaftliche Reihe 36). Berlin and New York: De Gruyter, 2011.

Literature

Arzhanov 2014: Yury Arzhanov, "Das Florilegium in der Hs. Vat. Sir. 135 und seine griechisch-arabischen Parallelen", in Martin Tamcke and Sven Grebenstein (eds), *Geschichte, Theologie und Kultur des syrischen Christentums* (Göttinger Orientforschungen, I. Reihe, Syriaca 46). Wiesbaden: Harrassowitz, 2014, pp. 41–56.

Arzhanov 2021: Yury Arzhanov, *Porphyry,* On Principles and Matter. *A Syriac Version of a Lost Greek Text with an English Translation, Introduction, and Glossaries* (Scientia Graeco-Arabica 34). Berlin and Boston, MA: De Gruyter, 2021.

Bagordo 2012: Andreas Bagordo, "Homer (Homerus)", in Christine Walde (ed.), *Brill's Encyclopedia of the Ancient World: New Pauly, Supplements 5: The Reception of Classical Literature*, transl. Duncan Smart and Matthijs H. Wibier. Leiden and Boston, MA: Brill, 2012, pp. 154–178.

Belfiore 1984: Elizabeth Belfiore, "A Theory of Imitation in Plato's *Republic*", *Transactions of the American Philological Association* 114 (1984), pp. 121–146.

Benevich 2022: Fedor Benevich, "A Case Study in Arabic Doxography: Šahrastānī's Account of Pythagoras and its Ismāʿīlī Background", in Andreas Lammer and Mareike Jas (eds), *Received Opinions: Doxography in Antiquity and the Islamic World* (Philosophia Antiqua 160). Leiden and Boston, MA: Brill, 2022, pp. 354–402.

Bhayro 2019: Siam Bhayro, "The Reception of Galen in the Syriac Tradition", in Bouras-Vallianatos and Zipser 2019a, pp. 163–178.

Blank 2017: David Blank, "Ammonius", in Edward N. Zalta (ed.), *The Stanford Encyclopedia of Philosophy*. Winter 2017 Edition, https://plato.stanford.edu/archives/win2017/entries/ammonius/.

Bonazzi et al. 2007: Mauro Bonazzi, Carlos Lévy, and Carlos Steel (eds), *A Platonic Pythagoras: Platonism and Pythagoreanism in the Imperial Age* (Monothéismes et philosophie 10). Turnhout: Brepols, 2007.

Bouras-Vallianatos 2019: Petros Bouras-Vallianatos, "Galen in Late Antique Medical Handbooks", in Bouras-Vallianatos and Zipser 2019a, pp. 38–61.

Bouras-Vallianatos and Zipser 2019a: Petros Bouras-Vallianatos and Barbara Zipser (eds), *Brill's Companion to the Reception of Galen* (Brill's Companions to Classical Reception 17). Leiden and Boston, MA: Brill, 2019.

Bouras-Vallianatos and Zipser 2019b: Petros Bouras-Vallianatos and Barbara Zipser, "Introduction", in Bouras-Vallianatos and Zipser 2019a, pp. 1–7.

Chandelier 2020: Joel Chandelier, "Medicine and Philosophy", in Henrik Lagerlund (ed.), *Encyclopedia of Medieval Philosophy: Philosophy between 500 and 1500*. Second edition. Dordrecht: Springer, 2020, pp. 1138–1146.

Chiaradonna 2016: Riccardo Chiaradonna, "Porphyry and the Aristotelian Tradition", in Andrea Falcon (ed.), *Brill's Companion to the Reception of Aristotle in Antiquity* (Brill's Companions to Classical Reception 7). Leiden and Boston, MA: Brill, 2016, pp. 321–340.

Collins 2008: Derek Collins, "The Magic of Homeric Verses", *Classical Philology* 103.3 (2008), pp. 211–236.

Daiber 1980: Hans Daiber, *Aetius Arabus. Die Vorsokratiker in arabischer Überlieferung* (Akademie der Wissenschaften und der Literatur, Veröffentlichungen der Orientalischen Kommission 33). Wiesbaden: F. Steiner, 1980.

D'Ancona 2020: Cristina D'Ancona, "Porphyry, Arabic", in Henrik Lagerlund (ed.), *Encyclopedia of Medieval Philosophy: Philosophy between 500 and 1500*. Second edition. Dordrecht: Springer, 2020, pp. 1576–1584.

Delatte 1915: Armand Delatte, *Études sur la littérature pythagoricienne* (Bibliothèque de l'École des Hautes Études, Sciences historiques et philologiques 217). Paris: É. Champion, 1915.

Delva 2017: Thijs Delva, "The Abbasid Activist Ḥayyān al-ʿAṭṭār as the Father of Jābir b. Ḥayyān: An Influential Hypothesis Revisited", *Journal of Abbasid Studies* 4.1 (2017), pp. 35–61.

De Piano 2014: Piera De Piano, "Porfirio e Proclo lettori di Omero. Esempi di esegesi omerica nella tradizione neoplatonica", *Incidenza dell'antico* 12 (2014), pp. 155–174.

Festugière 1963: André-Jean Festugière, "Modes de composition des Commentaires de Proclus", *Museum Helveticum* 20.2 (1963), pp. 77–100.

Forster 2018: Regula Forster, "Jābir b. Ḥayyān", in Kate Fleet et al. (eds), *Encyclopaedia of Islam Three Online*. Brill, 2018, https://doi.org/10.1163/1573-3912_ei3_COM_32665.
Garofalo 2019: Ivan Garofalo, "Galen's Legacy in Alexandrian Texts Written in Greek, Latin, and Arabic", in Bouras-Vallianatos and Zipser 2019a, pp. 62–85.
Ǧumʿa et al. 2020: Nashwa Ǧumʿa, Iman M. Hamed, and Peter E. Pormann, "Arabic Translation of Galen's *On the Affected Parts* and the Greek Textual Tradition", *The Classical Quarterly* 70.1 (2020), pp. 397–409.
Hadot 2015: Ilsetraut Hadot, *Athenian and Alexandrian Neoplatonism and the Harmonization of Aristotle and Plato*, transl. Michael Chase (Studies in Platonism, Neoplatonism, and the Platonic Tradition 18). Leiden and Boston, MA: Brill, 2015.
Halleux 1979: Robert Halleux, *Les textes alchimiques* (Typologie des sources du Moyen Âge occidental 32). Turnhout: Brepols, 1979.
Helmig and Steel 2021: Christoph Helmig and Carlos Steel, "Proclus", in Edward N. Zalta (ed.), *The Stanford Encyclopedia of Philosophy*. Fall 2021 Edition, https://plato.stanford.edu/archives/fall2021/entries/proclus/.
Horky 2013: Phillip Sidney Horky, *Plato and Pythagoreanism*. Oxford: Oxford University Press, 2013.
Izdebska 2020: Anna Izdebska, "Tetractys: A Pythagorean *terminus technicus* in the Process of Translation from Greek into Arabic", *Intellectual History of the Islamicate World* 9.1–2 (2020), pp. 140–168.
Johnston 2008: Sarah Iles Johnston, "Animating Statues: A Case Study in Ritual", *Arethusa* 41.3 (2008), pp. 445–477.
Kalvesmaki 2013: Joel Kalvesmaki, *The Theology of Arithmetic: Number Symbolism in Platonism and Early Christianity* (Hellenic Studies 59). Washington, DC: Center for Hellenic Studies, 2013.
Karamanolis 2006: George E. Karamanolis, *Plato and Aristotle in Agreement? Platonists on Aristotle from Antiochus to Porphyry*. Oxford: Oxford University Press, 2006.
Kraemer 1956: Jörg Kraemer, "Arabische Homerverse", *Zeitschrift der Deutschen Morgenländischen Gesellschaft* 106 (1956), pp. 259–316.
Kraus 1942–1943: Paul Kraus, *Jābir ibn Ḥayyān, Contribution à l'histoire des idées scientifiques dans l'Islam*, 2 vols (Mémoires de l'Institut d'Égypte 44–45). Cairo: Institut français d'archéologie orientale, 1942–1943.
Lamberton 1986: Robert Lamberton, *Homer the Theologian: Neoplatonist Allegorical Reading and the Growth of the Epic Tradition* (The Transformation of the Classical Heritage 9). Berkeley, CA: University of California Press, 1986.
Lorenz 2024: Hendrik Lorenz, "Ancient Theories of Soul", in Edward N. Zalta and Uri Nodelman (eds), *The Stanford Encyclopedia of Philosophy*. Summer 2024 Edition, https://plato.stanford.edu/archives/sum2024/entries/ancient-soul/.
Ludwich 1917–1918: Arthur Ludwich, "Plutarch über Homer", *Rheinisches Museum für Philologie* 72 (1917–1918), pp. 537–593.
Martelli 2019: Matteo Martelli, "Galen in the Late Antique, Byzantine, and Syro-Arabic Alchemical Traditions", in Bouras-Vallianatos and Zipser 2019a, pp. 577–593.
Martelli 2020: Matteo Martelli, "A Syriac Epitome of Galen's *On Simple Drugs*, Books 9–11: Text and Translation", *Galenos* 14 (2020), pp. 131–158.
Mavroudi 2020: Maria Mavroudi, "Homer in Greece from the End of Antiquity 1: The Byzantine Reception of Homer and His Export to Other Cultures", in Corinne Ondine Pache (ed.), *The Cambridge Guide to Homer*. Cambridge: Cambridge University Press, 2020, pp. 444–472.
Muftić 2018: Edin Muftić, "Homer: An Arabic Portrait", *Pro Tempore* 13 (2018), pp. 21–41.
Neuburger 1911: Max Neuburger, *Geschichte der Medizin*, vol. 2.1. Stuttgart: F. Enke, 1911.
Overwien 2003: Oliver Overwien, "Ḥunayn b. Isḥāq, *Ādāb al-falāsifa*: Griechische Inhalte in einer arabischen Spruchsammlung", in Rosa Maria Piccione and Matthias Perkams (eds), *Selecta colligere*, 1. Alessandria: Edizioni dell'Orso, 2003, pp. 95–115.
Overwien 2011: Oliver Overwien, "Das Bild des Kynikers Diogenes in griechischen, syrischen und arabischen Quellen", *Philologus* 155 (2011), pp. 92–124.

Overwien 2016: Oliver Overwien, "Eine spätantik-alexandrinische Vorlesung über Galens *De sectis* in al-Ruhāwīs *Bildung des Arztes* (*Adab al-ṭabīb*)", *Galenos* 10 (2016), pp. 195–206.

Overwien 2018: Oliver Overwien, "Der medizinische Unterricht der Iatrosophisten in der 'Schule von Alexandria' (5.–7. Jh. n. Chr.): Überlegungen zu seiner Organisation, seinen Inhalten und seinen Ursprüngen (zweiter Teil)", *Philologus* 162.2 (2018), pp. 265–290.

Overwien 2019: Oliver Overwien, *Medizinische Lehrwerke aus dem spätantiken Alexandria. Die* Tabulae Vindobonenses *und* Summaria Alexandrinorum *zu Galens* De sectis (Scientia Graeco-Arabica 24). Berlin and Boston, MA: De Gruyter, 2019.

Pietruschka 2014: Ute Pietruschka, "Gnomologia: Syriac and Arabic Traditions", in Houari Touati (ed.), *Encyclopedia of Mediterranean Humanism*. Spring 2014, http://www.encyclopedie-humanisme.com/?Gnomologia.

Pietruschka 2015: Ute Pietruschka, "Some Observations about the Transmission of Popular Philosophy in Egyptian Monasteries after the Islamic Conquest", in Damien Janos (ed.), *Ideas in Motion in Baghdad and Beyond: Philosophical and Theological Exchanges between Christians and Muslims in the Third/Ninth and Fourth/Tenth Centuries* (Islamic History and Civilization 124). Leiden and Boston, MA: Brill, 2015, pp. 81–108.

Pormann 2018: Peter E. Pormann, "Arabo-Islamic Tradition", in Peter E. Pormann (ed.), *The Cambridge Companion to Hippocrates* (Cambridge Companions to Philosophy). Cambridge: Cambridge University Press, 2018, pp. 340–361.

Raggetti 2015: Lucia Raggetti, "The 'Science of Properties' and Its Transmission", in J. Cale Johnson (ed.), *In the Wake of the Compendia: Infrastructural Contexts and the Licensing of Empiricism in Ancient and Medieval Mesopotamia* (Science, Technology, and Medicine in Ancient Cultures 3). Boston, MA and Berlin: De Gruyter, 2015, pp. 159–176.

Reid and DeLong 2018: Heather L. Reid and Jeremy C. DeLong (eds), *The Many Faces of* Mimēsis: *Selected Essays from the 2017 Symposium on the Hellenic Heritage of Western Greece* (The Heritage of Western Greece). Sioux City, IA: Parnassos Press, 2018.

Ruska 1923: Julius Ruska, "Über das Schriftenverzeichnis des Ǵābir ibn Ḥajjān und die Unechtheit einiger ihm zugeschriebenen Abhandlungen", *Archiv für Geschichte der Medizin* 15 (1923), pp. 53–67.

Sezgin 1971: Fuat Sezgin, *Geschichte des arabischen Schrifttums*, vol. 4: *Alchimie – Chemie, Botanik – Agrikultur bis ca. 430 H.* Leiden: Brill, 1971.

Sheppard 1982: Anne Sheppard, "Proclus' Attitude to Theurgy", *The Classical Quarterly* 32.1 (1982), pp. 212–224.

Steinschneider 1862: Moritz Steinschneider, *Zur pseudepigraphischen Literatur, insbesondere der geheimen Wissenschaften des Mittelalters.* Berlin: [Rosenthal & Comp.], 1862.

Steinschneider 1904: Moritz Steinschneider, "Zur alchimistischen Literatur der Araber", *Zeitschrift der Deutschen Morgenländischen Gesellschaft* 58.2 (1904), pp. 299–315.

Ullmann 1970: Manfred Ullmann, *Die Medizin im Islam* (Handbuch der Orientalistik, 1. Abt.: Der Nahe und der Mittlere Osten, Ergänzungsband 6.1). Leiden: Brill, 1970.

Ullmann 1972: Manfred Ullmann, *Die Natur- und Geheimwissenschaften im Islam* (Handbuch der Orientalistik, 1. Abt.: Der Nahe und der Mittlere Osten, Ergänzungsband 6.2). Leiden: Brill, 1972.

Wellmann 1895: Max Wellmann, *Die pneumatische Schule bis auf Archigenes, in ihrer Entwicklung dargestellt* (Philologische Untersuchungen 14). Berlin: Weidmann, 1895.

Wildberg 2024: Christian Wildberg, "Olympiodorus", in Edward N. Zalta and Uri Nodelman (eds), *The Stanford Encyclopedia of Philosophy*. Fall 2024 Edition, https://plato.stanford.edu/archives/fall2024/entries/olympiodorus/.

Zhmud 2017: Leonid Zhmud, "Греческая арифмология: Пифагор или Платон?" *ΣΧΟΛΗ* 11.2 (2017), pp. 428–459.
Published in Engl. as "Greek Arithmology: Pythagoras or Plato?", in Almut-Barbara Renger and Alessandro Stavru (eds), *Pythagorean Knowledge from the Ancient to the Modern World: Askesis, Religion, Science* (Episteme 4). Wiesbaden: Harrassowitz, 2016, pp. 321–346.

Athanasios Rinotas

Western Medieval Alchemical Influences in the *Anonymous of Zuretti*: With a Special Focus on the Case of Albertus Magnus

A couple of decades ago, historian of alchemy Stanton J. Linden argued – an assessment that remains valid today – that the study of alchemy is characterized by interdisciplinarity and, therefore, the understanding of alchemy does not depend only on the history of science but also on other disciplines such as philosophy, literature, history of art, etc. In this regard, it would be more appropriate to distinguish between *alchemies* rather than maintaining a *unitary and singular explanation of alchemy*.[1] This kind of methodological approach has significantly helped us organize and better study the alchemical systems met in different historical periods and thus understand their specific characteristics. In addition, we are now in a position to compare different alchemical traditions and opine either on their similarities or on their differences. With respect to the latter, modern scholarship has primarily focused on the relationship between medieval Latin and Arabic alchemy, while the relationship between medieval Latin and Byzantine alchemy has not been met with a similar interest.[2]

1 Bridging the Two Traditions

A reason which could justify the aforementioned "neglect" is to be found in the work of Maria Mavroudi, who has cogently shown that Byzantium tends to be neglected by modern scholarship and, thus, it is mostly absent in volumes dedicated either to medieval philosophy or medieval science.[3] That said, one can notice a similar attitude in books and general articles on medieval alchemy where, once again, Byzantium is the great absent.[4] This, however, does not mean that the relation between medieval Latin and Byzantine alchemy is completely unstudied. For instance, Lynn Thorndike has drawn some connections between the two traditions – even though his article is focused on the manuscript tradition – while Robert Halleux and Paul Mayvaert, in their study on

1 Linden 2003: 3–4.

2 For the relation between medieval Latin and Arabic alchemy, see Halleux 1996: 886–902 and Moureau 2020: 87–141.

3 Mavroudi 2015: 28–59.

4 For instance, see Newman 2013: 385–403; Principe 2013; Calvet 2018.

https://doi.org/10.1515/9783111589619-011

the *Mappae clavicula*, have shown that this Latin text has strong ties with the Greek Alchemical Corpus.[5]

This paper aims to further contribute to the aforementioned "neglected" relation. In particular, I will focus on the *Anonymous of Zuretti* and how it has been influenced by the alchemical thought of Albertus Magnus. To that end, I will first present in brief the alchemy of the *Anonymous of Zuretti* and that of Albertus Magnus, and then I will proceed by demonstrating how Albertus has influenced the text of *Anonymous*. In fact, my argument will follow the three topics met in *Anonymous*: (1) the art versus nature debate, which is apparent not only in the work of Albertus but also in the work of other medieval Latin scholars such as Roger Bacon and Thomas Aquinas; (2) the Albertian version of the three humidities of sulphur; and (3) the use of the Aristotelian *symbola* in Albertus' alchemy and their possible echoes in *Anonymous*.

Finally, my reading and association of the two authors are heavily based on philosophy, and thus I subscribe to the modern scholarly tradition, which deems that alchemy has used philosophical tenets and principles in order to entrench itself on "scientific" foundations.[6] In that way, it will also become clear that both medieval Latin and Byzantine "alchemists" used philosophy as an authoritative means to present this field of knowledge as legitimate.

2 The *Anonymous of Zuretti* and Albertus Magnus on Alchemy

Since its first edition by Carlo Oreste Zuretti,[7] *Anonymous* has attracted little attention in modern scholarship. Apart from the now-standard edition and a relevant study, both by Andrée Colinet,[8] one can hardly find any other scholarly work dedicated to it. Also, in the general accounts of alchemy, the text is either treated in passing or is completely absent.[9] However, despite this obvious "neglect", the aforementioned Byzantine text is worth examining since it offers us ample evidence regarding the relation between medieval Latin and Byzantine alchemy, whilst it also stands as a "bridge" between the two traditions.

5 Thorndike 1964: 1–48, Halleux and Meyvaert 1987: 7–58.

6 For the medieval Latin tradition, see especially Crisciani and Gagnon 1980. For the Byzantine tradition, see Magdalino and Mavroudi 2006: 13; Mavroudi 2006: 47; and Tatakis 1977: 225.

7 *Anonymi De arte metallica* (ed. Zuretti 1930).

8 The edition of the *Anonymus "Zuretti"* is Colinet 2000; the general study on the text is Colinet 2005.

9 For instance, Merianos and Viano treat the *Anonymous of Zuretti* in passing, while Martelli and Papathanassiou do not refer to it at all; see Merianos 2017: 236; Viano 2018: 474; Martelli 2016; 2019; Papathanassiou 2020: 486–495.

The *Anonymous of Zuretti* is a text that contains Latin alchemical texts translated into Greek. According to Forrai, the Latin-to-Greek translations were carried out in two phases; the first one was the period before the tenth century, and the second one was the period after 1261.[10] It is the second phase which is characterized by the translation and proliferation of many medieval Latin texts. In particular, it should be stated that the Byzantines felt that the Greek language and its literature were superior to the Latin ones. Thus, they considered the Latin representatives to belong to a subservient culture.[11] Likewise, the Byzantines did not hold the science of the Latins in high esteem, and, in general, they treated the science of the Westerners with great suspicion.[12] The situation changed after the conquest of Constantinople in 1204 by the Latins. The revival of learning after the Byzantine reconquest of the city in 1261 is still widely known as the "Palaeologan renaissance", during which the Byzantine scholars – as Efthymios Nicolaidis notes – reread the ancient sources and came into close communication with other traditions, such as the Arabic and the western Latin.[13] With respect to the medieval Latin tradition, Maximus Planudes (ca. 1255 – ca. 1305) translated Donatus' *Ars grammatica* and texts of Ovid, while he also translated St. Augustine's *De trinitate* and Thomas Aquinas' *Summa theologica*. Another Byzantine scholar of the same epoch, Manuel Holobolus (ca. 1245–1284), translated Boethius' logical works, and in the next century, Demetrius and Prochorus Cydones translated theological works of St. Augustine and Thomas Aquinas.[14] As Colinet has noticed, a similar efflorescence of alchemical texts had occurred during that period, since, apart from the *Anonymous of Zuretti*, others are also to be found, such as *Holkhamicus graecus* 290, the Byzantine Greek version of (Pseudo-) Albertus Magnus' *Semita recta*, and a fragment of Arnaldus de Villanova preserved in several manuscripts.[15]

Apparently, the text of *Anonymous* belongs to the second phase of the Latin-to-Greek translations and is mostly concerned with alchemical recipes. It is heavily influenced by medieval Latin authors such as Roger Bacon, (Pseudo-) Albertus Magnus, and Arnaldus de Villanova. According to Colinet, the oldest manuscript is dated to 1378 and was written in Calabria, while the original text was probably written around 1300 in southern Italy.[16] The fact that the first manuscript of *Anonymous* is of Calabrian origin can be justified and explicated by three points: firstly, up to the turn of the thirteenth century, the dominant language in Calabria was Greek, while after 1231, the Byzantine law began

10 Forrai 2021: 181.
11 Ebbesen 2002: 25.
12 Fisher 2012: 212.
13 Nicolaidis 2011: 109. On this cultural phenomenon, see, e.g., Fryde 2000. For skepticism concerning the appropriateness of the term "renaissance" with regard to Byzantium, see Cameron 2014: 20–21 and 121n25.
14 Ebbesen 2002: 25–26; Forrai 2021: 183.
15 Colinet 2000: XIV.
16 Colinet 2000: XXV and LXXXIV–LXXXVI.

to be abandoned and Latin started to gain more ground in the region.[17] By the end of the fourteenth century, Calabria was bilingual (Greek–Latin) and therefore offered the proper linguistic environment for the text of *Anonymous* to occur. Secondly, it is well-documented that Calabria had rich metal deposits of iron and copper since ancient times, and the region of Reggio was particularly rich in gold and iron.[18] This "metallic tradition" might have stood as a substantial reason for raising interest in alchemy. Finally, the region of Calabria was known for its literary and manuscript tradition since the copy center of Rossano retained an ongoing interest in copying manuscripts, albeit mostly religious.[19]

Concerning the content of *Anonymous*, it has already been mentioned that the text is heavily influenced by medieval Latin sources.[20] The extent of this influence is so great that it led Colinet to admit that there is barely any originality regarding the alchemical ideas contained in the work of *Anonymous*.[21] *Anonymous* generally adheres to the sulphur–mercury theory, which was very popular during the thirteenth century in medieval Latin Europe. According to *Anonymous*, all metals are composed of different ratios of sulphur and mercury; at the same time, their purity is also affected by the time needed for the metals to be formed. If the ratio of the two aforementioned ingredients is not right and the maturation time of metals is problematic and insufficient, then the metals which have been created are considered to be "sick" and therefore the alchemist intervenes in order to imitate the work of nature and "rectify" the problematic metals.[22] To that end, the alchemists employ elixirs which are used during the transmutation of metals. Apart from this brief presentation of *Anonymous*' general alchemy, a great – if not the greatest – deal of chapters is dedicated to alchemical substances. In fact, *Anonymous* divides alchemical substances into metals, spirits, and "auxiliaries", that is, substances that act as catalysts. In the category of metals, we find the six metals of gold, silver, iron, copper, tin, and lead, whereas spirits are divided into three categories, that is, mineral, animal, and vegetal ones.[23] With respect to mineral spirits, *Anonymous* names four: sulphur, mercury, arsenic, and sal ammoniac, while in the animal ones, he includes eggs, hair, and blood. However, he remains silent when it comes to vegetal spirits. Finally, the "auxiliaries", which are described as βοηθικά ("helpful substances"), contain such substances as borax, oil, vitriol, alums, etc., and they were used to facilitate and accelerate a transmutation. Alongside these substances, *Anonymous* provides us with a great number of alchemical recipes concerned with processing these substances.

---

17 Peters-Custot 2021: 243.
18 Noyé 2021: 439–442.
19 Zanetti 1999: 134–136.
20 Colinet 2000: XXXVII–LXVII.
21 Colinet 2005: 140.
22 Colinet 2005: 136.
23 Colinet 2000: LXXXVII–LXXXIX; 2005: 136.

Thus he offers us a description of such alchemical procedures as purification, sublimation, and calcination.

Turning to Albertus Magnus (ca. 1200–1280), the Dominican master was active during the thirteenth century; therefore, his alchemy is dated some decades before the work of *Anonymous*.[24] The alchemical ideas of Albertus are scattered all over his work, but the main book where he discusses alchemy is his *De mineralibus*. The *De mineralibus* was probably written between 1251 and 1254 and aimed to speak about minerals and thus fill the gap that existed in the Aristotelian corpus.[25] In general, Albertus regarded alchemy as something more than a mere art, and he was favorable to the possibility of transmutation of metals. Just like *Anonymous*, Albertus believed that metals were composed of sulphur and mercury, while the wrong ratio and impurities of these two were responsible for the "sickness" of metals. In this regard, the Dominican master considered skillful alchemists to act like skillful physicians, whose task was to purify the amount of sulphur and mercury contained within a metal and rearrange their ratio to accomplish a proper transmutation.[26] After performing this task, the alchemist should strengthen the elemental and celestial powers of the material, and then nature would perform the process and not the art of alchemy. Albertus' emphasis on the role of nature in alchemical transmutation was probably the reason for not having given a prominent place to elixir in his alchemy. In fact, Albertus often associated elixirs with a superficial kind of transmutation that was not able to bring about another specific form and thus did not seem to hold elixirs in high regard.[27] It should be stated, however, that the Dominican master showed a particular sensitivity towards the problem of the occurrence of a new specific form during alchemical transmutation, and he was more prone to accepting a transmutation in terms of matter rather than in terms of form.[28] Besides these general characteristics of his alchemy, Albertus, in his *De mineralibus*, gives us an account of the accidental properties of metals. At the same time, he also provides us with an account that is concerned with the characteristics of each metal. The last chapter of his book is dedicated to the discussion of the "intermediaries", that is, minerals that seem to be between stones and metals. The latter category of minerals is very close to what *Anonymous* has described as "auxiliaries".

---

24 Partington 1937: 3–20; Halleux 1982: 57–80; Anawati 1981: 126–133; Kibre 1980: 187–202.

25 Weisheipl 1980: 13–51.

26 Albertus Magnus, *De mineralibus*, Lib. III, tr. i, cap. 9 (ed. Borgnet 1890–1899: 5, col. 71a–b); Engl. transl. Wyckoff 1967: 177–179.

27 Rinotas 2022a: 173–193.

28 Rinotas 2022b: 57–82.

3 Albertian Influences in the *Anonymous of Zuretti*

Trying to establish a connection between the text of *Anonymous* and Albertus Magnus is not a novel attempt. In fact, Colinet has already portrayed the *Semita recta* as a source for *Anonymous*' text, even though the authenticity of this Albertian work is seriously questionable.[29] The fact that the *Semita recta* is likely not a genuine text of Albertus makes the connection between the two parts fragile, since Colinet pays little attention to the relation between the text of *Anonymous* and the genuine texts of Albertus. Yet, pursuing the endeavor of strengthening the *liaison* between the two authors has both a historical and a contextual basis. On the one hand, Albertus was a Dominican, and it has already been well attested that the Dominican order was actively present in the intellectual life of Byzantium. According to Ebbesen, the Dominicans were settled in Constantinople in 1232, and it is probably due to them that the works of Vincent of Beauvais (*Speculum doctrinale*) and Thomas Aquinas were translated and disseminated during the thirteenth and fourteenth centuries.[30] Given this "Dominican influence", it would not be too perilous to surmise that the Albertian works might have started getting known during the aforementioned centuries. Besides, the Greek translation of the *Semita recta*, albeit written around 1462, favors such a hypothesis. In addition, Colinet has examined the two Greek texts for possible influences or a relation between each other, but she found that they are not connected in any way.[31] On the other hand, it should be noted that Colinet creates *liaisons* between the text of *Anonymous* and Albertus' *De mineralibus*, even though to a lesser extent. That said, the rest of this paper aims to further connect the text of *Anonymous* and the *De mineralibus*. These connections are indicative and far from being exhaustive. In what follows, I focus on the subjects of (1) art versus nature debate, (2) the three humidities of sulphur, and (3) the role of *symbola* in alchemy.

Art versus Nature Debate

Aristotelian philosophy played a crucial role in associating alchemy with natural procedures and connecting it to the general notion of nature. In particular, the theory of the two exhalations, which is contained in Aristotle's *Meteorologica*, provided the proper theoretical framework for sulphur–mercury theory, while in *Physica*, one may find Aristotle's claim that art may successfully reproduce nature's products or even surpass them.[32] The latter proposition stood as a legitimate argument for alchemists in order to express themselves on the validity of alchemy and its potential accomplishments.

29 Colinet 2000: XLVIII–LII; regarding *Semita recta*'s authenticity, see p. XLVIIIn166; Newman 1991: 193–194; Calvet 2012: 129–133.

30 Ebbesen 2002: 26; Forrai 2021: 183.

31 Colinet 2000: LI–LII.

32 *Meteorologica* 3.6.378a26; *Physica* 2.8.199a15–17.

Yet, a similar dictum was found in Avicenna's *De congelatione et conglutinatione lapidum*, which triggered a fervent discussion on art and nature in the medieval Latin world. Avicenna wrote this text between 1020 and 1027, and it was translated into Latin by Alfred of Sareshel around 1200. Alfred's translation was added at the end of the Latin translation of Aristotle's *Meteorologica*, which was translated then by Gerard of Cremona (Books I–III) and Henricus Aristippus (Book IV). As a result, the *De congelatione* was regarded as a genuine Aristotelian text, and thus, enjoyed for some time great authoritative status.[33] The importance of the *De congelatione* lies in the fact that it contains two arguments on the impossibility of alchemical transmutation: (1) art is inferior to nature, and therefore its products are not equal to those of nature, and (2) alchemical transmutation is impossible because the true specific characteristics of metals are to be found beneath the level of human sense, and, as a result, the alchemists cannot manipulate something they cannot perceive.[34] The former argument is especially present in the alchemical debate of the thirteenth century and it is expressed by the maxim "let the practitioners of alchemy know that species cannot be transmuted",[35] implying in this manner the superiority of nature over art. Such medieval authors as Roger Bacon, Thomas Aquinas, Albertus Magnus, and Aegidius Romanus were involved in the debate and took different positions with respect to nature's superiority over art.

Before we turn to the actual analysis of *Anonymous* and Albertus Magnus' text, it should be mentioned that the connection between the two texts should not be regarded as a direct one; Albertus was not the main influence on *Anonymous* in this matter. This subsection aims to show that one may find in *Anonymous* the echoes of this debate, and Albertus is used as an exemplary thinker on the matter for the sake of the analysis.

The text of *Anonymous* contains some excerpts which echo the debate mentioned above. In general, its author seems to hold art's power and effectiveness in high regard; therefore, *Anonymous* does not endorse Avicenna's position on the superiority of nature. *Anonymous*' first statement on the relation between art and nature is found in chapter thirty-nine, where he says:

> Εἰδέναι τοιγαροῦν χρὴ ὅτι, ὥσπερ ἐν τῷ τῶν σωμάτων λόγομεν μιμεῖσθαι τόν τεχνίτην φύσιν, καὶ οὕτω προβιβάζειν τὰ εὐτελῆ μέταλλα εἰς χρυσοῦ καί ἀργύρου τελείωσιν [...].
>
> It should be known that inside the part of metals, we say that the artisan imitates nature and leads base metals towards the perfection of gold and silver.[36]

Concerning this excerpt, Colinet does not provide any explanatory notes or commentary. Yet, one may detect the influence from Aristotle's previous statement in his

33 Newman 1989: 423–445.
34 Newman 1989: 427–429; 1991: 3–4.
35 Rubino and Pagani 2016: 43.
36 *Anonymus "Zuretti"* 39.3 (ed. Colinet 2000: 56, 17–20). Unless otherwise noted, translations are by the author.

Physica, where art imitates or even surpasses nature. The author of *Anonymous*, at least according to this passage, seems to adhere to the first part of Aristotle's proposition since he asserts that an artisan imitates nature and leads metals towards their perfection. Moreover, the Greek word τελείωσις, which is used to denote the notion of perfection, reveals a possible Aristotelian influence, given that the term is reminiscent of the Aristotelian term of τέλος. Further on, in chapter forty-eight, where the preparation of sulphur is discussed, the author of *Anonymous* expresses a similar idea stating:

> Τῆς καθάρσεως δὲ ταύτης ληψόμεθα παράδειγμα φυσικόν, διότι ἡ τέχνη ὀπαδός ἐστι τῆς φύσεως διό, ὥσπερ τελειοῖς αὐτὸ ἡ φύσις διὰ χρόνου, οὕτω ποιεῖται ἡ τέχνη διά τινων ἀλόγων ἐφόδων.
>
> But for this purification we will take an exemplar based on nature since art is an adherent of nature, because just like nature perfects something throughout time, so does art through some methods that do not rely on reason.[37]

Likewise, in this excerpt, the author of *Anonymous* reiterates the concept we came across in the previous passage: the imitation of nature by art. However, in this case, *Anonymous* also offers some insight regarding the methodology used in alchemy, which does not depend on *logos*, that is, reason. It should be stressed here that Aristotle does not exclude the notion of reason in arts; thus, *Anonymous* does not speak in an Aristotelian sense.[38] One plausible interpretation would be that the author of the text implies the difference between the seven liberal arts and the mechanical arts. In fact, liberal arts were meant to liberate a person's mind, while mechanical arts were treated as inferior and banausic, aiming to cover bodily and material needs. That said, *Anonymous* might have tried to convey the meaning that the art of alchemy and its method is not that pertinent to mind.[39] Finally, *Anonymous*' references to the art–nature relation culminate in chapter fifty-eight, where he speaks about sal ammoniac and expresses a quite favorable approach towards art. Specifically, he says:

> Εἰδέναι δὲ χρὴ ὅτι τὸ ἀμμωνιακὸν ἅλας διπλοῦν ἐστι, φυσικὸν καὶ τεχνικόν. [...] Τεχνικὸν δέ ἐστιν ὃ γίγνεται ἀπὸ τέχνης, καὶ ἐκεῖνός ἐστι κρεῖττον ἐν τῷ ἡμετέρῳ ἔργῳ ἤπερ τὸ φυσικόν [...].
>
> It should be known that sal ammoniac is of two kinds, natural and artificial. [...] The artificial is the one which is made by art, and it is better for our work than the natural one [...].[40]

37 *Anonymus "Zuretti"* 48.1 (ed. Colinet 2000: 80, 26–29).
38 For instance, in the *Nicomachean Ethics* (6.4.1140a) Aristotle defines art as "a reasoned capacity to make", while in his *Metaphysics* (981a–b) the artist is presented as one "who knows the reason why he does what he does".
39 Whitney 1990: 28; Weisheipl 1965: 54–90.
40 *Anonymus "Zuretti"* 58.2 (ed. Colinet 2000: 97, 1–9).

As Colinet argues, the distinction of sal ammoniac into natural and artificial is met in Arabic alchemy, particularly in al-Rāzī's and Jābir's work.[41] Yet, the author of *Anonymous* follows the *Semita recta* and its narrative, a pseudo-Albertian text. The important thing here is that we have the first acknowledgement in *Anonymous* that a work of art can be better than that of nature, and thus, it is the first time that art appears superior to nature. Obviously, this excerpt of *Anonymous* addresses the second part of Aristotle's statement in *Physica*, according to which art can surpass nature. In addition, the text seems to be in indirect conversation with Avicenna's text since the author of *Anonymous* takes the opposite position. Even though one cannot tell whether the anonymous author knew the Avicennan text, Colinet has provided us with some direct and indirect connections between the two texts. Accordingly, relating this text to the Avicennan art–nature debate is not far-fetched.[42]

Turning to Albertus Magnus, it has been already mentioned that the Dominican master was well aware of the Avicennan *De congelatione*, and therefore, he took a position with respect to the art–nature debate. In this regard, in Albertus' works one may often find the Avicennan proposition *sciant artifices alkimie species metallorum mutare non posse* ("let the artisans of alchemy know that the species of metals cannot be transmuted"), which denoted art's limitations in terms of bringing about a proper alchemical transmutation. However, Albertus' use of the Avicennan dictum does not imply that he always had a negative attitude towards art's efficacy. In fact, as will be shown shortly, Albertus was seriously concerned with the problem of alchemical transmutation for a long time, and therefore, his opinion was neither fixed nor uniform throughout his scholarly career. In what follows, I will present some indicative excerpts from his work that clearly show the aforementioned shift in his opinion.[43]

In 1246, Albertus Magnus finished his commentary on the second book of Peter Lombard's *Sententiae*. When Albertus treated the subject of whether a demon can induce a new substantial form in matter, he rejected this possibility by evoking the *Sciant artifices*. He notes:

> Item, Ars non transmutat a forma substantiali in formam: quia dicit Aristoteles in IV Meteororum: "Sciant artifices alchimiae species transmutari non posse": ergo nec daemones, quia ipsi non operantur nisi per modum artis.

> Likewise, art does not transmute a substantial form into [another substantial] form, because Aristotle says in *Meteorology* IV that "the artificers of alchemy should know that species cannot

41 Colinet 2000: 289n432.

42 Colinet 2000: 240n17, 244n47, 281n363, and 288n430.

43 For a first but not thorough account on Albertus' use of the *Sciant artifices*, see Newman 2004: 38–44. I have dealt with the subject extensively in Rinotas 2022b.

> be transmuted"; therefore, demons cannot [transmute them], because they work only by means of art.[44]

In this passage, Albertus is in communication with Avicenna's *De congelatione* and its first argument, according to which art is inferior to nature. In fact, Albertus rejects the demonic ability to induce a new form in matter because demons work by means of art, which has limited capacity and cannot bring about what nature can. Yet, Albertus' seemingly rigid opinion on art and its capacity changed in his *De mineralibus*, where he discussed the possibility of metallic transmutation. In the ninth chapter of the third book, Albertus treats the topic of the possibility of the transmutation of metals. Having referred to the *Sciant artifices*, he further supplements the quotation by saying:

> propter quod et ipse subjungit, quod non permutantur species, nisi forte in primam materiam et in materiam metallorum reducantur, et sic juvamine artis deducantur in speciem metalli quod voluerint.

> And for this reason, he himself (Avicenna) adds that "specific forms are not transmuted, unless perhaps they are first reduced to prime matter (*materia prima*)" – the [indeterminate] matter of [all] metals – and then, with the help of art, developed into the specific form of the metal they want.[45]

In this excerpt, Albertus presents himself as more open to the possibility of metallic transmutation and, thus, to art's actual ability to bring about such an effect. For Albertus, a metal can be transmuted if it is first turned to its prime matter, and then, through art, a new form may be induced. This statement by Albertus clearly shows that he is favorable towards a transmutation based on matter, which is further suggested in his *Quaestiones super de animalibus*, written around 1260. There, Albertus poses the question, "Whether an aquatic animal, once its form is changed, can become terrestrial", and initially rejects this view by invoking the *Sciant artifices* and the incapacity of art to transmute species. However, the Dominican master creates a strong analogy and connection between animals and metals. When he finally settles the matter, he distinguishes between a transformation in terms of matter, which is possible, and one in terms of form, which is impossible.[46] In this regard, Albertus confirms his previous statement in his *De mineralibus*, and he seems to give more credit to art than he did in his commentary on the *Sententiae*.

All in all, both the author of *Anonymous* and Albertus Magnus show a positive attitude towards art and its ability – at least – to complement nature. As has already been

---

44 Albertus Magnus, *Super IV Libros Sententiarum*, Lib. II, dist. 7, art. 8 (ed. Borgnet 1890–1899: 27, col. 154b); Engl. transl. Newman 2004: 46.

45 Albertus Magnus, *De mineralibus*, Lib. III, tr. i, cap. 9 (ed. Borgnet 1890–1899: 5, col. 71a); Engl. transl. Wyckoff 1967: 177–178.

46 Rinotas 2022b: 77–81.

mentioned, the anonymous author has consulted the *Semita recta*, which was by then circulated under the name of Albertus Magnus; this indicates that Albertus was already known and of interest to the author of *Anonymous*. An indirect influence cannot be ruled out concerning the art–nature debate.

The Three Humidities of Sulphur

The three humidities of sulphur is a topic that has already been illustrated by Colinet,[47] who has stressed that the author of *Anonymous* follows the *Breve breviarum* of Pseudo-Roger Bacon on the matter but has also mentioned that a similar theory is found in Albertus Magnus' *De mineralibus*.[48] That said, Colinet's presentation highlights how the *Breve breviarum* influenced the text of *Anonymous* with respect to the topic of the three humidities of sulphur, while the role of Albertus' *De mineralibus* remains undetermined. However, in 2014, William Newman published a paper that showed how the *De mineralibus* had influenced the *Breve breviarum* on the matter of the three humidities of sulphur, and we are now in a better position to reevaluate and reassess Albertus' influence over the text of *Anonymous*.[49] In this regard, this subsection aims to clarify Albertus' role and contribution to the matter.

The author of *Anonymous* refers to the topic of the three humidities of sulphur again in chapter forty-eight, where he speaks about the preparation of sulphur. When it comes to sulphur's humidities, he says:

> Διότι, ὡς λέγεται, τρεῖς ὑγρότητας ἔχει, ὧς αἱ δύο εἰσὶν περισσοί, ἡ δὲ τρίτη ἀναγκαία, ὅθεν ἀναγκαῖόν ἐστι τὸ καθαρθῆναι αὐτὸ ἀπὸ τῶν δύο περισσῶν ὑγροτήτων, τῆς τρίτης ἐναπομεινούσης, καὶ ἐναπομένει ἡ οὐσία αὐτοῦ καθαρή, ἰσχυρὴ ἑνωμένη, ἐξ οὐσίας γεώδους καὶ ὕδατος κεκαθαρμένου, συνκεκραμένη.
>
> Because, as it is said, [sulphur] has three humidities, of which two are superfluous and the third is necessary; whence it is necessary to purify it from its two superfluous humidities while the third one remains [intact] and its substance stays pure, strongly unified, purified from its earthy substance and from its water, balanced.[50]

As has already been mentioned, Colinet informs us that the theory of the three humidities of sulphur is taken from Pseudo-Roger Bacon's *Breve breviarum*, yet she stresses that the description of *Anonymous* is quite summarized and, therefore, it poses difficulties in its understanding.[51] What should be kept in mind before proceeding with our

47 Colinet 2005: 147.
48 Colinet 2000: 254n128.
49 Newman 2014.
50 *Anonymus "Zuretti"* 48.1 (ed. Colinet 2000: 80, 20–25).
51 Colinet 2005: 147.

analysis is that sulphur consists of three humidities of which the two are superfluous, and the third one is necessary. Apart from this, there is no further information on which way a humidity can be superfluous or necessary; Albertus Magnus' account on the matter will be examined to shed some light on this topic.

The reason for focusing on Albertus' account of the three humidities is that Newman has shown that the *Breve breviarum* derives from Albertus' *De mineralibus*. Therefore, it is proper to show how Albertus has expressed himself on this topic and then how Pseudo-Roger Bacon treated it.[52] Albertus refers to the topic in the fifth chapter of the third book, where he refers to the "ability or inability of metals to be consumed by burning". There he says:

> Scimus igitur adustibilitatem metallorum esse ex sulphure, et non ex argento vivo secundum se. Adhuc autem scimus, quod omne quod habet cum terrestri unctuosam valde humiditatem mixtam, tres invenitur habere humiditates, quarum una valde est aerea et ignea exterius partibus adhaerens, propter horum elementorum elevationem qua semper evolant ad superficies rerum in quarum veniunt mixtionem et complexionem. Aliam autem habet magis aqueam sub illa proxime natantem in partibus rei. Tertiam autem habet humiditatem radicitus infixam et immersam partibus quae terminata est ad complexionem, et ideo sola complexionalis non facile separabilis ab ipsa sine totius rei destructione. Oportet igitur et in sulphuris natura sic esse.

> We know, therefore, that the ability of metals to be burnt is [due to] the Sulphur, and not to the Quicksilver by itself. Furthermore, we also know that in anything that contains very unctuous moisture mixed with earthiness, the moisture is of three kinds. One of these is extremely airy and fiery, adhering to the surface, as a consequence of the [upward] motion of those elements [Fire and Air], so that they always rise to the surface of things in which they are mixed and combined. The second, close beneath this, contains more wateriness floating about among the parts of the thing. The third has its moisture firmly rooted and immersed in the parts and bounded in the combination; and therefore, this is the only one that is not easily separated from the combination, unless the thing is totally destroyed. And therefore, this must be the nature of Sulphur.[53]

This source of Albertus provides insight into the three humidities and serves as an explanatory aid for interpreting the text of *Anonymous*. In fact, Albertus distinguishes two humidities that reside at the surface of the metal. The first one is extremely airy and fiery, and its position on the surface is due to the upward motion of the elements; it is followed by the second one, which is watery. The third humidity is deeply rooted in the metal, not easily separated, and denotes the nature of sulphur. Albertus' description of the third kind of humidity alludes to the notion of *humidum radicale* ("radical

52 Throughout his paper, Newman repeats in several instances that the *Breve breviarum* depends on, or derives from, Albertus' *De mineralibus*. Newman argues that the content and the terminology used by Pseudo-Roger indicate that the *De mineralibus* is the source of the *Breve breviarum*. For more, see Newman 2014: 329 and 337–338.

53 Albertus Magnus, *De mineralibus*, Lib. III, tr. ii, cap. 5 (ed. Borgnet 1890–1899: 5, col. 80a); Engl. transl. Wyckoff 1967: 197.

moisture"), a kind of moisture in the depths of an organism, and any lack of it is connected with the end of life.[54] We can now understand why the absence of this humidity in the combination denotes its destruction. Furthermore, Albertus' text helps us explicate the source of *Anonymous*. Specifically, the "two superfluous humidities" are the airy–fiery and the watery ones which Albertus mentions, while the third humidity, which is "strongly unified" and "balanced", is what Albertus describes as "firmly rooted and immersed in the parts and bounded in the combination". Consequently, Albertus' theory of sulphur's three humidities was further accommodated by Pseudo-Roger Bacon in his *Breve breviarum*, where he says:

> Sulphur also has three humidities, one of which is airy and unctuous, and consumes, burns, blackens, and breaks all metals [...]. But the second humidity of sulphur is phlegmatic and neither burning nor blackening but evaporating in the fire [...]. The third humidity of sulphur is radical and natural, through which the substantial parts of it are conjoined, and this is the water of the sulphur that binds its earth and through which the pure substance of sulphur is joined together and has its existence [*constat* & *conjungitur*].[55]

In Pseudo-Roger's passage Albertus' influence is easily detected.[56] To be precise, Albertus' airy–fiery humidity becomes airy and unctuous in Pseudo-Roger. At the same time, the second one, described as "watery" in Albertus, is denoted as "phlegmatic" in Pseudo-Roger. Yet, this phlegmatic humidity is also watery since it evaporates in the fire. Finally, the third humidity in Pseudo-Roger is named "radical and natural" and has the same role as in Albertus: binding the parts of the combination.

To sum up, it has been shown, on the one hand, that Newman's recent study indicates that Albertus' *De mineralibus* was the actual source for Pseudo-Roger's *Breve breviarum* with respect to the theory of the three humidities, and, on the other, that Albertus deserves further credit on the influence he exerted on *Anonymous* on the matter at hand. In this regard, even though *Anonymous* follows the *Breve breviarum*, we

---

54 McVaugh claims that Albertus takes the notion of radical moisture from Avicenna and his *Canon*; for more, see McVaugh 1974: 268–269.

55 The English translation is taken from Newman 2014: 338.

56 It should be stressed that there is another source in Albertus' *De mineralibus* that is closer to the one of the *Breve breviarum*, which has already been adduced. For the sake of brevity, I will just give its translation without any comments: "And we know from what has been said that both Sulphur and Quicksilver – and anything else at all which, because of its constitution (*complexione*), is moist and liquefiable by Fire – contains three [kinds of] moisture, just like living things, plants and animals. One [kind of] moisture is thick and undigested (*crassus et indigestus*), rising to the surface like grease, fat, or oil; and it is this that makes things inflammable. And the second is like the phlegmatic humour (*phlegmaticus humectans*) that moistens the parts of things, but does not contribute to generation or growth. And the third is the radical (*radicalis*) moisture saturating the essential parts of a thing; and because of this moisture the parts [of things] are firm, and grow, and are nourished. We have found that silver is hard and dry. And therefore it must have been thoroughly cleansed of the two superfluous [kinds of] moisture, and the subtlety of the third is responsible for its excellent mixture" (transl. Newman 2014: 336).

should – at least – acknowledge that Albertus' *De mineralibus* should be considered an indirect influence.

The Role of *Symbola* in Alchemy

I will now deal with a topic that has not attracted much attention in modern scholarship: the role of the Aristotelian *symbolon* in alchemy. My aim is to show, first, how Albertus uses the notion of *symbolum* in his alchemy and, second, how this theory can be regarded as a possible influence in the text of *Anonymous*. To this end, I will start my analysis with Albertus' work, and then I will proceed with *Anonymous*, as the former can serve both as an explanatory tool and a possible source for the latter.

The notion of *symbolon* is found in Aristotle's second book of *De generatione et corruptione* (2.4.331a20–32). According to Aristotle, transmutation between two elements can be performed easier and faster due to their *symbola*, that is, due to a shared quality that acts as a facilitator between the two elements. For instance, the element of fire, which has the qualities of hot and dry, can be transformed or transmuted to air, which in turn has the qualities of hot and moist, if the quality of moist overpowers the one of dry. In that case, the common quality of hot acts as a "symbol" (σύμβολον), which facilitates the transformation. Concerning the meaning of the Aristotelian *symbolon*, Harold Joachim offers some insight: "Aristotle uses the term here and elsewhere to mean a part of one whole, which is capable of fitting in with a complementary part so as to constitute another whole".[57] Likewise, Williams also gives an account of *symbola*:

> *Symbola* are produced by breaking in two some object, e.g., a bone or a coin, each of a pair of friends or contracting parties, A and B, taking one part, so that at some future date A may prove his association with B by producing his part of the object, which can be seen to be the counterpart of B's. Here Aristotle uses it to refer to the quality that is shared by two "consecutive" (cf. 331b4) simple bodies, e.g., the coldness which is a differentia of both water and earth. The coldness that is in water has its "counterpart" in the coldness that is in earth.[58]

What we should keep in mind from these accounts is the fact that a *symbolon* has the capacity to fit in with another complementary part to create a whole, while the common quality which is shared by two elements can be seen in terms of being "counterparts". That said, a *symbolon* allows an element to "come in touch" with another and transform itself into another element through the shared quality.

Turning to Albertus, the Dominican master was acquainted with the Aristotelian *symbola*, and there are two instances in his *De mineralibus* where he uses them within an alchemical context. In the first chapter of the second tractate of the third book of the

57 Joachim 1922: 220–221.
58 Williams 1982: 162.

De mineralibus, Albertus discusses "The solidification and liquefaction of metals" and states, while referring to mercury:

> quod ideo fit, quia argentum vivum quod in plumbo est, habet humidum unctuosum, sicut saepius dictum est, et ideo non adhaeret superficiei, nisi per aliquod habens symbolum in natura.
>
> And this is because the Quicksilver in the lead contains unctuous moisture, as has often been said; and therefore it will not adhere to a surface except through having something in common (*symbolum*) with its nature.[59]

In this passage, Albertus refers to *symbolum* without giving any introduction or acknowledging his source for the concept. As Wyckoff notes, Albertus treats *symbola* as a "common factor" that provides a thing with the ability either to transmute itself to something else or to unite itself with another thing.[60] In this instance, the second case applies, since the quicksilver of lead may adhere to a surface only if these two share something in common: a *symbolon*. A bit further in the text, in the sixth chapter, Albertus discusses the "cyclical production of metals from each other" and states:

> His autem adjiciendum est, commune omnibus metallis esse, quod propinqua est valde materia eorum. Scimus autem ex his quae in scientia Peri geneseos determinata sunt, quod inter habentia symbolum in materia et virtutibus et potentiis naturalibus, facilis est transmutatio ad invicem. Propter quod et multorum Philosophorum, quorum tamen pater est Hermes Trismegistus, qui Propheta Philosophorum vocatur, assertio est, circularem esse metallorum generationem, et ex se invicem, sicut et icircularis est generatio elementorum: quod etiam mihi videtur verissimum.
>
> An additional statement should now be made: one thing that is common to all metals is that their materials are closely related. We know, from what has been determined in the science of Generation and Corruption, that among [things] having a common property (*symbolum*) in their material, powers, and potentialities, the transmutation of any one into another is easy. And this is the reason for the assertion of many philosophers – whose father is Hermes Trismegistus, called the prophet of philosophers – that the production of the metals is cyclical, from each other, just as the production of the elements is cyclical. And this seems to me very true.[61]

This text is much more revealing than the previous one. First of all, Albertus refers to the *De generatione et corruptione* as his source of *symbolum*, while he also acknowledges that transmutation is easier due to them. Then, after accepting that the material of metals is closely related, he asserts that "the production of the metals is cyclical, from each other, just as the production of the elements is cyclical". This statement is

---

59 Albertus Magnus, *De mineralibus*, Lib. III, tr. ii, cap. 1 (ed. Borgnet 1890–1899: 5, col. 76a); Engl. transl. Wyckoff 1967: 188.
60 Wyckoff 1967: 188.
61 Albertus Magnus, *De mineralibus*, Lib. III, tr. ii, cap. 6 (ed. Borgnet 1890–1899: 5, col. 81b); Engl. transl. Wyckoff 1967: 200.

essential because Albertus alludes to an alchemical transmutation of metals by means of *symbola*, and thus, he bases alchemy on natural processes and natural philosophy. It should also be mentioned that the theory of *symbola* conforms with the sulphur–mercury theory, since the latter is likewise based on the elements and their qualities.[62]

After this short analysis of Albertus' use of *symbola* in his alchemy, we are now in a position to turn to the text of *Anonymous* and examine a source that might also seem to use the same theory with Albertus. Specifically, in chapter forty, the anonymous author discusses mercury and states:

> Καὶ διὰ τοῦτο ποιοῦμεν τὸ ἐλεξίρ, ἤγουν τὴν ζύμην τῶν μετάλλων, δι' οὗ βάπτομεν πάντα τὰ μέταλλα εἰς χρυσὸν καὶ ἄργυρον, καὶ μάλιστα διὰ τῆς μεσυμβολήσεως τοῦ ἑρμοῦ, διότι αὐτός ἐστι πηγὴ καὶ θεμέλιον πάντων.
>
> And for that we make the elixir, that is, the ferment of metals, by means of which we tinge all metals gold and silver and especially through the intervention [μεσυμβολήσεως] of mercury because it [mercury] is the source and foundation of everything.[63]

To better understand this excerpt, we need some context that precedes it. Previously, the author of *Anonymous* speaks about the four "metallic spirits" (μεταλλικὰ πνεύματα), that is, sulphur, arsenic, sal ammoniac, and mercury, all of which have the ability to tinge metals. Then, he adduces several examples of tingeing metals, such as tin and lead, and asserts that the natural species of metals can take three to four different forms as they are tinged.[64] What follows is the above excerpt, which states that the elixir is made for reasons of coloration, helping metals to be tinged gold and silver, while the role of mercury is highlighted as the "source and foundation of everything".

Colinet has commented on this source in two different instances. In the first one, in her edition of the text, she follows mainly a textual approach and indicates the sources upon which the text of *Anonymous* is based. Particularly, according to Colinet, the reference of *Anonymous* to the elixir as a ferment of metals is also found in al-Rāzī's *Alums and Salts*, while the one to mercury as the source and foundation of everything is a redaction of sources and probably comes from Pseudo-Albertus' *Semita recta*.[65] In the second instance, in her article on *Anonymous*, Colinet has a more contextual approach, since she focuses on the sulphur–mercury theory which exists in *Anonymous*, and points towards the superiority and importance of mercury over sulphur as a metallic "ingredient".[66] However, in either case, Colinet does not comment upon the use of the

---

62 Obrist 1996: 233.

63 *Anonymus "Zuretti"* 40.1 (ed. Colinet 2000: 60, 10–13).

64 *Anonymus "Zuretti"* 40.1 (ed. Colinet 2000: 59, 13–60, 9).

65 Colinet 2000: 272n285 and n286.

66 Colinet 2005: 141–142. The importance of mercury in medieval Latin alchemy has been stressed by Newman who has shown that in Pseudo-Geber's *Summa perfectionis* one can find the "Mercury alone"

word μεσυμβολήσεως, and therefore, she does not seem to suspect a connection with the Aristotelian theory of *symbola*. That said, it is worth examining if such a connection is plausible.

First of all, the word μεσυμβόλησις is a *hapax*, and thus we do not have examples from other sources that could shed some light on the use of the term. It should be mentioned, however, that in the Latin text of the *Semita recta*, which is the main work that influenced the excerpt at hand, there is no Latin equivalent for μεσυμβόλησις. In fact, the *Semita recta* reads *per mercurium*, with no Latin noun corresponding to the Greek word μεσυμβόλησις.[67] Thus, we may assume that the author of *Anonymous* has added the noun μεσυμβολήσεως on purpose to denote a special meaning, given that the text could simply read διὰ τοῦ ἑρμοῦ, without altering the sense of the phrase.

From my research, the word μεσυμβόλησις is not attested in any major dictionary, while the orthographically closest term to it is the word μεσεμβόλησις. The latter has the following meanings; the Liddell–Scott–Jones *Lexicon* defines it as "*intervention* of a planetary influence, *interception*", thus within an astrological context, while Trapp explains it as *Einwirkung* ("effect", "impact", or "influence").[68] In this regard, the term μεσυμβόλησις could be seen either as a word that does not really add any particular meaning to the phrase διὰ τοῦ ἑρμοῦ, and therefore it raises questions about its use by the author of *Anonymous*, or else as a word that implies a specific meaning, not apparent at first glance.

The necessary examination of the text's manuscripts reveals the following. The older witness, which is dated to 1377/1378, *Vaticanus graecus* 1134, f. 26 verso, reads μεσυμβολίσεως. However, the sixteenth-century *Neapolitanus* III D 19, which is most likely copied from the *Vaticanus*, reads μεσεμβολήσεως, according to the *apparatus criticus* in Zuretti's edition.[69] It is also noteworthy that Zuretti has chosen to render the term as μετασυμβολήσεως. Colinet, in her turn, evidently equates μεσυμβόλησις to μεσεμβόλησις, not only because she translates μεσυμβόλησις as *intervention* in French,[70] but primarily because in the "Index grec" of her edition, she clearly identifies μεσυμβόλησις with μεσεμβόλησις.[71] This is a bizarre editorial choice, since she

theory, according to which all metals come from mercury alone. This theory has an Albertian influence too. For more, see Newman 1991: 204–208; 2013: 395.

67 See *Anonymus "Zuretti"* 40.1 (ed. Colinet 2000: 60, *apparatus fontium* to lines 10–13): "[...] ita massae [...] metallorum transubstantiantur in album uel rubeum [in rubeum uel album *M*] et maxime per mercurium quia ipse est fons et origo omnium metallorum".

68 See Liddell–Scott–Jones and the *Lexikon zur byzantinischen Gräzität*, s.v. "μεσεμβόλησις".

69 *Anonymi De arte metallica* (ed. Zuretti 1930: 98, 8, *apparatus criticus*). Unfortunately, I did not have access to *Neapolitanus* III D 19. For *Vaticanus graecus* 1134 and *Neapolitanus* III D 19, see Zuretti 1927: 153–192 and 231–260 respectively; Colinet 2000: xxv–xxxv; Pausillo 2020: 141–159.

70 *Anonymus "Zuretti"* 40.1 (ed. Colinet 2000: 60 [Fr. transl.]): "[...] grâce à l'intervention du mercure [...]".

71 Colinet 2000: 362.

otherwise does not provide an *apparatus criticus* for the reading of the word within the text, where any relevant comment is entirely absent. The reason why the copyist of the Neapolitanus converted μεσυμβολίσεως to μεσεμβολήσεως is related, in all likelihood, to Colinet's assessment of his working method; it seems that he was not content with just copying the manuscript, but he corrected the spelling, the accentuation, as well as the grammatical errors. As Colinet further comments, "pour lui, il n'existe qu'un grec, le grec classique, et il fait la chasse à toutes les 'fautes', éliminant quand il le peut, tout particularité dialectale, démotique ou d'époque différente".[72] As for the astrological connotations of μεσεμβόλησις, they could be related to "mercury" (written with the symbol ☽ in *Vaticanus graecus* 1134) that follows, which denotes both the metal and the planet. On the other hand, by the time the anonymous author was using the term, it might have acquired a much broader sense. Therefore, it is not implausible that the phrase διὰ τῆς μεσυμβολήσεως τοῦ ἑρμοῦ could simply mean "through the intervention [that is, the effect] of mercury".

Be that as it may, we could also examine the possibility that the word μεσυμβόλησις holds a special meaning, based on the notion of *symbolum* in Albertus Magnus, that we discussed above. Although it is hard to deduce a direct influence of Albertus' *De mineralibus* on *Anonymous*, this cannot be ruled out, as argued earlier. In addition, Wyckoff has noticed that the notion of *symbolon* is also found in Pseudo-Avicenna's *De anima*, a text extensively used in Colinet's comments, and therefore a closer relation between *Anonymous* and the *De anima* is implied.[73] Yet, my actual argument on the connection between the word μεσυμβόλησις with the notion of *symbolon* is bolstered upon examination of further textual evidence.

As we have already seen, in Albertus, the mercury of lead requires a *symbolum* in order to adhere to a metallic surface.[74] This is something we should bear in mind because it will help us to better understand the source of *Anonymous* and the meaning of μεσυμβόλησις. In *Anonymous*, an elixir is used to tinge metals gold and silver; this process is particularly enhanced with the intervention of mercury. In the same text, mercury is associated with dyeing methods in other instances too, since it belongs to the metallic spirits which have the capacity to tinge bodies (metals) white and red.[75] Now, it is worth examining how *Anonymous* conceives of and defines the notion of tincture. In the third chapter, the author speaks of the lime of metals and states:

> Βαφή ἐστι πρόθεσίς τις χρωματίζουσα τὴν τίτανον καὶ ἀχωριστέραν ταύτῃ προσμένουσαν· ἡ βαφή ἐστιν ὕδωρ διβάφον τὴν τίτανον καὶ ταύτῃ ἀχωρίστως προσκείμενον [...].

72 Colinet 2000: xxxv.

73 Wyckoff 1967: 188. For Colinet's references to Pseudo-Avicenna's *De anima*, see for instance, Colinet 2000: 240n17, 242n27 and n28, 246n59, 256n147, 257n152, 258n157, 260n175, 267n231.

74 See note 59 above.

75 *Anonymus "Zuretti"* 40.1 (ed. Colinet 2000: 59, 15–20).

> Tincture is an addition [πρόθεσις] which colours the lime and which stays with it (lime) without being separated; tincture is a water which completely tinges the lime and is inseparably attached to it.[76]

In this text, we have two definitions of tincture, according to which, it is defined as (1) an "addition" (πρόθεσις), which has the ability to color, and (2) as a water, which has the ability to tinge and to adhere to lime. The word πρόθεσις also indicates that the act of coloration is performed on the surface of the metal, and that is why it acts as an "addition". In the text of *Anonymous*, we saw that μεσυμβόλησις and mercury are connected to tingeing metals. In this regard, we may be able to associate the two sources. There is yet another excerpt in *Anonymous* which defines mercury as "a composite body which is composed of subtle sulphureous earth and of water which has been strongly mixed in".[77] This excerpt alludes to a liquid-like kind of mercury, thus, we can assume that the water, which has the ability to tinge and to adhere to lime or any metallic surface, could be mercury as well. It should be recalled that, in Albertus, mercury's attachment to a surface is effected through *symbola*; therefore, this text may be the key to understanding the meaning of μεσυμβόλησις. To put it plainly, when the author of *Anonymous* speaks of dyeing metals specifically through the "intervention" (μεσυμβολήσεως) of mercury, he may intend to convey the meaning of attaching mercury, and thus a tincture, to a metallic surface by means of *symbola*.

4 Conclusion

The primary aim of this paper is to create a dialogue between two seemingly distinct alchemical traditions, that is, the Byzantine and the medieval Latin one, and show that they share some common traits which allow us to infer that there is evidence for actually relating them. To this end, the text of *Anonymous* has been placed in dialogue with Albertus Magnus' *De mineralibus*, and we came up with the ensuing results. First, in the text of *Anonymous* one can find echoes of the art versus nature debate which was launched in Latin Europe during the thirteenth century. The author of *Anonymous* does not only accept that art imitates nature, but also that art surpasses nature, and thus he exposes a more favorable stance towards art. Second, it was shown that on the topic of the three "humidities", *De mineralibus* could be regarded as an indirect influence on *Anonymous*, since Newman's study has shown that Albertus' book strongly influenced the *Breve breviarum*, which in turn was the main influence on *Anonymous* on this topic.

76 *Anonymus "Zuretti"* 3 (ed. Colinet 2000: 9, 20–23).

77 *Anonymus "Zuretti"* 100.27 (ed. Colinet 2000: 217, 17–18): Ἑρμῆς οὖν ἐστι σωματικὸν σύνθετον ἐκ γῆς λεπτῆς θείωδου καὶ ὕδατος ἰσχυρῶς προσμεμιγμένου.

In this regard, we are now in a position to draw a connection between the two texts. Finally, I have attempted to shed some light on the obscure meaning of the term μεσυμβόλησις, which is found in *Anonymous*, offering an interpretation that is based on the Aristotelian theory of *symbola* and found in Albertus' *De mineralibus* too.

As a final remark, it should be stated that the two texts – and, by extension, the two traditions – not only share common ground in terms of content, but it can also be ascertained that one is influenced by the other and that one can be used as an exegetical tool for explicating obscure themes of the other.

Bibliography

Sources

Albertus Magnus, *De mineralibus*: Auguste Borgnet (ed.), *B. Alberti Magni Opera omnia*, 38 vols. Paris: L. Vivès, 1890–1899, vol. 5.
Engl. transl.: Dorothy Wyckoff, *Albertus Magnus, Book of Minerals*. Oxford: Clarendon Press, 1967.

Albertus Magnus, *Super IV Libros Sententiarum*: Auguste Borgnet (ed.), *B. Alberti Magni Opera omnia*, 38 vols. Paris: L. Vivès, 1890–1899, vol. 27.

Anonymi De arte metallica: Carlo Oreste Zuretti (ed.), *Catalogue des manuscrits alchimiques grecs*, vol. 7: *Anonymi de arte metallica seu de metallorum conversione in aurum et argentum*. Brussels: M. Lamertin, 1930.

Anonymus "Zuretti": Andrée Colinet (ed. and Fr. transl.), *L'Anonyme de Zuretti ou l'Art sacré et divin de la chrysopée par un anonyme* (Les Alchimistes grecs 10). Paris: Belles Lettres, 2000.

Literature

Anawati 1981: Georges C. Anawati, "Albert le Grand et l'alchimie", in Albert Zimmermann (ed.), *Albert der Grosse. Seine Zeit, sein Werk, seine Wirkung* (Miscellanea Mediaevalia 14). Berlin and New York: De Gruyter, 1981, pp. 126–133.

Calvet 2012: Antoine Calvet, "L'alchimie du Pseudo-Albert le Grand", *Archives d'histoire doctrinale et littéraire du Moyen Âge* 79.1 (2012), pp. 115–160.

Calvet 2018: Antoine Calvet, *L'alchimie au Moyen Âge: XIIe–XVe siècles* (Études de philosophie médiévale 107). Paris: J. Vrin, 2018.

Cameron 2014: Averil Cameron, *Byzantine Matters*. Princeton, NJ: Princeton University Press, 2014.

Colinet 2000: Andrée Colinet, *L'Anonyme de Zuretti ou l'Art sacré et divin de la chrysopée par un anonyme* (Les Alchimistes grecs 10). Paris: Belles Lettres, 2000.

Colinet 2005: Andrée Colinet, "*L'Anonyme de Zuretti*, un traité alchimique italo-grec de 1300", In Cristina Viano (ed.), *L'alchimie et ses racines philosophiques. La tradition grecque et la tradition arabe* (Histoire des doctrines de l'Antiquité classique 32). Paris: Librairie Philosophique J. Vrin, 2005, pp. 135–152.

Crisciani and Gagnon 1980: Chiara Crisciani and Claude Gagnon, *Alchimie et philosophie au Moyen Âge: perspectives et problèmes*. Montreal: L'Aurore/Univers, 1980.

Ebbesen 2002: Sten Ebbesen, "Greek-Latin Philosophical Interaction", in Katerina Ierodiakonou (ed.), *Byzantine Philosophy and Its Ancient Sources*. Oxford: Oxford University Press, 2002, pp. 15–30.

Fisher 2012: Elizabeth A. Fisher, "Manuel Holobolos and the Role of Bilinguals in Relations between the West and Byzantium", in Andreas Speer and Philipp Steinkrüger (eds), *Knotenpunkt Byzanz. Wissensformen und kulturelle Wechselbeziehungen* (Miscellanea Mediaevalia 36). Berlin and Boston, MA: De Gruyter, 2012, pp. 210–222.

Forrai 2021: Réka Forrai, "Translations I: From Other Languages into Greek: Section I. Latin", in Stratis Papaioannou (ed.), *The Oxford Handbook of Byzantine Literature*. Oxford: Oxford University Press, 2021, pp. 180–188.

Fryde 2000: Edmund Fryde, *The Early Palaeologan Renaissance (1261–c. 1360)* (The Medieval Mediterranean 27). Leiden and Boston, MA: Brill, 2000.

Halleux 1982: Robert Halleux, "Albert le Grand et l'alchimie", *Revue des Sciences philosophiques et theologiques* 66 (1982), pp. 57–80.

Halleux 1996: Robert Halleux, "The Reception of Arabic Alchemy in the West", in Roshdi Rashed (ed.), *Encyclopedia of the History of Arabic Science*, vol. 3: *Technology, Alchemy and Life Sciences*. London: Routledge, 1996, pp. 886–902.

Halleux and Mayvaert 1987: Robert Halleux and Paul Mayvaert, "Les origines de la *Mappae clavicula*", *Archives d'histoire doctrinale et littéraire du Moyen Âge* 54 (1987), pp. 7–58.

Joachim 1922: Harold H. Joachim, *Aristotle, On Coming-to-Be and Passing-Away (De generatione et corruptione). A Revised Text with Introduction and Commentary*. Oxford: Clarendon Press, 1922.

Kibre 1980: Pearl Kibre, "Albertus Magnus on Alchemy", in James A. Weisheipl (ed.), *Albertus Magnus and the Sciences: Commemorative Essays 1980* (Studies and Texts 49). Toronto: Pontifical Institute of Mediaeval Studies, 1980, pp. 187–202.

Lexikon zur byzantinischen Gräzität: Erich Trapp et al., *Lexikon zur byzantinischen Gräzität, besonders des 9.–12. Jahrhunderts* (Österreichische Akademie der Wissenschaften, Philosophisch-Historische Klasse, Denkschriften 238, 250, 276, 293, 326, 352, 417, 494; Veröffentlichungen der Kommission für Byzantinistik 6.1–8). Vienna: Verlag der Österreichischen Akademie der Wissenschaften, 1994–2017.

Liddell–Scott–Jones: Henry George Liddell, Robert Scott, and Henry Stuart Jones, *A Greek-English Lexicon. Ninth Edition with a Revised Supplement*. Oxford: Clarendon Press, 1996.

Linden 2003: Stanton J. Linden, *The Alchemy Reader: From Hermes Trismegistus to Isaac Newton*. Cambridge: Cambridge University Press, 2003.

Magdalino and Mavroudi 2006: Paul Magdalino and Maria Mavroudi, "Introduction", in Paul Magdalino and Maria Mavroudi (eds), *The Occult Sciences in Byzantium*. Geneva: La Pomme d'or, 2006, pp. 11–37.

Martelli 2016: Matteo Martelli, "Greco-Egyptian and Byzantine Alchemy", in Georgia L. Irby (ed.), *A Companion to Science, Technology, and Medicine in Ancient Greece and Rome* (Blackwell Companions to the Ancient World). Chichester: Wiley Blackwell, 2016, pp. 217–231.

Martelli 2019: Matteo Martelli, *L'alchimista antico. Dall'Egitto greco-romano a Bisanzio*. Milan: Editrice Bibliografica, 2019.

Mavroudi 2006: Maria Mavroudi, "Occult Science and Society in Byzantium: Considerations for Future Research", in Paul Magdalino and Maria Mavroudi (eds), *The Occult Sciences in Byzantium*. Geneva: La Pomme d'or, 2006, pp. 39–95.

Mavroudi 2015: Maria Mavroudi, "Translations from Greek into Latin and Arabic during the Middle Ages: Searching for the Classical Tradition", *Speculum* 90.1 (2015), pp. 28–59.

McVaugh 1974: Michael McVaugh, "The 'Humidum Radicale' in Thirteenth-Century Medicine", *Traditio* 30 (1974), pp. 259–283.

Merianos 2017: Gerasimos Merianos, "Alchemy", in Anthony Kaldellis and Niketas Siniossoglou (eds), *The Cambridge Intellectual History of Byzantium*. Cambridge: Cambridge University Press, 2017, pp. 234–251.

Moureau 2020: Sébastien Moureau, "*Min al-Kīmiyā' ad Alchimiam*. The Transmission of Alchemy from the Arab-Muslim World to the Latin West in the Middle Ages", *Micrologus* 28 (2020), pp. 87–141.

Newman 1989: William Newman, "Technology and Alchemical Debate in the Late Middle Ages", *Isis* 80.3 (1989), pp. 423–445.

Newman 1991: William R. Newman, *The* Summa Perfectionis *of Pseudo-Geber. A Critical Edition, Translation and Study* (Collection de travaux de l'Académie Internationale d'Histoire des Sciences 35). Leiden: Brill, 1991.

Newman 2004: William R. Newman, *Promethean Ambitions: Alchemy and the Quest to Perfect Nature*. Chicago, IL: University of Chicago Press, 2004.

Newman 2013: William R. Newman, "Medieval Alchemy", in David C. Lindberg and Michael H. Shank (eds), *The Cambridge History of Science*, vol. 2: *Medieval Science*. Cambridge: Cambridge University Press, 2013, pp. 385–403.

Newman 2014: William R. Newman, "Mercury and Sulphur among the High Medieval Alchemists: From Rāzī and Avicenna to Albertus Magnus and Pseudo-Roger Bacon", *Ambix* 61.4 (2014), pp. 327–344.

Nicolaidis 2011: Efthymios Nicolaidis, *Science and Eastern Orthodoxy: From the Greek Fathers to the Age of Globalization*, transl. Susan Emanuel (Medicine, Science, and Religion in Historical Context). Baltimore, MD: The Johns Hopkins University Press, 2011.

Noyé 2021: Ghislaine Noyé, "Byzantine Calabria", in Salvatore Cosentino (ed.), *A Companion to Byzantine Italy* (Brill's Companions to the Byzantine World 8). Leiden and Boston, MA: Brill, 2021, pp. 434–452.

Obrist 1996: Barbara Obrist, "Art et nature dans l'alchimie médiévale", *Revue d'histoire des sciences* 49 (1996), pp. 215–286.

Papathanassiou 2020: Maria Papathanassiou, "The Occult Sciences in Byzantium", in Stavros Lazaris (ed.), *A Companion to Byzantine Science* (Brill's Companions to the Byzantine World 6). Leiden and Boston, MA: Brill, 2020, pp. 464–495.

Partington 1937: James R. Partington, "Albertus Magnus on Alchemy", *Ambix* 1 (1937), pp. 3–20.

Pausillo 2020: Giorgia Pausillo, "Nuove considerazioni sui manoscritti alchemici di Domenico Pizzimenti", *Scripta* 13 (2020), pp. 141–159.

Peters-Custot 2021: Annick Peters-Custot, "Greek Communities in Post-Byzantine Italy", in Salvatore Cosentino (ed.), *A Companion to Byzantine Italy* (Brill's Companions to the Byzantine World 8). Leiden and Boston, MA: Brill, 2021, pp. 225–251.

Principe 2013: Lawrence Principe, *The Secrets of Alchemy* (Synthesis). Chicago, IL: University of Chicago Press, 2013.

Rinotas 2022a: Athanasios Rinotas, "Elixir as Means of Contrasting with Nature in Albert the Great's Alchemy", in Mattia Cipriani and Nicola Polloni (eds), *Fragmented Nature: Medieval Latinate Reasoning on the Natural World and Its Order* (Studies in Medieval History and Culture). London: Routledge, 2022, pp. 173–193.

Rinotas 2022b: Athanasios Rinotas, "The *Sciant artifices* in the Work of Albert the Great: Towards Two Kinds of Transmutation?" *Early Science and Medicine* 27.1 (2022), pp. 57–82.

Rubino and Pagani 2016: Elisa Rubino and Samuela Pagani, "Il *De mineralibus* di Avicenna tradotto da Alfredo di Shareshill", *Bulletin de philosophie médiévale* 58 (2016), pp. 23–87.

Tatakis 1977: Basileios N. Tatakis, *Η Βυζαντινή Φιλοσοφία*. Athens: Moraitis School. Society for the Study of Modern Greek Culture and General Education, 1977.

Thorndike 1964: Lynn Thorndike, "Relation between Byzantine and Western Science and Pseudo-Science before 1350", *Janus* 51.1 (1964), pp. 1–48.

Viano 2018: Cristina Viano, "Graeco-Egyptian Alchemy", in Alexander Jones and Liba Taub (eds), *The Cambridge History of Science*, vol. 1: *Ancient Science*. Cambridge: Cambridge University Press, 2018, pp. 468–482.

Weisheipl 1965: James A. Weisheipl, "Classification of the Sciences in Medieval Thought", *Mediaeval Studies* 27 (1965), pp. 54–90.

Weisheipl 1980: James A. Weisheipl, "The Life and Works of St. Albertus Magnus", in James A. Weisheipl (ed.), *Albertus Magnus and the Sciences: Commemorative Essays 1980* (Studies and Texts 49). Toronto: Pontifical Institute of Mediaeval Studies, 1980, pp. 13–51.

Whitney 1990: Elspeth Whitney, "Paradise Restored: The Mechanical Arts from Antiquity through the Thirteenth Century", *Transactions of the American Philosophical Society* 80.1 (1990), pp. 1–169.
Williams 1982: Christopher J. F. Williams, *Aristotle's* De generatione et corruptione. *Translated with Notes* (Clarendon Aristotle Series). Oxford: Clarendon Press, 1982.
Zanetti 1999: Ugo Zanetti, "L'Italie de tradition byzantine et sa culture religieuse", *Studia Orientalia Christiana. Collectanea* 32 (1999), pp. 113–140.
Zuretti 1927: Carlo Oreste Zuretti, *Catalogue des manuscrits alchimiques grecs*, vol. 2: *Les manuscrits italiens*. Brussels: M. Lamertin, 1927.

Vangelis Koutalis

Χημεία, *ars sacra*, *alchymia*, *philosophia naturalis*: *Marcianus graecus* 299 (coll. 584) among the Early Modern *literati*

The manuscript nowadays identified as *Marcianus graecus* 299 (coll. 584; hereafter: **M**), dated to the late tenth century, has been an object of study and a point of reference for the historians of alchemy and, to a much lesser extent, for the historians of chemistry, during the last two centuries.[1] In one very important respect, it could not have been otherwise: it is the earliest surviving witness to the corpus of the Greek treatises on chrysopoeia, or on the "sacred art", which circulated among scholars and alchemical adepts in Constantinople during the Late Middle Ages, and throughout Europe during the Early Modern Era. Any discussion which deals with the emergence of what now is subsumed under the general rubric of "alchemy" in the Graeco-Roman Egypt and its transformations in Byzantium necessitates revisiting the texts included in **M** and reassessing its significance as a historical document. Nevertheless, little attention so far has been paid to the well-known fact that this valuable primary source was part of the collection of Greek and Latin manuscript and printed books that Cardinal Bessarion (ca. 1403–1472), a high-profile early modern *literatus*, donated to the Republic of Venice in 1468.

1 A Codex Unworthy to Occupy the Attention of "New Chemists"

One of the oldest library institutions in Italy, the resplendent Biblioteca Nazionale Marciana, owes its origin, in large part, to the addition of Bessarion's collection to the acquisitions of the Basilica of St Mark.[2] In the very act of donation, signed in Viterbo, on 14 May 1468, it was stipulated that (1) an adequate storage space should be provided by the Venetian authorities, (2) the manuscripts should be accessible to everyone wishing to consult them, (3) no item in the collection should be taken away from Venice, and (4) the lending out of manuscripts would be permitted on the condition that a refundable

1 For this codex, see Kopp 1869–1875: 2.257–262; M. Berthelot 1885: 347–352 ["Appendice D: 2"]; 1889: 173–186; Reitzenstein 1919; Zuretti et al. 1927: 1–22 (no. 1); Lagercrantz 1927; 1932; Festugière 1939: 80–84; Rehm 1939; Mioni 1981: 427–433 (no. 299); Letrouit 1995: 11–14; 2002: 85–104; Saffrey 1995; Schütt 1996; Martelli 2011: 5–10; Roberts 2019; 2022; Carlotta 2023. See also the contribution of Michèle Mertens to this volume, which re-examines Zosimus of Panopolis' writings, as transmitted in **M**, from a compilatory viewpoint.

2 For what follows, see Labowsky 1966; Malone-Lee 2024: 176–182.

https://doi.org/10.1515/9783111589619-012

security deposit would be paid amounting to twice the value of the item borrowed. The construction of the building which eventually housed the collection began seven decades later, in 1537. When Pietro Bembo (1470–1547), in 1530, was appointed official historian and librarian of the Republic of Venice, the bequest passed on by Bessarion still remained in the Ducal Palace packed up in closed chests. Owing to the reorganization then attempted, the collection was transferred to the Basilica of St Mark sometime before February 1565. The chests were unpacked, the codices were placed on benches, and their potential readers were even able to borrow them in a regular way, by leaving the deposit required.[3]

Even in the period before Bembo's term in office, well-connected scholars could gain access to Bessarion's codices. Regardless, both before and after the implementation of standard procedures for making the manuscript holdings of St Mark available to the public, not all types of users could effectively be given the opportunity to consult, copy, or borrow a codex. At least until the nineteenth century, we may quite plausibly assume that the pool of actual readers barely expanded beyond the educated upper echelons of society. Juan Andrés y Morell (1740–1817), an exiled Jesuit priest from Valencia, enjoyed such an elevated social status. He was a scion of the Aragonese nobility, and he had also been a professor of rhetorics at the University of Gandía and a professor of philosophy at the Jesuit College of Ferrara. Then he became protégé and tutor of the children of the Marquis Bianchi in Mantua, and a decade later (in 1799) he was appointed *Riformatore*, or Reformer of Studies, and Director of the Library at the University of Pavia.[4] During his visit to Venice in 1788, he was shown around, over the course of a few days, the manuscript volumes, which Bessarion had bequeathed more than three centuries earlier, by the historian of literature and bibliographer Jacopo Morelli (1745–1819), then librarian of the Biblioteca Marciana.[5]

In one of the "family letters" that the Jesuit scholar addressed to his brother, Carlos, and published in Madrid, in 1790, as part of his five-volume *Cartas familiares*,[6] several pages are dedicated to this antiquarian excursion. After his relatively terse remarks concerning the biblical and ecclesiastical codices, Juan Andrés transiently mentions the prevalence of original works and commentaries associated with Plato in the philosophical subdivision of Bessarion's collection, extolling the latter as "a treasure of Greek erudition and philosophy" (*un tesoro de erudicion y filosofía griega*). He also notices, the presence of some unpublished "opuscules" attributed to ancient and more recent philosophers, such as Sextus Empiricus, Symeon Seth, and Michael Psellus.[7] At this point,

3 Zorzi 1987: 87–119; Kidwell 2004: 277–280.

4 Gabaldón 2017; Giménez López 2021.

5 Burigana 2012.

6 This book is a series of travel memoranda from the Italian cities. For the five-volume edition, see the entry in Sommervogel et al. 1890–1932: 1, col. 346 (no. 10).

7 Andrés 1786–1793: 3.62. Unless otherwise noted, translations are by the author.

we are reminded of the fact that the former possessor and donor of the manuscripts had a passionate commitment to Plato and had publicly defended his "venerated master" (*venerado maestro*) with ardor.[8] But no further analysis is offered.

The sender of the letter does not seem to have been much interested in this strictly philosophical territorial sea. He is more willing to enter into details when he goes on to speak about the Greek mathematical and astronomical primary sources then enshrined in the Venetian library.[9] At the end of this part of his report he concludes that the Marciana "in the natural sciences, holds many and good codices to study, equally as in the sacred sciences" (*igualmente que en las ciencias sagradas, tenia en las naturales muchos y buenos códices que estudiar*).[10] And it is at this point that he recommends to his fellow men-of-letters one such codex that he himself examined with particular interest. This codex, which preserves writings of the "Greek chemists" (*químicos griegos*), is no other than **M**.[11] According to Juan Andrés, it would be an indispensable asset in any future endeavor to get the, hitherto unpublished, "corpus of the Greek chemists" (*cuerpo de los químicos griegos*) printed. Among its contents there are plenty of things that can incite curiosity and are "worthy to occupy the attention of our chemists and naturalists" (*dignas de ocupar la atencion de nuestros químicos y naturalistas*).[12]

This late eighteenth-century eyewitness testimony has not passed unnoticed by later historians. In 1869, Herman Kopp, in his meticulous disquisition "Zur Kenntniss der Sammlungen griechischer alchemistischer Aufsätze" (On the Knowledge of the Collection of the Greek Alchemical Treatises), which forms the largest portion of the second volume of his *Beiträge zur Geschichte der Chemie*, quoted, in one of his footnotes, a passage from the relevant section of the German translation of *Cartas familiares*.[13] Andrés' rather extensive allusion to the version of the "corpus of the Greek chemists" transmitted by **M** is simply used by Kopp to underscore how blatantly incomplete was the respective description given in the first printed catalogue of the Greek codices of the Marciana, compiled by Anton Maria Zanetti, the immediate predecessor of Jacopo Morelli as librarian, with the collaboration of the Hellenist and Hebraist Antonio Bongiovanni.[14] Far more intriguing, however, is the fact that a Jesuit antiquary, connoisseur, and educator, in a period sometimes termed the "High Enlightenment", singled out a codex containing chrysopoetic texts as a precious specimen – not of medieval esoteric thought, but rather of the Byzantine natural-philosophical tradition in general, and, more specifically, of the Greek "chemical" literature.

---

8 Andrés 1786–1793: 3.62.
9 Andrés 1786–1793: 3.63–65.
10 Andrés 1786–1793: 3.65–66.
11 Andrés 1786–1793: 3.66.
12 Andrés 1786–1793: 3.67.
13 Kopp 1869–1875: 2.257–258n44.
14 Zorzi 1987: 262–264.

A similar evaluation was made in the fifth of the seven volumes of Andrés' mammoth survey *Dell'origine, de' progressi e dello stato attuale d'ogni letteratura* (On the Origin, Progress and Actual State of All Literature), published in Parma between 1782 and 1799.[15] The opening chapter of this volume, which is the third chapter of the section focused on physics, bears the title "Della chimica" (On Chemistry) and it recounts the historical development of this discipline. Erroneously invested with an aura of great antiquity, through "beautiful dreams" (*belli sogni*) and "slender lines of reasoning" (*sottili ragionamenti*),[16] "chemistry" had only very recently been lifted up in prestige. Notwithstanding this delay, it had become "the dominant science and landlady in the broad field of the natural sciences" (*la scienza dominante, e padrona nello spazioso campo delle scienze naturali*).[17] From very early on in his narrative, the Jesuit scholar makes clear that the history of "chemistry" should be reconstructed as a history of principles and theories, not as the history of practical operations.[18] He also admits that in the then-current state of historical knowledge it would be impossible to determine, with any satisfying precision, when the nascent phase of "chemistry" began and, even less so, to name the first writers who begun to uncover the principles and to establish the rules. But he tentatively situates the earliest stages of this theoretical development "in the school of Alexandria and the times of the decline of Greek philosophy" (*nella scuola d'Alessandria, e ne' tempi della decadenza della greca filosofia*).[19] This newly sprung branch was not a field of practical knowledge. It was a distinct field of theoretical reflection. "Chemistry" emerged when the practical operations carried out in this area of inquiry were no longer being designed and performed at random, but "under fixed rules and for well-founded reasons" (*per costanti regole, e con fondate ragioni*).[20]

For Andrés, the earliest form of "chemistry" was more strongly linked to the late antique and Early Byzantine variants of Greek philosophy than to the Egyptian esotericism or to the Egyptian craftmanship. He even unabashedly downplays the oft-cited assertion that the Roman Emperor Diocletian (r. 284–305) ordered the burning of the ancient Egyptian chrysopoetic books as merely an "opinion" (*opinione*) of low credibility, since it was first uttered only by a much later source, the tenth-century Byzantine *Suda* lexicon.[21] Then again, he thinks that there is a kernel of historical truth beneath this dubious evidence. The same kernel of truth, in his view, underlies the widespread

15 Sommervogel et al. 1890–1932: 1, cols 343–346 (no. 9).

16 Andrés 1782–1799: 5.1.

17 Andrés 1782–1799: 5.30.

18 Andrés 1782–1799: 5.2.

19 Andrés 1782–1799: 5.6–7.

20 Andrés 1782–1799: 5.6.

21 Andrés 1782–1799: 5.3. Today, there is unanimous scholarly agreement that the reference to Diocletian's suppression of the Egyptian chrysopoeia was initially made in the chronicle of John of Antioch, written in the late sixth or early seventh century. The compilers of the *Suda* lexicon reproduced, at this instance, an excerpt from this chronicle. See Ioannes Antiochenus, *Fragmenta* (ed. Roberto 2005:

myths exalting the remote Egyptian origins of "chemistry" and the questionable ascription of authorship of the earliest "chemical" texts, which were infested with "arcane and useless doctrines" (*misteriose ed inutili dottrine*), to Alexandrian or Egyptian authors. This kernel is the conjecture which Andrés has already uttered. In the late antique school of Alexandria "an arcane and obscure, sacred and theological philosophy" (*una filosofia misteriosa ed oscura, sacra e teologica*) was taught and a proneness to boast of understanding what elsewhere would remain incomprehensible did triumph. It was in this intellectual environment that "chemistry" was born in the form of an "art always called great, holy, mystical, sacred, and divine by the first authors" (*arte chiamata sempre da' primi autori grande, santa, mistica, sacra, e divina*).[22]

Andrés acknowledges that various manuscript collections of the "Greek chemists" do survive, scattered over different libraries and deserving scholarly scrutiny. They contain texts written by a multitude of authors, some of whom bear famous and trusted names, such as Heliodorus, Synesius, Zosimus, Olympiodorus. Those expounders of the "divine art" are probably posterior to the homonymous ancient scholars, or at least, in the cases of Zosimus and Olympiodorus, the "chemical" writings which they have passed down to us must have been reworked by more recent hands.[23] Other "Greek chemists", such as Petasius, Christianus, and Sergius, or Stephanus, the "prince of Greek chemists" (*principe de' greci chimici*),[24] have not masked their identities through the appropriation of illustrious names. At any rate, this is a promising field for future research. But the sole "chemical" codex which Andrés has personally inspected is **M**. Apart from being a splendid example of the Byzantine manuscript production, as a physical object, the Venetian codex is a historical document that demonstrates the "great abundance of chemical writers and writings among the diligent and inquisitive Greek scholars" (*copia grande, che v'era di scrittori, e di scritti chimici presso gli studiosi e curiosi greci*), especially through the insertion of a very long list of chemical authors, at the beginning, and the appearance of a number of treatises composed by authors up to then unknown to modern chemists and unregistered by bibliographers.[25]

Only a few pages later, in the same chapter on the history of "chemistry", the late medieval Latin "chemistry" is summarily discarded. It is deemed to have been thwarted from the outset by futile "mechanical research on the making of gold" (*meccaniche ricerche di fare l'oro*), without any serious attempt to delve into how the natural bodies can be decomposed and recomposed. It is also seen as tainted to the very core by speculations obscured "with the unintelligible jargon of odd and meaningless words"

frag. 248, p. 428 = Mariev 2008: frag. 191, p. 348). For the *Suda* entry under discussion, as well as for other entries pertaining to alchemy, see Alexandre Roberts' contribution to this volume.

22 Andrés 1782–1799: 5.7.

23 Andrés 1782–1799: 5.7–8.

24 Andrés 1782–1799: 5.8.

25 Andrés 1782–1799: 5.8–9.

(*coll'inintelligibile gergo di strane ed insignificanti parole*).[26] Prior to this overall decline, "chemistry" in the medieval Islamic world had been assigned broader and higher tasks. It had been cultivated conjointly with mineralogy and put to the service of medicine.[27] The Arab "chemists" had both made some "chemical" discoveries and exposed as fictions the theories of "false chemists" (*falsi chimici*).[28] By contrast, in the Latin West, there was an abundance of chemical literature, but what has not been justifiably buried in oblivion is only a small number of written documents which, despite their historical significance, were absolutely valueless for the "advancement of true physics" (*avanzamento della vera fisica*).[29]

Instead of outright hostility and contempt, Andrés shows curiosity and ambivalence towards the "divine art" of Greek "chemists". In all likelihood, chrysopoeia, in the narrow sense of the artificial production of gold, was their exclusive goal. More mildly put, the question of how to turn base metals into precious ones, "which is only one of the many problems in chemistry" (*che non è che un problema della chimica*), was the center around which their theoretical speculations gravitated.[30] Yet, Greek "chemistry", although not itself a familiar topic for present-day chemists and bibliographers, "can form a distinct phase in the history of the sciences" (*può fare un' epoca nella storia delle scienze*) and it should also be restored to its proper rank as a segment in the array of Greek intellectual contributions. Its position would be rather lower and less luminous than the position bestowed to other branches of knowledge, but it would be "nonetheless discernible among the sciences appropriate to the Greeks" (*osservabile nondimeno fra le scienze proprie de' greci*).[31]

This presumption was destined to be harshly refuted in the short run. Towards the end of the eighteenth century, a process of sharp differentiation between the present and the past of chemistry was initiated. In 1800, Antoine François de Fourcroy (1755–1809) – one of the closest collaborators of Antoine Laurent Lavoisier during the 1780s and a zealous champion of the new, "antiphlogistic", or oxygen-based, chemistry[32] – drew up a sketch of the history of chemistry, in the first, introductory section of his *Systême* [*sic*] *des connaissances chimiques*.[33] Here, he had no hesitation in stating that, despite the involvement of many Greeks in the *mystères de l'alchimie*, and despite the propensity, shared by all Greek philosophers, to treat extensively and methodically the problem of the primary elements of natural bodies, or even that of the conversion of one element into another, their arts were not more advanced than those of the Egyptians.

26 Andrés 1782–1799: 5.11–12.
27 Andrés 1782–1799: 5.10.
28 Andrés 1782–1799: 5.11.
29 Andrés 1782–1799: 5.12.
30 Andrés 1782–1799: 5.7.
31 Andrés 1782–1799: 5.9–10.
32 Smeaton 1962; Kersaint 1966.
33 Smeaton 1962: 225–227 (nos. 64–71).

"Their physics was systematic" (*Leur physique étoit systématique*), but "their chemistry was null" (*leur chimie* [*étoit*] *nulle*).[34] No consideration whatsoever is given by Fourcroy to Byzantine "chemistry". From his perspective, from the seventh century, when the Arabs [*sic*] destroyed the library of Alexandria,[35] to the middle of the seventeenth century, the sciences in general, and chemistry in particular, made but little progress.[36]

Fourcroy concedes that he has faithfully followed the historical exposition of Torbern Bergman (1735–1784), professor of chemistry, metallurgy, and pharmacy at the University of Uppsala.[37] In fact, the text alluded to is a graduation dissertation, under the title *Historiae chemiae medium seu obscurum aevum a medio saeculo VII. ad medium saec. XVII* (Middle of the History of Chemistry or Dark Age, from the Middle of the Seventh Century to the Middle of the Seventeenth Century), defended in 1782 by Pehr Afzelius Arvidsson (1760–1843), a student and assistant of Bergman.[38] This dissertation, in 1787, was republished in the fourth volume of Bergman's collected works, his *Opuscula chemica et physica*.[39] In the very same volume, which was translated, no more than a year later, into German, and in 1791, into English,[40] Afzelius Arvidsson's text was preceded by another historical graduate dissertation, *De primordiis chemiae* (On the Origins of Chemistry). Presented in 1779 at the University of Uppsala by Jacob Paulin (1752–1797), a chemistry student also supervised by Bergman,[41] the second dissertation had already been separately reprinted once, in 1780.[42]

In both these historical accounts, which were based on the inaugural lecture delivered by Bergman, on the occasion of his becoming professor, in 1769,[43] no role is reserved for Byzantine "chemistry", and the manuscript collections of the "Greek chemists". In the dissertation of Paulin, readers are reassured, in passing, that the Greek treatises attributed to Moses "on chemical composition and on duplication" (*de compositione chemica* & *de duplicatione*), which one may come across in the Royal Library of Paris, as well as those of Mary the Jewess, on the "Practice of chemical art" (*Practica*

34 Fourcroy 1800: 1.14. The British chemist William Nicholson, who translated Fourcroy's book into English, in this passage mitigates the force of the concluding sentence (1804: 1.17): "[...] their philosophy was systematic, and their chemistry had no scientific existence". This does not occur in the German and Spanish translations: Veith and Wiedemann 1801: 1.146: [...] *ihre Physik war systematisch; Chemie hatten sie gar nicht*; Olive and González Azaola 1803–1809: 1.154: [...] *su fisica era sectaria, y ninguna su química*.

35 The belief that the destruction of the Great Library of Alexandria took place after the Arab conquest of the city is a legend which was initially disseminated by Muslim sources and must have become current in the late twelfth century. See Lewis 2008; Qassem 2008.

36 Fourcroy 1800: 1.15–16.

37 Lennartson 2020.

38 Moström 1957: 50 (no. 150.4) and 65 (no. 200); Lennartson 2020: 182–183.

39 Moström 1957: 48–51 (no. 150).

40 Moström 1957: 67–68 (no. 204.4) and 93–94 (no. 290).

41 Lennartson 2020: 179.

42 Moström 1957: 45 (no. 139), 50 (no. 150.4), and 57 (no. 170).

43 Lennartson 2020: 125–126.

in artem chemicam) and on the "Practice of the philosophical stone" (*Practica lapidis philosophici*), betray later authorship.[44] Some pages before, there is also a second-hand enumeration of names of authors whose Greek manuscript treatises on chrysopoeia lurk, unedited, in libraries.[45] Even worse, this cursory listing is topped off with the proclamation that "day after day the alchemical hodgepodge was exacerbated, crammed with Pythagorean-Kabbalistic and Scholastic follies" (*quotidie in pejus ruebat farrago alchemistica, dementiis Pythagorico-caballisticis & Scholasticis referta*) up to the seventh and eight centuries. By then, the mystic and scholastic shrouds had become so dense that "the chemical studies and the other disciplines were covered by the thickest mist and everything was overwhelmed by horrid barbarity" (*chemicas & alias disciplinas involveret caligo, horridaque barbaries omnia obrueret*).[46] In the dissertation of Afzelius Arvidsson, we are told, already in the first paragraph, that "the darkness with which this interval in the history of chemistry is wrapped" (*tenebrae, quibus Historia Chemiae hoc intervallo involvitur*) is comparable to the darkness of the Middle Ages in the political history. The character of this long chemical "dark age" was itself hardly chemical. It should be more properly called "Hermetic" (*Hermeticum*) or "alchemistic" (*Alchemisticum*).[47]

2 A Codex Which May Conduce to the Renovation of Early Modern *philosophia naturalis*

After the outburst of the political revolution of 1789 in France, the late eighteenth-century reforms in chemical theory, practice, and terminology were rhetorically depicted by some of their pioneers as facets of an abrupt, radical break with the preceding chemical traditions. A singular all-embracing "revolution" in chemistry was ostensibly waged, casting off the dead weight of the past "metaphysical preconceptions".[48] In the aftermath of this "chemical revolution", what Andrés, less than a decade before the publication of Fourcroy's *Système*, had called "Greek chemistry" was relegated to the same precarious status as the late medieval Latin *alchemia*. The profound debts that the revolutionary "new chemistry", whether unwittingly or not, owed to the late medieval alchemical theoretical constructions and experimental methods were consigned to oblivion and erased.[49] In the same vein, the Graeco-Egyptian and the Byzantine versions

44 Bergman and Paulin 1787: 68.

45 Bergman and Paulin 1787: 60–61.

46 Bergman and Paulin 1787: 62.

47 Bergman and Afzelius Arvidsson 1787: 85–86.

48 For a comprehensive outline, see McEvoy 2010.

49 An example of alleged radical novelty is the conception of the chemical element as the end-point of chemical analysis, developed by Lavoisier. It has been shown that this conception was, to a considerable extent, built upon earlier alchemical corpuscular matter theories. See Newman 2009.

of χημεία have been displaced outside the history of natural philosophy, medicine, and chemistry since then. It is only within the last fifty years that there have been efforts to countermand this exclusion, due to a gradual turn in the historiography of alchemy towards a critical reappraisal of the dichotomy between chemistry and alchemy.[50]

Recent historiography has highlighted the fact that, even until the middle of the eighteenth century, alchemy – in contrast to what was the norm in the next two and a quarter centuries – was not dissociated from other, well established, disciplines studying natural phenomena. In the Early Modern period, quite often, the scholarly assessment of alchemy was directly proportional to the assessment of the experimental natural philosophy in general, notwithstanding the occasional vitriolic denunciations of folk chrysopoeia by prestigious polymaths or alchemists with academic credentials.[51] Could this divergence between the early modern and the late eighteenth-century scholarly attitudes towards pre-modern χημεία, or towards the Byzantine "sacred art" specifically, be evinced also in the case of Bessarion himself? The Byzantine émigré donor of **M** was a high-ranking Church official, initially in the East and later in the West, a patron of humanists, and an esteemed scholar himself, actively engaged in theological and philosophical debates which were current at the time.[52] His own contribution to the resurgence of interest in classical philosophy, and most of all, his involvement in the controversy that erupted among Byzantine scholars, on Italian soil, in the middle of the fifteenth century, over the differences between the Aristotelian and the Platonic philosophy, have been thoroughly explored and extensively written about.[53] On the other hand, his contribution to the dissemination of Byzantine alchemical literature in the Latin West is usually viewed from a rather narrow perspective.

It is only incidentally that Bessarion's relation to **M** is brought forward. Almost always, his involvement in the history of alchemy is, tacitly and explicitly, construed as a collateral consequence of his humanist devotion to the task of rescuing the written monuments of the Greek and Byzantine past. Are there, however, indications that the presence of **M** in his personal library signified something more than mere possession of such a written monument? Or that the former Orthodox Metropolitan of Nicaea, who later became a Roman Catholic Cardinal, might have found this codex appealing, not as a bizarre remnant of the past, but as a vestige of the past of an "art", or field of knowledge, that he and his contemporaries considered intrinsically interesting?[54] One such indication is provided, indeed, by the position of **M** in the division in which he classified his collection of manuscripts. Here, the detailed information contained in the

---

50 For a concise overview of this historiographical turn, see Principe 2011.

51 See Joly 2013: 85–186; Principe 2013: 83–210; Kahn 2016.

52 For the life and work of Bessarion, see Coccia 1973; Monfasani 1986; Mioni 1991; Coluccia 2009; Kennedy 2018; Malone-Lee 2024.

53 For some relevant studies published during the last decade, see Del Soldato 2012; 2020: 25–31; Monfasani 2013; Mariev 2013; Malone-Lee 2019.

54 This question is discussed at length in Koutalis 2022: 223–320.

monograph written by Lotte Labowsky on the inventories of Bessarion's library[55] will be a valuable guide.

In the lower margin of f. 8 recto in **M** the following note is discernible: τόπος π΄ ("locus 80"). It has been written by the hand of Bessarion himself.[56] This is the number which represents the exact place of **M** in his library, in keeping with his personal system of classification. Six other codices, now held at the Biblioteca Nazionale Marciana, in Venice, were subsumed under the same "locus". Four of them preserve pharmacological and medical works of Galen, such as the *Περὶ* [*κράσεως καὶ*] *δυνάμεως τῶν ἁπλῶν φαρμάκων βιβλία ια΄* (Eleven Books [on the Mixtures] and Powers of Simple Drugs) and the *Ἱπποκράτους ἀφορισμοὶ καὶ Γαληνοῦ εἰς αὐτοὺς ὑπομνήματα* (Commentary on the Aphorisms of Hippocrates):

1) *Marcianus graecus* 286 (coll. 626), dated to the fifteenth century;[57]
2) *Marcianus graecus* 277 (coll. 630), of the thirteenth century;[58]
3) *Marcianus graecus* 278 (coll. 873), of the fifteenth century;[59]
4) *Marcianus graecus* 275 (coll. 893), of the fifteenth century.[60]

The remaining two include works on moral philosophy. In the one, the *Ἐπικτήτου διατριβαί* (The Teachings of Epictetus) of Flavius Arrianus (ca. 86 CE–160 CE) are handed down, whereas the other is one out of a total of twenty-six, either complete or fragmentary, codices owned by Bessarion, in which texts of the Middle Platonic philosopher Plutarch of Chaeronea (first century CE) are copied; at this instance, a section from the *Ἠθικά* (Morals): (5) *Marcianus graecus* 251 (coll. 723), dated to the fifteenth century,[61] and (6) *Marcianus graecus* 249 (coll. 287), of the twelfth century respectively.[62]

Bessarion, therefore, had joined together **M**, a chrysopoetic codex of the late tenth century, with four codices, extending in date from the thirteenth to the fifteenth century, which transmit Galenic, mostly theoretical, pharmacological and medical treatises, and two codices, extending from the twelfth to the fifteenth century, which transmit ethical treatises. It also must be pointed out that all the principal texts in these six codices go back to the first two centuries CE. Even so, as Lotte Labowsky has warned us,[63] the classification made by Bessarion, the division of his personal manuscript and printed book collection in eighty-nine "loci", may have been influenced by criteria of demarcation other than those supplied by a thematic or a chronological approach. The provenance

---

55 Labowsky 1979.
56 See Mioni 1981: 427 (no. 299).
57 Labowsky 1979: 227, 255, 295, 439; Mioni 1981: 409–410 (no. 286); Touwaide 2016: 332 (no. [1684]).
58 Labowsky 1979: 165, 242, 254, 295, 438; Mioni 1981: 400–401 (no. 277); Touwaide 2016: 330 (no. [1672]).
59 Labowsky 1979: 165, 228, 255, 295, 438; Mioni 1981: 402 (no. 278); Touwaide 2016: 330 (no. [1673]).
60 Labowsky 1979: 165, 227, 255, 294, 438; Mioni 1981: 399 (no. 275); Touwaide 2016: 330 (no. [1669]).
61 Labowsky 1979: 176, 192, 252, 299, 438; Mioni 1981: 366–367 (no. 251).
62 Labowsky 1979: 176, 200, 252, 302, 438; Mioni 1981: 364 (no. 249). See also Manfredini 1994.
63 On this point, and what follows in this paragraph, see Labowsky 1979: 20–23.

of each codex or the time of acquisition, perhaps, had also played a non-negligible role. The very designation τόπος or "locus" may well have corresponded to an actual location in space, and, thus, may well have been embedded in the spatial structure of Bessarion's library in Rome, but it would not be prudent to infer that a uniform and consistent system of bookshelves or benches had been in operation, similar to that of an organized public library. In thematic terms, roughly half of the eighty-nine "loci" seem to have been occupied by theological and ecclesiastical works. With no sharp dividing lines drawn between different thematic areas, the rest "loci", up to the eightieth, were populated by legal, philosophical, and scientific works. The last nine, from the eighty-first to the eighty-ninth, were dedicated mainly to grammar and poetry.

Keeping these cautions in mind, we should not overlook an important conclusion which can be deduced from the evidence gathered by Labowsky's study. In 1469, after the ratification of his donation, Bessarion sent to the Venetian authorities 466 volumes packed in thirty boxes. One of the many codices and books which was not enclosed in these boxes was **M**. In the first inventory of the collection bequeathed, which was compiled in Viterbo, in 1468, and accompanied the Act of Donation, **M** is numbered as 440. On account of the delay in its transfer, in the new inventory of 1474, after Bessarion's death, it is numbered as 584, which means that it was transposed after the number 466, the upper limit of the sum of the volumes sent first to Venice.[64] Of the other six codices that had been assorted with **M** in "locus 80", by Bessarion, only the four codices containing works of Galen were also left aside from this first selection and transfer. *Marcianus graecus* 275, in the inventory of 1474, was allocated the number 653; *Marcianus graecus* 277, the number 993; *Marcianus graecus* 278, the number 662; and *Marcianus graecus* 286, the number 654.[65] The two codices from "locus 80" in which the *Ἐπικτήτου διατριβαὶ* of Arrianus and the *Ἠθικὰ* of Plutarch survive, *Marcianus graecus* 249 and *Marcianus graecus* 251, in 1474 were numbered anew as 162 and 21 respectively.[66] They had been put inside the boxes of the first consignment.

The volumes that had not been acquired in the years following the official act of donation and had not been consigned to Venice in 1469 were those that Bessarion was less willing to give away, because he intended to employ them, one way or the other, in his own final intellectual pursuits and in the endeavours of the intellectual milieu around him. On the contrary, the codices from his collection that were considered as more valuable, in consonance with the antiquarian standards then existing, were those that were dispatched to Venice before the death of Bessarion, as a splendid sample in advance. The famous philosopher, mentor, humanist, and Church official was by no means indifferent to the short-term glow and the social esteem of his contemporaries.[67]

64 Labowsky 1979: 176 and 223.
65 Labowsky 1979: 227–228, 242.
66 Labowsky 1979: 192 and 200.
67 Labowsky 1979: 46–47.

The fact is that **M** was among the manuscript and printed volumes that he chose to keep for his own use, in the final years of his life. If anything, this permits the assumption that the Byzantine émigré scholar might have seen in **M**, not only a codex precious in terms of its antiquarian value, but also a codex precious in terms of its content and worthy of closer examination.

In the case of Bessarion and his colleagues, the closer examination of a manuscript often entailed the multiplication of its copies. Very revealing, in this regard, is the example of the codices from Bessarion's library that hand down works of Galen.[68] Maria Rosa Formentin, half a century ago, in her monograph on the Venetian Greek medical codices, got impressed, not so much with the great number of the Galenic texts that one could run across in the library of Bessarion, as with the number of the new manuscript copies of these texts that the circle of scholars and scribes around Bessarion had brought forth. Over a hundred treatises of Galen are reposited in fourteen codices owned by Bessarion, now registered in Biblioteca Nazionale Marciana in a series, as *Marciani graeci* 275–288. But only five among these are of a date anterior to his own time: *Marciani graeci* 276, 277, 278, 283, 288. The remaining nine were penned by scribes working for him.[69] Two of these earlier codices, *Marcianus graecus* 277 and *Marcianus graecus* 278, had been placed to "locus 80", together with **M**, and had been excepted from the first consignment in 1469. Yet, the available evidence shows that **M** was far from being a top priority for the copyists Bessarion had recruited.

A perusal in the catalogues of the Greek alchemical manuscripts that have been published leaves little doubt that at the end of the fifteenth century and at the first half of the sixteenth century the circulation of **M** was very limited. Among the present holdings of the Italian libraries, the alchemical codices which derive provably from **M** are seven:[70]

1) Città del Vaticano, Biblioteca Apostolica Vaticana, MS *Barberinianus graecus* 513, *saec.* XVII.
2) Città del Vaticano, Biblioteca Apostolica Vaticana, MS *Vaticanus graecus* 2259, *saec.* XVII.
3) Milano, Biblioteca Ambrosiana, MS A 57 *inf.*, *saec.* XVI (2nd half).
4) Milano, Biblioteca Ambrosiana, MS A 193 *inf.*, *saec.* XVI (2nd half; *terminus ante quem*: 1594).
5) Napoli, Biblioteca Nazionale "Vittorio Emanuele III", MS *Neapolitanus* III D 17, *saec.* XVI (year: 1565).

---

68 Formentin 1978: 48–85.

69 Formentin 1978: 48.

70 See the recent catalogue of Giorgia Pausillo (Pausillo 2022: 56–61, 140–143, 157–171, 180–186, 210–218). In the volume of the *Catalogue des manuscrits alchimiques grecs* for the Italian libraries (Zuretti et al. 1927), the corresponding number is lower by one, since *Vaticanus graecus* 2259 was not included there at all; see Zuretti et al. 1927: 23–37 (no. 2), 71–91 (nos. 6–7), 103–120 (no. 16), 194–207 (no. 35), 217–224 (no. 43).

6) Venezia, Biblioteca Nazionale Marciana, MS *graecus* Appendix IV. 28 (coll. 543), *saec.* XVI (2nd half).
7) Venezia, Biblioteca Nazionale Marciana, MS *graecus* 598 (coll. 909), *saec.* XVI (2nd half).

None of these manuscripts were copied during the lifetime of Bessarion or in the decades following his death. Most of them were copied in the second half of the sixteenth century.

According to the manuscript descriptions in the volumes of the *Catalogue des manuscrits alchimiques grecs* covering other European countries, only three Greek alchemical codices deriving from **M** were copied in the fifteenth and the early sixteenth centuries:

1) Kassel, Universitätsbibliothek Kassel, Landesbibliothek und Murhardsche Bibliothek der Stadt Kassel, MS *Chemicus* 2° 1, *saec.* XV/XVI (1st half).[71]
2) Oxford, Bodleian Library, MS *Canonicus graecus* 95, *saec.* XV.[72]
3) San Lorenzo de El Escorial, Real Biblioteca del Monasterio de San Lorenzo de El Escorial, MS Φ.I.11, *saec.* XV.[73]

The codex in Escorial came to be owned by the poet, historian, and collector of Greek manuscripts, Diego Hurtado de Mendoza (1503/1504–1575), who had stayed for a long time (1539–1546) in Venice, as imperial ambassador.[74] The codex in Kassel, several years after it was first inscribed, passed into the possession of John Dee (1527–1609), the mathematician, astrologer, alchemist, and spiritual advisor to Queen Elizabeth I (r. 1558–1603).[75] Mendoza was a key figure in Spanish humanism and a bibliophile who had the greatest share in the growth of the Greek manuscript and printed-book collection of the library of Escorial.[76] Dee, especially through his 1564 occult philosophical tract *Monas hieroglyphica* (The Hieroglyphic Monad)[77] played an influential role in the re-evaluation of alchemy, during the second half of the sixteenth century, as a renewed and promising version of natural philosophy, which could be appended to astronomy and mathematics.[78]

Dee owned yet another Greek alchemical manuscript dated to the early sixteenth century, which today forms the fourth part of a convolute classified as Archbishop

---

71 Goldschmidt 1932: 117–139 ("Codices germanici", no. 1); Abraham 1997: LXIII–LXIV.
72 Singer et al. 1924: 3–12 (no. 2).
73 Zuretti 1928: 4–39 (no. 3).
74 González Palencia and Mele 1941–1943; Spivakovsky 1970.
75 Parry 2011.
76 Graux 1880: 163–273; Martínez Manzano 2018; 2021.
77 Josten 1964: 84–111.
78 Clulee 1988: 75–142; Rampling 2012.

Selden B 18, at the Bodleian Library, in Oxford.[79] Both the Bodleian manuscript and the one which now resides in Kassel were purchased by Dee, three years after the publication of his *Monas*, in the very same day, 4 April 1567. But, in his *Monas* he had already explicitly mentioned certain names of purported Greek alchemical authors. He had also quoted, in Greek, although to a considerable extent freely, a passage from the tract *Περὶ πορφύρας καὶ χρυσοῦ ποιήσεως· φυσικὰ καὶ μυστικά* (On the Making of Purple and Gold: Natural and Secret Questions), attributed to Democritus, combined with a passage from Stephanus of Alexandria's "Ἐπιστολὴ πρὸς Θεόδωρον" (Letter to Theodorus).[80] The only source cited by Dee, on this occasion, is the "great Democritus":

> [...] ut facilè iam magno Democrito aurem praebeant: τὸ τῆς Ψυχῆς praedicenti ἴαμα, καὶ πάντος [*sic*] μόχθου λυτήριον κατασκσευᾶσαι [*sic*] βουλομένοις, non Μυθικὸν esse hoc Dogma, sed Μυστικὸν et Arcanum: ut et illi, qui asseruit, quod "λόγω δημιούργου [*sic*] κόσμου μεθοδεύεται: ἵνα ὁ θεόφρων καὶ ὁ θεογενὴς ἄνθρωπος, διὰ τῆς εὐθείας ἐργασίας: καὶ θεολογικῶν, καὶ μυστικῶν λόγων μαθῇ".
>
> so that they may indeed readily lend their ear to the great Democritus, announcing to those who wish to affect a healing of the soul and a deliverance from all distress that this doctrine is not mythical, but mystical and arcane; as also [they may listen] to that [author] who has asserted that the logos of the creative universe works by rules so that man, godly-minded and born of God, may learn by straightforward work and by theological and mystical language.[81]

The sentence quoted from Pseudo-Democritus[82] in the original text reads as follows:

> Οὗτοι δὲ ἀκρίτῳ καὶ ἀλόγῳ ὁρμῇ τὸ τῆς ψυχῆς ἴαμα καὶ παντὸς μόχθου λύτρον κατασκευάσαι βουλόμενοι, οὐκ αἰσθήσονται βλαβησόμενοι. Δοκοῦντες γὰρ ἡμᾶς μυθικόν, ἀλλ' οὐ μυστικὸν ἀπαγγέλειν λόγον, οὐδεμίαν ἐξέτασιν ποιοῦνται τῶν εἰδῶν· [...].
>
> On the contrary, these novices who rashly and without reasoning want to prepare a medicine for the soul and for relieving any distress are not aware that they will fail. In fact, they believe that we are presenting a legendary rather than a secret discourse, so they do not carry out any close examination of the species [...].[83]

Among the Greek words reproduced by Dee the variant reading λυτήριον appears. As reported in the apparatus provided by Mateo Martelli, in his critical edition of Pseudo-Democritus' treatise, this version differs from the version found in **M** (λύτρον). It occurs in two other witnesses to the Greek chrysopoetic corpus, which are both held by the Bibliothèque nationale de France: *Parisinus graecus* 2325, dated to the thirteenth

---

79 Singer et al. 1924: 17–22 (no. 4); Hutter 1982: 281–283 (no. 188).

80 See Matton 1995: 337–338; Koutalis et al. 2018: 30n67, 43.

81 Dee 1564: f. 23 recto–verso; Engl. transl. Josten 1964: 198–201. Sylvain Matton (Matton 1995) did not identify Stephanus' text as a source tacitly quoted by Dee in this passage. The same omission is also made by Conrad Josten (Josten 1964).

82 For Democritus the alchemist, see Martelli 2011: 125–172; 2013: 1–73.

83 [Democritus], *Physica et mystica* 16.166–170 (ed. Martelli 2013: 96 [text], 97 [Engl. transl.]).

century, and *Parisinus graecus* 2327, copied in the year 1478.[84] In the first, the word used is λυτήριον, in the second, λητήριον. Therefore, if Dee had access to any of the surviving Greek alchemical manuscripts, before or during the composition of his *Monas*, the most probable candidate is *Parisinus graecus* 2325.[85]

A second very interesting detail here is the divergence, this time in terms of a difference in meaning, between Stephanus' original text and Dee's wording. Stephanus of Alexandria draws a distinction between two types of χημεία, the legendary, on the one side, and the hidden and mystical or secret, on the other:

> Καὶ ἄλλο ἐστὶν ἡ μυθικὴ χημεία καὶ ἄλλο ἡ μυστικὴ καὶ κρυπτή· καὶ ἡ μὲν μυθικὴ πολυπληθίᾳ λόγων συγχέεται, ἡ δὲ μυστικὴ λόγῳ δημιουργίας κόσμου μεθοδεύεται, ἵνα ὁ θεόφρων καὶ ὁ θεογενὴς ἄνθρωπος διὰ τῆς εὐθείας ἐργασίας καὶ θεολογιῶν καὶ μυστικῶν λόγων μάθῃ περὶ τοῦ ἐνύλου κόσμου, [...].
>
> For the legendary chemistry is one thing, and the mystical and hidden is another. For the legendary chemistry is confounded in a multitude of words, but the mystical is operated by the word of the Creator of the world, that the man who is holy and born of God may learn by the direct operation and by theological and mystical words.[86]

In the paraphrasis of this passage in the *Monas*, the readers become acquainted, not with any "chemistry" (χημεία), but with a "dogma", a teaching, that is, not with a discipline investigating concrete objects, nor with a field of knowledge in which theory and practice are inseparably interwoven, but rather with an all-embracing theoretical contemplation and instruction.

Likewise, it is not of χημεία or of *chemia*, but of a "rambling" (*verbosa*) *alchymia* and an "arcane and hidden" (*arcana et abdita*) *alchymia*[87] that we read in the first published Latin translation of Stephanus' *Ἐπιστολή*. Its translator was the Calabrian humanist Domenico Pizzimenti (1550–1592), who consulted two surviving Greek codices, now held by the Biblioteca Nazionale "Vittorio Emanuele III" in Naples: MS *Neapolitanus* III D 18 and MS *Neapolitanus* III D 17.[88] The latter is one of the above-discussed seven codices, in the Italian libraries, which derive from **M**. It was copied in 1565 by the Greek professional scribe Cornelius Murmuris. *Neapolitanus* III D 18 was copied in Rome by another Greek émigré scribe, John Mavromates, sometime between 1555 and 1573. It is an apograph of a codex dated to the end of the fourteenth century or the beginning of the fifteenth century (Biblioteca Apostolica Vaticana, Città del Vaticano, *Vaticanus graecus*

84 See Lebègue 1924: 1–17 (no. 1), 17–62 (no. 2).
85 For the relation of this codex to **M**, see Lagercrantz 1932: 399–432; Rehm 1939: 399–415; Letrouit 1995: 11–14; Martelli 2011: 14–43.
86 Stephanus Alexandrinus, *De magna et sacra arte*, Theod. 21–25 (ed. Papathanassiou 2017: 167); Engl. transl. Taylor 1938: 39.
87 Pizzimenti 1572: f. 30 verso.
88 Formentin 2004.

1174).[89] In 1570, the Latin versions of the alchemical treatises attributed to Democritus, Synesius, and Pelagius, which he had drafted, were printed in Padua under the title *Democritus Abderyta De arte sacra, sive de rebus naturalibus* (Democritus of Abdera On the Sacred Art, or On the Natural Things). Then, in 1572, this volume was reprinted in Cologne, bound together with another book, Antoine Mizauld's *Memorabilium sive arcanorum omnis generis, per aphorismos digestorum, centuriae IX* (Nine Hundreds of Memorable Things or Secrets of Every Kind, Expounded through Aphorisms). In the same year, a second, enlarged edition, including the treatises of Stephanus and Michael Psellus, was prepared and printed in Padua.[90]

The very title of Pizzimenti's book illustrates how difficult it was to recast Greco-Egyptian and Byzantine χημεία as a recognizable piece in the historical background of the established branches of natural philosophy, or of natural philosophy itself. The singling out of Democritus as the primary actor in this "sacred art" apparently aims at circumventing this difficulty; all the more so in a period when corpuscular and atomist theories of matter were beginning to spread rapidly throughout West European scholarly circles. By the last quarter of the sixteenth century, there were natural philosophers who openly extolled Democritus as the original inventor of physics, to the detriment of the authority of Aristotle, and, hence, of the late medieval Aristotelian natural philosophical tradition.[91] But in the same historical conjuncture, a variety of symbolic personae were imposed on the ancient Atomist philosopher of Abdera. Christoph Lüthy has recorded at least four such fictitious Democriti, active in the early modern scholarly imagination: Democritus the Atomist, Democritus the laughing philosopher, Democritus the moralizing anatomist, and Democritus the alchemist.[92]

The laughing philosopher, which was paired by his weeping colleague, Heraclitus, cropped up much earlier than the other three personae of Democritus. His laughter, in the early modern enactment of this topos, was reinterpreted as "a philosophical and even melancholy response to the world", as the proper stance of a "Demo-criticus", who sarcastically criticizes both the existing reality and the common representations of this reality.[93] Underlying the topos of Democritus' laughter and Heraclitus' tears, as Vincent Robert-Nicoud recently noticed, was the assumption that the world is "upside down" and should be restored to its previous order or to its pristine state.[94] Pizzimenti, according to Lüthy, was the early modern scholar who brought Democritus the alchemist to the stage, due to the publication of his Latin translations of the Greek chrysopoetic

89 Pausillo 2020: 141–159.

90 For these editions, see Ferguson 1894; Manzi 1974: 40 (no. 19); Neville 2006: 2.176.

91 Lüthy and Nicoli 2023.

92 Lüthy 2000.

93 Lüthy 2000: 455–461.

94 Robert-Nicoud 2018: 52–57.

texts.[95] And behind that new symbolic persona lurked something more than a mere need for restoration. It was a promise of restoration.

In actual life, the late medieval and the early modern variants of *alchymia* had been fields in which corpuscular theories proliferated, paving the way for the emergence of modern *philosophia experimentalis.*[96] On an imaginary level, the encounter, emblematically heralded in the title-page of Pizzimenti's anthology of translations, between Democritus the pioneer in the "sacred art" and Democritus the pioneer in the philosophical understanding of "natural things" transformed Democritus the inventor of natural philosophy into Democritus the renovator of natural philosophy, by means of a restoration of *ars chymica*. In a dedicatory letter prefixed to his book, addressed to the Cardinal Antoine Perrenot de Granvelle (1517–1586), diplomat of the Spanish court and art collector, who was also interested in alchemy,[97] Pizzimenti stressed that

> hanc philosophiae partem chymicorum ignorantium culpa infimam & sordidissimam omnium in pristinam dignitatem reuocandam existimaui, [...].
>
> this part of philosophy, which, through fault of the ignorant chymists, is the most sordid and vile of all, should be restored to its pristine dignity.[98]

He further explained that if there is anything true in the *ars chymica*, "it is all enclosed within these writings of Democritus" (*id totum in his Democritii scriptis contineri*).[99]

3 A Codex "Not Unpleasant to Those Who Have Dedicated Themselves to the Chymical Art"

But how could the content of "these writings of Democritus" be described, lest it be assimilated into the *alchymia* of the *chymici ignorantes*? Bessarion and the later compilers of the inventories of his library had come up against analogous questions of classification with regard to **M**. In the inventory of 1468, while Bessarion was still alive, **M** was placed at the end of a subsection containing works of Platonic and Neoplatonic literature. It was vaguely labelled as "Stephani et aliorum". In the inventory of 1474, the label of **M** was enriched with the qualifiers "philosopher" and "of Alexandria", as well as with information concerning the medium used in its production: "Stephani philosophi alexandrini et aliorum, in pergameno". In the inventory of 1524, when the

95 Lüthy 2000: 471–473.
96 Newman 2006.
97 Kahn 2007: 149–150; Legnani 2013.
98 Pizzimenti 1572: f. 4 recto.
99 Pizzimenti 1572: f. 4 verso.

humanist Andrea Navagero (1483–1529) was completing his term as librarian, five years before the transfer of Bessarion's collection to the Basilica of St. Mark, **M** was relocated to a division containing mathematical works and copies of Aristotle's works on natural philosophy. A new label was attached to it, too: "Stephani philosophi et aliorum complurium de sacra arte, in pergameno". Finally, in the inventory of 1543, when Pietro Bembo was passing the baton to his successor, Benedetto Ramberti (1503–1547), **M** was classified as part of a set of various works on moral philosophy, natural philosophy (e.g., Timaeus Locrus, *Περὶ φύσιος κόσμω καὶ ψυχᾶς*; On the Nature of the World and the Soul), and astronomy, the common feature of which seems to be no other than the difficulty itself of their classification in an unmistakably demarcated field.[100]

A century later, Anton Maria Zanetti and Antonio Bongiovanni, in their catalogue of the Greek manuscripts in the Marciana, numbered **M** among the medical codices, placing it at the end of the respective thematic section, immediately after a codex in which four medical treatises of the Byzantine physician John Zacharias Actuarius (ca. 1275–ca. 1330) are copied, and began their description of its contents with a general characterization of it as "collection of Greek chemists" (*chemicorum graecorum collectio*).[101] Morelli, in his own catalogue, published in 1802, did not retain the division into thematic groups employed by Zanetti and Bongiovanni, but left their general characterization of **M** intact.[102] When Juan Andrés spoke of "Greek chemists", he was actually using the same words as those used by the librarians of the Marciana, during the second half of the seventeenth century. With exactly the same words, one of the leading figures of the early modern European *respublica literaria*, the highly reputed bibliographer Johann Albert Fabricius (1668–1736), headed his Latin translation of the report prepared by the German physician and philologist Thomas Reinesius (1587–1667)[103] about a Greek alchemical manuscript compilation, preserved in the codex now registered in Forschungsbibliothek Gotha of the University of Erfurt as *Chartaceus* A 242, in the twelfth volume of his *Bibliotheca graeca*, printed in 1724.[104] This compilation was copied in 1623 from another manuscript witness to the Greek Alchemical Corpus that was then kept in the library of Augsburg and has now been lost.[105]

Originally written in German, on behalf of the Duke of Saxony, Johann Wilhlem, Reinesius' report was completed in 1634, but it became available in printed form many decades later, in 1714, when it was inserted in Ernst Salomon Cyprian's catalogue of the

---

100 Labowsky 1979: 176, 223, 247, 294.
101 Zanetti and Bongiovanni 1740: 140–141 (no. CCXCIX).
102 Morelli 1802: 173–178 (no. CCXCIX).
103 See Hase 1858.
104 This volume of *Bibliotheca graeca* was one of the three volumes of its first edition (1705–1728) that was not reworked and republished by Gottlieb Christoph Harles in the *editio nova* of 1790–1809. See Petzholdt 1866: 679–681. For Fabricius, see Petersen 1998.
105 Reinesius 1724. For a more recent description of the codex of Gotha see Goldschmidt 1932: 145–172 (no. 3).

manuscripts of the Ducal Library at Gotha.[106] Both in the first lines of the original report drawn up by Reinesius and in the sentence that Cyprian selected as a general heading for this entry in his catalogue – intended to characterize the contents of the Gotha codex – instead of a "collection of Greek chemists", we are told of a volume in which a variety of treatises "on chrysopoeia" (περὶ χρυσοποιΐας)[107], or "on the divine art, as they call it" (*de divina arte, wie sie die nennen*), are assembled, some genuinely produced by famous and non-famous Greek authors, others of spurious authorship.[108] The same general heading, "auctores graeci περὶ χρυσοποιΐας", was applied to the contents of a codex now enshrined in the Kongelige Bibliothek in Copenhagen, numbered as Thott 207,2º. This codex was copied from *Chartaceus* A 242 of Gotha by the historian of literature and philosophy Daniel Georg Morhof (1639–1691),[109] and later became the property of the Danish statesman and manuscript collector Otto Thott (1703–1785).[110] In the first folia of Thott 207,2º, the report of Reinesius was added, serving as an introduction to the following "chrysopoetic" treatises and their authors.[111]

A summary of Reinesius' report was also included in the chapter "de libris Physicis secretoribus, praecipuè Chemicis" (On the More Secret Books on Physics, Especially on Chemistry) of the first volume of the monumental *Polyhistor sive de notitia auctorum et rerum commentarii* (Polyhistor, or Commentaries for the Notice of Authors and Things Studied), which Morhof published in 1688. There, the erudite historian of literature, who five years later would author an *Epistola de transmutatione metallorum*, accepting the tenability of his contemporary chrysopoetic undertakings,[112] repeated what had been hypothesized by Reinesius in relation to the route of transmission of Greek chrysopoetic literature. These treatises, Reinesius had alleged, were originally written by "some monks and other scholars" (*etliche münche und andere gelehrte*) in Alexandria. Then, in Constantinople, they were brought together in "one unitary body" (*ein corpus*), and afterwards they arrived to Italy through the exiled scholars fleeing there after the conquest of Constantinople, along with a series of other valuable Greek books, unknown to the West up to then, such as Quintus Smyrnaeus' Ὁμήρου παραλειπόμενα, the poems of Tryphiodorus and Colluthus, or the Ἀργοναυτικὰ of Orpheus of Croton. The first name in the corresponding list of emigré scholars – given in parentheses by Reinesius but omitted by Morhof in his summarized version – is that of Bessarion.[113]

106 Cyprian 1714: 87–99. Reinesius' role in bringing to the fore the Greek Alchemical Corpus as a valuable primary source is touched upon also by Matton 1995: 329–331.
107 Cyprian 1714: 87: *Autores Graeci* περὶ χρυσοποιΐας.
108 Cyprian 1714: 88.
109 Kern 1928; Waquet 2000.
110 Suarez and Woudhuysen 2010: 1203.
111 Schartau 1994: 243–254.
112 Principe 2000.
113 Cyprian 1714: 90; Morhof 1688–1692: 1.102.

Reinesius for the first time publicly referred to the same codex in a book published in 1640, his *Variarum lectionum libri III priores.*[114] In this lengthy miscellany, made up of chapters on specific topics whose only common denominator is that they fall within the province of the textual and source criticism of classical literature, the authors that Fabricius and Juan Andrés would later tag as "chemists" were described as φιλόσοφοι practising and teaching the "sacred and great art" of χημευτικὴ or χρυσοποιΐα.[115] The question with which Reinesius deals at that particular point in his book is the meaning of the appellation of "poet by profession, according to his own words" (ποιητής, ὡς αὐτός φησι, τὸ ἐπιτήδευμα) ascribed to the early fifth-century pagan Greek historian and Roman imperial official Olympiodorus of Egyptian Thebes in the *Βιβλιοθήκη* (*Bibliotheca*) of the ninth-century Patriarch of Constantinople Photius,[116] the primary source thanks to which the only surviving fragments of Olympiodorus' *Ὕλη ἱστορίας* (Material for History) have come down to us. There is no evidence, Reinesius observes, that Olympiodorus, composed works in verse or that he was counted among the writers who were specialized in this literary genre, the "poets" in the narrowest and most popular sense of the term.[117] But beyond its common signification, the term ποιητὴς was also used to designate those who delved into the "sacred" chrysopoetic art and explicated it through the written word, the interpreters of "the accomplishment of the mystery" (τοῦ μυστηρίου τὴν ποίησιν), as indicated by a list of such authors, inscribed in the codex of Gotha, in which they are given the attribute of being ποιηταί. Reinesius identifies Olympiodorus who is mentioned in this list with Olympiodorus the historian, vindicating in this way the claim of Photius.[118] *Apud Romanos verò* ("assuredly in the Romans"), he elaborates further, among whom the art of χημεία itself was either utterly ignored or considered deceitful, "no distinct peculiar name had been conferred on these gold-making philosophers" (*aurifices isti philosophi peculiare nomen tum non habuêre*).[119]

A few decades later, Ole Borch (1626–1690), royal physician, poet, lecturer on philology, botany, and chemistry at the University of Copenhagen, and also one of the forerunners of the modern historiography of chemistry, author of the first separately printed monograph in Latin on the origins and progress of chemistry,[120] took up again the problem of the antiquity of the art of χημεία, in his book on the "wisdom"

114 *VD17* 39:154190G.

115 Reinesius 1640: 154.

116 Photius, *Bibliotheca*, cod. 80, 56b.13–14 (ed. Henry 1959–1977: 1.166).

117 Reinesius 1640: 154. For Olympiodorus see Rohrbacher 2002: 73–81. Only one hexameter verse written by him survives, interpolated into his history. It has been suggested that Olympiodorus might have been the author of *Βλεμυομαχία* (*Blemyomachia*), a Homerizing poem known only in fragments. See Livrea 1976.

118 Reinesius 1640: 155.

119 Reinesius 1640: 155.

120 See Koch 1866; Schepelern 1983; Shackelford 2004: 344–350. Factually, Borch was the first early modern scholar that became principally known as a historian of chemistry. From a present-day standpoint,

that Hermes Trismegistus and the ancient Egyptians supposedly upheld and propagated in common with the ancient "chemists" (*Hermetis, Aegyptiorum, et chemicorum sapientia*, 1674).[121] Arguing against the polemical refutation that Hermann Conring (1606–1681) – professor of natural philosophy, rhetoric, medicine, and politics at the University of Helmstedt[122] – had launched more than fifteen years earlier against the Paracelsian genealogical account of iatrochemistry (*De Hermetica Aegyptiorum vetere et Paracelsicorum nova medicina*, 1648),[123] Borch rejects Reinesius' identification of "Olympiodorus the chemist" (*Olympiodorus chemicus*) with Olympiodorus the historian. He does so on the grounds that the former neither hailed from Thebes nor adhered to paganism, as the latter reportedly did. "Olympiodorus the chemist", as attested by his surviving manuscript chrysopoetic treatise, came from Alexandria and was committed to the Christian faith. Accordingly, there is no advantage in making conjectures about the possible alternative meanings of the word ποιητὴς and the proclivity of Olympiodorus the historian to verse-writing.[124]

The identity match that was first suggested by Reinesius has been dismissed by the historians of alchemy in the late twentieth and early twenty-first centuries along similar lines. What remains a matter of debate[125] is the possible connection between Olympiodorus the alchemist, credited with a commentary on a treatise of Zosimus known only through the indirect tradition, and Olympiodorus the philosopher, an early sixth-century pupil of Ammonius in the school of Alexandria, whose extant works include three commentaries on Platonic dialogues and two on Aristotelian treatises.[126] Be that as it may, Borch's reproach against Reinesius' assumption about the identity of Olympiodorus the "poet" of chrysopoeia was prompted by his eagerness to challenge Reinesius' scholar authority in the public discussions on the antiquity of "chemistry". The main target of his attack was Hermann Conring's book, in which "chemistry" was stripped of its primeval aura.

Conring[127] had maintained that "all the hitherto extant chemical writers are posterior to Constantine" (*hodie exstant scriptores Chemici omnes infra Constantinum sunt*), the Roman Emperor of the early fourth century, and he invoked, among many others, the respected opinion of the "most learned" (*doctissimus*) Reinesius on Olympiodorus,

---

however, the primacy in the formation of the history of chemistry as a distinct field of modern scholarly research can more plausibly be granted to Torbern Bergman. See Nielsen 2001.

121 Neville 2006: 1.183–184.

122 Stolleis 1983; Fasolt 2004: 46–91.

123 Kelly and Stolleis 1983: 545 (nos. 89–89a).

124 Borch 1674: 76.

125 Letrouit 1990; 1995: 49–56; Viano 2021: 20. Cf. Viano 2006: 199–206; 1995: 99–102.

126 Tarrant 1998.

127 The first and second editions (in 1648 and 1669 respectively) of Conring's book have been discussed also by Matton 1995: 330–331.

the "chemical writer", to corroborate his thesis.[128] Incidentally, in the same year that Borch's book was published, Peter Lambeck (1628–1680), imperial librarian in Vienna since 1662,[129] put into print his own, very copious, deliberations on the relation between Olympiodorus the chemist, Olympiodorus the historian, and Zosimus. These appeared in the fifty-first entry of his catalogue of the Greek medical codices of the Bibliotheca Caesarea in Vienna, where he provides the description of a codex now residing in Vienna, at the Österreichische Nationalbibliothek, classified as MS *medicus graecus* 2, copied from **M**, in 1564.[130] The outcome of Lambeck's analysis was that Olympiodorus the historian, Olympiodorus the "chymical practitioner and writer" (*Operator ac Scriptor Chymicus*), and Olympiodorus the Alexandrian philosopher are different profiles of one and the same person, whereas Zosimus the *Historicus et Chymicus* was elder than this single, but multi-faceted, Olympiodorus.[131] Still unaware of Lambeck's contribution, Borch demurs about Conring's general inference, contending that many chemical writings are to be found in the Vatican Library, the Royal Library in Paris, the Library of St Mark in Venice, the Bavarian Library, the Library of Cologne, and elsewhere, which Conring should have taken into serious consideration, before laying down, "dictatorily" (*dictatoriè*), his pronouncement.[132]

Only more thorough, unprejudiced studies on the Greek "chemical" manuscripts, Borch insists, could allow more reliable conclusions on the depth of the past of "chemistry". His opponent, nevertheless, had not ignored the amount and the value of the manuscript tradition. In the first edition of his book, for example, it was mentioned that the works of Zosimus, together with those of other "Greek chemists", remain unedited, stored in Italian, French, and German libraries.[133] In the second, revised and enlarged edition, of 1669, which was, actually, the edition that sparked the polemical response of Borch, since it contained, as an appendix, a public "apologetic discourse" directed against the criticisms voiced in Borch's *De ortu et progressu chemiae dissertatio* (Dissertation on the Birth and Progress of Chemistry, 1668),[134] Conring had also interposed a piece from a recently published summary catalogue of the Greek manuscripts kept at the library of the royal palace-convent of the Escorial.[135]

Authored by the Jesuit Alexandre Barvoets (1619–1654),[136] this catalogue had been printed at Antwerp in 1648.[137] It had been inserted immediately after the "epistle to

128 Conring 1648: 22.
129 König 1979.
130 Goldschmidt 1932: 3–24 (no. 1).
131 Lambeck 1674: 184.
132 Borch 1674: 79.
133 Conring 1648: 17.
134 Neville 2006: 1.183.
135 Conring 1669: 33–34.
136 Dehergne 1973: 26 (no. 85).
137 Sommervogel et al. 1890–1932: 1, col. 995 (no. 2).

the reader" with which Balthasar Cordier opened his bilingual edition of nineteen sermons delivered by Origen on the Book of Jeremiah, falsely presented as written by Cyril of Alexandria.[138] In 1653, Barvoets' catalogue was reprinted, as one out of a set of ten supplements, in the first volume of the vast repository of bibliographies and chronicles that was crafted by another Jesuit, the erudite humanist Philippe Labbé (1607–1667),[139] which bore the title *Nova bibliotheca MSS. Librorum.*[140] Not much later, in 1671, a fire would consume almost half of the Greek codices of the Escorial.[141]

In Barvoets' catalogue, it is explicitly stated that in the library of the Spanish royal palace "in around seven or eight codices" (*in septem aut octo fere Codicibus*) Greek treatises could be encountered "on the philosophers' stone, on the sun and the moon, namely on the production of gold and silver, on the preparation of metals, chymical salts, etc." (*de lapide Philosophorum, de sole ac lunâ, id est, auri ac argenti confectione, praeparatione metallorum, salis chymici, etc.*), written by *auctores chymici* who all appear under lofty names, in most cases spuriously attached to them.[142] Furthermore, the Jesuit compiler points out that many of these manuscripts seem to coincide with the Greek manuscripts on the same topics procured by Maximilian I, Duke and later Elector of Bavaria (r. 1597–1651). And he quotes a passage from a letter sent by the professor of eloquence, and expert on ancient coins, Simon Wangnereck (1605–1657),[143] on behalf of Maximilian, listing the names of the authors encountered "in this most precious and entirely unheard-of chrysopoetic book repository" (*in hac pretiosissima & planè inaudita Chrysopaeorum Bibliotheca*).[144]

The only such codex, now held by the Bayerische Staatsbibliothek, is the one registered as MS *graecus* 112.[145] It is dated around 1620 and inscribed in the same hand as the codex of Gotha, *Chartaceus* A 242, about which Reinesius had written his 1634 report. These two codices derive from **M**. This holds also for yet another Greek alchemical codex copied in the same hand, which now belongs to the Universitätsbibliothek in Leipzig (*Lipsiensis Universitatis graecus* 66).[146] An obvious conclusion would be that in the mid-seventeenth century **M** attracted much more scholarly attention than was the case in the previous century. But many dilemmas still lingered as for the proper answer to the question of how its contents could be best described. The controversy between Conring and Borch over whether or not the origins of "chemistry" reach back to a remote antiquity was not erupted solely on the basis of philological and historical

138 Sommervogel et al. 1890–1932: 2, col. 1441 (no. 10).
139 Duclos 2001.
140 Sommervogel et al. 1890–1932: 4, cols 1315 (nos. 48 and 54), 1322–1323 (no. 72); Cochetti 1992: 54–57.
141 Graux 1880: 319–320.
142 Barvoets 1648: XXVI.
143 Stöger 1855: 391.
144 Barvoets 1648: XXVI.
145 Goldschmidt 1932: 246–272 (no. 10); Hajdú 2003: 33–42.
146 Goldschmidt 1932: 222–246 (no. 9).

questions. Above anything else, at stake was the endorsement by the European scholarly milieus of Paracelsian "chemical philosophy", which combined alchemical theory with medical practice, as a legitimate branch of natural philosophy. Robert Halleux has noted that the scrutiny of how and when "chemistry" emerged should be seen as the scholarly counterpart of an "epidemic of transmutations which inflamed Europe" (*épidémie de transmutations qui enfièvre l'Europe*), during the same period, allegedly furnishing empirical proofs for the Paracelsian theories on metals.[147] Both Conring and Borch were willing to apply the terms "chemists" and "chemistry" to the authors and the treatises found in **M**. Conring, however, did not subscribe to the Paracelsian ambitions – expressed in a polemical thesis resembling the promise made by Pizzimenti in 1570 with regard to the *ars sacra* of Democritus: the partisans of Paracelsus portrayed their own version of *alchymia* as the remedy capable of renovating natural philosophy by reviving a primordial tradition.

The Chian scholar Leone Allacci (1588–1669), who lived in Rome for most of his career and rose to great prominence as a theologian, philologist, manuscript copyist, editor, and librarian of the Biblioteca Vaticana,[148] stayed away from the controversy over the ancient roots of Paracelsianism, without staying away from **M**. In his 1634 historical diatribe *De Psellis et eorum scriptis* (On the Authors Bearing the Name Psellus and Their Writings),[149] he announced that he was planning to edit and publish a comprehensive collection of Greek texts on the "chymical art", as part of a larger collection of various ancient texts (*Varia antiqua*) which he was preparing. The Greek originals in this collection would be complemented with their Latin versions, and it was implied that for the accomplishment of this particular task the Latin translations issued by Pizzimenti would also be used. The plan of Allacci was to publish a single-topic volume, "not unpleasant to those who have dedicated themselves to the chymical art" (*Chymicae artis deditis non iniucundum*), in which, apart from the treatises translated by Pizzimenti, the unedited Greek texts of Zosimus, Olympiodorus, Christianus, and other authors of this kind would be included, along with relevant short treatises of his own.[150]

The first two volumes of Allacci's promised "Varia antiqua" found their way into print in 1653, under the title of *Σύμμικτα, sive Opusculorum Graecorum et Latinorum, vetustiorum et recentiorum libri duo* (Collection, or Two Books of Older and More Recent Greek and Latin Opuscules),[151] edited by Barthold Nihus and printed by Joan Blaeu in Amsterdam, with false printer's name (Jost Kalckhoven) and place of

147 Halleux 1980: 812.

148 Papadopoulos 1987; 1989; Cerbu 1986; Hartnup 2004: 53–104; Surace 2014: 199–204.

149 Papadopoulos 1984–1986: 1.22 (no. 246); Jacono 1962: 12–13 (no. 10).

150 Allacci 1634: 45–46.

151 Papadopoulos 1984–1986: 1.25 (no. 280); Jacono 1962: 27 (no. 36).

publication (Cologne) on the title-page.[152] None of them contained any alchemical text. When, though, Ole Borch visited Rome, from October 1665 to March 1666, Allacci confided to him, during one of their casual conversations, that he still intended "to bring the Greek chemical writers out into the public light" (*Graecos scriptores Chemicos publicae luci* [...] *exponere*). Recalling in his *Conspectus scriptorum chemicorum illustriorum* (Overview of the More Distinguished Chemical Writers), published posthumously in 1697,[153] the oral assurance given back then, Borch regretted the fact that the completion of this edition was impeded by the growing physical feebleness, and finally by the death of the aged scholar.[154] Allacci died in 1669, and none of the remaining eight volumes of his *Σύμμικτα* was ever published. His plan to publish the Greek Alchemical Corpus was not executed, but the evidence supporting the possibility that he was indeed working on it is substantial.[155] Among the relevant codices that he consulted one derives from **M**: a codex of the Biblioteca Apostolica Vaticana, registered as *Barberinianus graecus* 197, which has been copied by Allacci himself and Lorenzo Porzio (1604–1676), a younger *scriptor graecus* who since 1654 collaborated with him in their shared workspace of the Biblioteca Vaticana.[156] In preparing his "chymical" or "chemical" anthology, Allacci browsed through Latin "alchymical" manuscripts too. He had started recording the bibliographical information required for mapping competently the literature on *alchymia* and χημεία which was then accessible.[157]

As the end of the Chian scholar was approaching, the table of contents of the complete series of *Σύμμικτα*, projected at a total of ten volumes, was found among his papers and was published in 1668 by the poet and humanist Andrea Peschiulli (1601–1691), who had become one of his most intimate friends.[158] The collection of the Greek alchemical texts was envisaged to form the ninth volume of *Σύμμικτα*.[159] Incidentally, Peschiulli was himself one of the *Chymicae artis dediti*, at whom this collection was principally aimed. If we are to believe his younger contemporary biographer, Domenico de Angelis, he was engrossed "much more than what would be deemed befitting to a wise man" (*più di quell, che ad uom saggio si convenia*) in the study and application of alchemy

152 Sterck 1936.
153 Neville 2006: 1.183.
154 Borch 1697: 10.
155 Koutalis 2022: 377–380.
156 For Portius, who was also of Chian descent on his father's side, see Ciaramitaro 2006.
157 A. Berthelot 1887: 848–854.
158 Peschiulli signed, as one of the witnesses, the two *codicilli*, the testamentary supplements, dictated by Allacci on his deathbed. In the formal testament of the Chian scholar, Peschiulli is addressed as "suo amico" (his friend) and is granted a considerable amount of money (five *scudi*) per month for the rest of his life. See the texts of the testament and the two *codicilli* as published by Legrand 1894–1896: 3.447–459; Peschiulli is mentioned on pp. 450, 458, and 459.
159 Allacci 1668: 21–23.

(*alchimia*).[160] Among his unedited works, as enumerated by de Angelis, there is a Latin five-part book entitled *Lusus chymici* (A Play of the Chymist), the fourth section of which is an "oration" (*declamatio*) against the "pseydo-chymists" (*pseydochymici*), whereas the fifth is devoted to a novel kind of alchemy, a "chymistry hitherto unheard of" (*Chymia hactenus inaudita*).[161] The preoccupation of Peschiulli with alchemy struck many of those "who deemed him to be a man of consummate prudency and of solid and mature apprehension" (*che lo stimarono per uomo di consumata prudenza, e di saldo, e maturo intendimento*) as being somewhat "odd" (*strana*).[162] It did not tarnish his scholarly reputation nonetheless. Two other late-seventeenth- and early-eighteenth authors speak of Peschiulli as a "man excellently versed in all erudition" (*huomo versatissimo in ogni erudition*)[163] and "worthy of the highest praise from the Republic of the Erudite Scholars" (*meritevole di ogni eccelso encomio nella Repubblica degli Eruditi*).[164]

Many years later, in 1745, excerpts of **M** were published for the first time as an appendix to an edition prepared by the physician and philologist Johannes Stephanus Bernard (1718–1793).[165] This edition contained the Greek original text and a Latin translation of *Περὶ πυρετῶν σύντομος σύνοψις* (Synopsis on Fevers), a work dubiously attributed to the Alexandrian medical commentator Palladius. The texts published were a lexicon on chrysopoeia, under the Latin title *Lexicon alphabeticum chemicum* and the poems of Heliodorus, Theophrastus, Hierotheus, and Archelaus, as *Excerpta ex scriptoribus chemicis*, preceded by an introduction in which the contents of **M** were described and a list of the "philosophers of the sacred science" was reproduced.[166] They were transcribed from **M** and edited by Jacques Phillipe d'Orville (1696–1751), a book-collector and former professor of history, eloquence, and Greek at Amsterdam (1730–1742).[167] By the mid-eighteenth century, **M** was regarded as a chemical codex, with hardly any ambiguities remaining.

In the ninth chapter of the third book ("De l'imperfection des mots"; On the Imperfection of Words) of his polemical philosophical dialogue against John Locke's empiricism, *Nouveaux essais sur l'entendement humain* (New Essays on Human Understanding) – completed in 1705 but published posthumously in 1765 – Gottfried Wilhelm Leibniz (1646–1716), speaking through his fictional mouthpiece Theophilus, laments that ancient Greek medical practice was transmitted to the early modern West indirectly, through second-hand Arabic reports. "The spring-water was muddied in the Arab rivulets" (*l'eau de la source a eté troublée dans les ruisseaux des Arabes*) and thus

160 de Angelis 1710: 162.
161 de Angelis 1710: 163.
162 de Angelis 1710: 162.
163 Tasselli 1693: 515.
164 Scarfò 1720: 169.
165 Douma 1939.
166 Douma 1939: 52–53.
167 Singer et al. 1924: 12–16 (no. 3); Madan et al. 1895–1953: 4.37.

the recourse to the original Greek texts is indispensable for any proper understanding of the procedures followed, and the treatments proposed, by the ancient Greek physicians. Medicine, Theophilus argues, "as it is empirical" (*entant qu'elle est Empirique*), is the third most profitable area of knowledge, second only to religion and history, from the repository of the ancient traditions which are preserved in written form.[168] It is for this reason that he stresses the importance of the physicians who, besides being properly trained in current medicine, are also acquainted with antiquity. And for this same reason he goes on to add:

> j'ai eté bien faché que Reinesius excellent dans l'un et l'autre genre, s'etoit tourné plutot à eclaircir les rites et Histoires des Anciens, qu'à retablir une partie de la connoissance, qu'ils avoient de la nature, où il a fait voir qu'il auroit encore pû reussir à merveille.
>
> it was most vexing to me that Reinesius, who excelled in both fields, preferred to work on explaining the rituals and the history of the ancients, rather than on recovering some part of their knowledge of nature – a task which he has shown that he could also have performed with marvellous success.[169]

Leibniz's comment on Reinesius' reluctance to study the written monuments of ancient medicine as monuments of medicine practised by the ancients, transmitting "some part of their knowledge of nature", resonates to some extent with Borch's strictures against Reinesius concerning the chemical quality of the contents of **M**.

Not too long thereafter, chemists such as Bergman and Fourcroy would enter into the historical scene, and the history of **M** among the early modern *literati* would be transformed into a history indwelling in a non-chemical universe. This history from the outset had certain serious "blind spots". It revolved around a Byzantine codex, preserving a florilegium formed in Constantinople, as Reinesius in 1634 had assumed, but it repressed and gradually erased the characteristically Byzantine aspects of the contents of this codex, and more broadly of χημεία (e.g., those arising from the Eastern Christian assimilation of pagan intellectual traditions). The *literati* who were the principal actors in this history tended also to repress the aspects of χημεία that were grounded in the experience of working. Their debates or quarrels centered, mostly, on the possible associations to Greek philosophy. The technical-operational texts of **M** were seen as irrelevant to the philosophical texts, and vice-versa. Ironically, the historical discussions on pre-modern chemistry have been framed by both those "blind spots" up until our days.

168 Leibniz 1765: 298; ed. Robinet and Schepers 1962: 336.
169 Leibniz 1765: 299; ed. Robinet and Schepers 1962: 336; Engl. transl. Remnant and Bennett 1996: 336.

Acknowledgement: I would like to thank Matteo Martelli for his comments and suggestions and Mina Markoula for her language editing.

Bibliography

Sources

Allacci 1634: Leonis Allatii *De Psellis et eorum scriptis diatriba*. Rome: Mascardi, 1634.

Allacci 1668: Leonis Allatii *Συμμίκτων sive Opusculorum Graecorum, et Latinorum vetustiorum ac recentiorum libri X*, ed. Andreas Peschiullius. Rome: Mascardi, 1668.

Andrés 1782–1799: Juan Andrés, *Dell'origine, de' progressi e dello stato attuale d'ogni letteratura*, 7 vols. Parma: Stamperia Reale, 1782–1799.

Andrés 1786–1793: Juan Andrés, *Cartas familiares del Abate D. Juan Andrés a su hermano D. Carlos Andrés*, 5 vols. Madrid: A. de Sancha, 1786–1793.

Barvoets 1648: Alexandre Barvoets, "Catalogus praecipuorum auctorum ineditorum graece M.SS. qui in Bibliotheca Scorialensi asservantur. Opera Alexandri Barvoetii è Societate Iesu", in Balthasar Cordier (ed.), *S. P. N. Cyrilli Archiepiscopi Alexandrini Homiliae XIX in Ieremiam Prophetam, Hactenus ineditae, ac nunc demum ex antiquissimo codice M. S. Regiae Bibliothecae Scorialensis descriptae, & latinitate donatae à Balthasare Corderio Antuerpiensi, Soc. Iesu Theologo*. Antwerp: Plantin – Moretus, 1648, pp. XXI–XXXVIII.

Bergman and Afzelius Arvidsson 1787: Torbern Bergman and Pehr Afzelius Arvidsson, *Historiae chemiae medium seu obscurum aevum a medio saeculo VII. ad medium saec. XVII*, in Torberni Bergman *Opuscula physica et chemica pleraque seorsim antea edita nunc collecta et revisa*, vol. 4, ed. Ernestus Benjamin Gottlieb Hebenstreit. Leipzig: J. G. Müller, 1787, pp. 85–141.

Bergman and Paulin 1787: Torbern Bergman and Jacob Paulin, *De primordiis chemiae*, in Torberni Bergman *Opuscula physica et chemica pleraque seorsim antea edita nunc collecta et revisa*, vol. 4, ed. Ernestus Benjamin Gottlieb Hebenstreit. Leipzig: J. G. Müller, 1787, pp. 1–84.

Borch 1674: *Hermetis, Aegyptiorum, et chemicorum sapientia ab Hermanni Conringii animadversionibus vindicata per Olaum Borrichium*. Copenhagen: P. Haubold, 1674.

Borch 1697: Olai Borrichii *Conspectus scriptorum chemicorum illustriorum, libellus posthumus cui praefixa Historia vitae ipsius ab ipso conscripta*. Copenhagen: S. Garmann, 1697.

Conring 1648: Hermanni Conringii *De Hermetica Aegyptiorum vetere et Paracelsicorum nova medicina liber unus. Quo simul in Hermetis Trismegisti omnia, ac universam cum Aegyptiorum tum Chemicorum doctrinam animadvertitur*. Helmstedt: H. Müller, 1648.

Conring 1669: Hermanni Conringii *De Hermetica medicina libri duo. Quorum Primus agit de Medicina, pariterque de omni Sapientia veterum Aegyptiorum: Altero non tantùm Paracelsi, sed etiam Chemicorum, Paracelsi laudatorum aliorumque, potissimum quidem Medicina omnis, simul vero & reliqua universa doctrina examinatur. Editio secunda infinitis locis emendatior & auctior*. Helmstedt: H. Müller, 1669.

Cyprian 1714: Ernst Salomon Cyprian, *Catalogus codicum manuscriptorum Bibliothecae Gothanae*. Leipzig: J. F. Gleditsch, 1714.

de Angelis 1710: Domenico de Angelis, "Vita di Andrea Peschiulli da Corigliano", in *Le vite de letterati Salentini scritte da Domenico de Angelis uno de'dodici Colleghi d'Arcadia*, vol. 1. Florence: [s.n.], 1710, pp. 137–164.

Dee 1564: Ioannis Dee *Monas Hieroglyphica*. Antwerp: G. Silvius, 1564.

[Democritus], ***Physica et mystica***. Matteo Martelli (ed. and Engl. transl.), *The* Four Books *of Pseudo-Democritus* (Sources of Alchemy and Chemistry 1; Ambix 60, Supplement 1). Leeds: Maney Publishing, 2013, pp. 79–103.

Fourcroy 1800: Antoine François Fourcroy, *Système des connaissances chimiques, et de leurs applications aux phénomènes de la nature et de l'art*, 10 vols. Paris: Baudouin, 1800.
Engl. transl.: William Nicholson, *A General System of Chemical Knowledge; and Its Application to the Phenomena of Nature and Art. By A. F. Fourcroy*, 11 vols. London: Cadell and Davies et al., 1804.

Germ. transl. (vols 1 and 5): David Joseph Veith and Christian R. W. Wiedemann, *Fourcroy's System der chemischen Kenntnisse und Darstellung ihrer Anwendung auf die Erscheinungen der Natur und zu den Zwecken der Kunst*, 2 vols. Braunschweig: F. Vieweg, 1801.
Span. transl.: Pedro María Olive and Gregorio González Azaola, *Sistema de los conocimientos químicos y de sus aplicaciones a los fenomenos de la naturaleza y del arte: Obra escrita en frances por A. F. Fourcroy*, 10 vols. Madrid: Imprenta Real, 1803–1809.

Ioannes Antiochenus, *Fragmenta*: Umberto Roberto (ed. and It. transl.), *Ioannis Antiocheni fragmenta ex Historia chronica* (Texte und Untersuchungen zur Geschichte der altchristlichen Literatur 154). Berlin and New York: De Gruyter, 2005.
Sergei Mariev (ed. and Engl. transl.), *Ioannis Antiocheni fragmenta quae supersunt omnia* (Corpus Fontium Historiae Byzantinae 47). Berlin and New York: De Gruyter, 2008.

Lambeck 1674: Petri Lambecii *Sacrae Caesareae maiestatis Consiliarii, historiographi ac bibliothecarii, Commentariorum de Augustissima Bibliotheca Caesarea Vindobonensi liber sextus*. Vienna: M. Cosmerovius, 1674.

Leibniz 1765: Gottfried Wilhelm Leibniz, *Nouveaux essais sur l'entendement humain*, in *Œuvres philosophiques latines & françoises de feu Mr. de Leibnitz. Tirées de ses manuscrits qui se conservent dans la bibliothèque royale à Hanovre*, et publiées par Mr. Rud. Eric Raspe. Amsterdam and Leipzig: J. Schreuder, 1765, pp. 1–496.
Edition: André Robinet and Heinrich Schepers (eds), *Gottfried Wilhelm Leibniz, Nouveaux essais* (Gottfried Wilhelm Leibniz, Sämtliche Schriften und Briefe, Sechste Reihe, Philosophische Schriften 6). Berlin: Akademie Verlag, 1962.
Engl. transl.: Peter Remnant and Jonathan Bennett, *G. W. Leibniz: New Essays on Human Understanding* (Cambridge Texts in the History of Philosophy). Second edition. Cambridge: Cambridge University Press, 1996.

Morelli 1802: Iacobi Morellii *Bibliothecae Regiae Divi Marci Venetiarum custodis bibliotheca manuscripta graeca et latina: Tomus primus*. Bassano: Remondini, 1802.

Morhof 1688–1692: Danielis Georgi Morhofi *Polyhistor sive de notitia auctorum et rerum commentarii. Qvibus praeterea varia ad omnes disciplinas consilia et subsidia proponuntur*, 2 vols. Lübeck: P. Böckmann, 1688–1692.

Photius, *Bibliotheca*: René Henry (ed. and Fr. transl.), *Photius, Bibliothèque*, 8 vols (Collection byzantine). Paris: Belles Lettres, 1959–1977.

Pizzimenti 1572: *Democritus Abderita De arte magna, sive de rebus naturalibus. Nec non Synesii, & Pelagii, & Stephani Alexandrini, & Michaelis Pselli in eundem commentaria. Dominico Pizimentio Vibonensi interprete*. Second edition. Padua: S. Galignano, 1572.

Reinesius 1640: Thomae Reinesii *Variarum lectionum libri III priores. In quibus de scriptoribus sacris & profanis classicis plerisque disseritur* […]. Altenburg: O. Michael, 1640.

Reinesius 1724: Thomae Reinesii "De Collectione MSS. Chemicorum Graecorum, qua exstat in Biblioth. Gothana", transl. Johann Albert Fabricius, in Johann Albert Fabricius (ed.), *Bibliotheca graeca, sive notitia scriptorum veterum graecorum*, vol. 12: *In quo post elenchum situs episcopatuum orbis Christiani ultra quater-mille, et scriptorum historiae ecclesiasticae notitiam, de jctis & medicis Graecis disseritur*. […]. Hamburg: T. C. Felginer, 1724, pp. 748–760.

Scarfò 1720: Grisocarpo Pegeo P. A. della Colonia Sebezia [Grisostomo Scarfò], "Andrea Peschiulli", in Giovanni Mario Crescimbeni (ed.), *Notizie istoriche degli Arcadi morti*, vol. 1. Rome: A. de'Rossi, 1720, pp. 167–170.

Stephanus Alexandrinus, *De magna et sacra arte*: Maria K. Papathanassiou (ed.), *Stephanos von Alexandreia und sein alchemistisches Werk. Die kritische Edition des griechischen Textes eingeschlossen*. Athens: Cosmosware, 2017.
Partial Engl. transl.: Frank Sherwood Taylor, "The Alchemical Works of Stephanos of Alexandria: Translation and Commentary: Part II", *Ambix* 2.1 (1938), pp. 39–49.

Tasselli 1693: Luigi Tasselli, *Antichità di Leuca: Città già posta nel campo salentino. De' luoghi, delle terre, e d'altre città del medesimo promontorio* [...]. *Opera del M. R. P. Luigi Tasselli di Casarano Predicatore, e Teologo Capuccino, già Diffinitore de' Padri dello stesso Ordine nella Provincia di Otranto* [...]. Lecce: Micheli, 1693.

Zanetti and Bongiovanni 1740: Anton Maria Zanetti and Antonio Bongiovanni, *Graeca D. Marci Bibliotheca codicum manu scriptorum per titulos digesta. Praeside et moderatore Lavrentio Theupolo equite ac D. Marci Proc. iussu senatus*. Venice: S. Occhi, 1740.

Literature

Abraham 1997: Lyndy Abraham, "Introduction", in Lyndy Abraham (ed.), *Arthur Dee*, Fasciculus chemicus, *Translated by Elias Ashmole* (English Renaissance Hermeticism 6). New York: Garland, 1997, pp. XI–LXXXI.

A. Berthelot 1887: André Berthelot, "Rapport sur les manuscrits alchimiques de Rome", *Archives des missions scientifiques et littéraires*, 3ème série, 13 (1887), pp. 819–854.

M. Berthelot 1885: Marcellin Berthelot, *Les Origines de l'alchimie*. Paris: G. Steinheil, 1885.

M. Berthelot 1889: Marcellin Berthelot, *Introduction à l'étude de la chimie des anciens et du moyen âge*. Paris: G. Steinheil, 1889.

Burigana 2012: Riccardo Burigana, "Morelli, Jacopo", in Benvenuto Bertoni (ed.), *Dizionario Biografico degli Italiani*, vol. 72. Rome: Istituto della Enciclopedia Italiana, 2012, pp. 628–631.

Carlotta 2023: Vincenzo Carlotta, "Translating Alchemical Practice into Symbols: Two Cases from Codex *Marcianus graecus* 299", in Katja Krause, Maria Auxent, and Dror Weil (eds), *Premodern Experience of the Natural World in Translation*. New York: Routledge, 2023, pp. 217–229.

Cerbu 1986: Thomas Cerbu, *Leone Allacci (1587–1669): The Fortunes of an Early Byzantinist*. PhD Dissertation. Harvard University, 1986.

Ciaramitaro 2006: Giuliana Ciaramitaro, "Lorenzo Porzio (1604–1676): cenni biografici", *Schede umanistiche* NS 20.1 (2006), pp. 151–164.

Clulee 1988: Nicholas H. Clulee, *John Dee's Natural Philosophy: Between Science and Religion*. London: Routledge, 1988.

Coccia 1973: Antonio Coccia, "Vita e opere del Bessarione", *Miscellanea Franciscana* 73.3–4 (1973), Special Issue: *Il Cardinale Bessarione nel V centenario della morte 1472–1972*, pp. 265–293.

Cochetti 1992: Maria Cochetti, "Philippe Labbe", *Il bibliotecario* 33–34 (1992), pp. 49–61.

Coluccia 2009: Giuseppe L. Coluccia, *Basilio Bessarione: lo spirito greco e l'occidente* (Accademia delle Arti del Disegno, Monografie 15). Florence: Leo S. Olschki, 2009.

Dehergne 1973: Joseph Dehergne, *Répertoire des Jésuites de Chine de 1552 à 1800* (Bibliotheca Instituti Historici S. I. 37). Rome: Institutum Historicum S. I.; Paris: Letouzey & Ané, 1973.

Del Soldato 2012: Eva Del Soldato, "*Illa litteris Graecis abdita*: Bessarion, Plato, and the Western World", in Marco Sgarbi (ed.), Translatio Studiorum*: Ancient, Medieval and Modern Bearers of Intellectual History* (Brill's Studies in Intellectual History 217). Leiden and Boston, MA: Brill, 2012, pp. 107–122.

Del Soldato 2020: Eva Del Soldato, *Early Modern Aristotle: On the Making and Unmaking of Authority*. Philadelphia, PA: University of Pennsylvania Press, 2020.

Douma 1939: Sjoerd Douma, *Johannes Stephanus Bernard, medicus en philoloog, 1718–1793*. Assen: Van Gorcum & Comp., 1939.

Duclos 2001: Paul Duclos, "Labbe, Philippe", in Charles E. O'Neill and Joaquín M. Domínguez (eds), *Diccionario histórico de la Compañía de Jesús: biográfico-temático*, vol. 3. Rome: Institutum Historicum S. I.; Madrid: Universidad Pontificia Comillas, 2001, pp. 2250–2251.

Fasolt 2004: Constantine Fasolt, *The Limits of History*. Chicago, IL: University of Chicago Press, 2004.

Ferguson 1894: John Ferguson, *On the First Edition of the Chemical Writings of Democritus and Synesius: Four Papers Read to the Philosophical Society of Glasgow, November, 1884–October, 1894*. Glasgow: R. Anderson, 1894.

Festugière 1939: André-Jean Festugière, "Alchymica", *L'antiquité classique* 8 (1939), pp. 71–95.

Formentin 1978: Mariarosa Formentin, *I codici greci di medicina nelle tre Venezie* (Università di Padova, Studi bizantini e neogreci 10). Padua: Liviana, 1978.

Formentin 2004: Maria Rosa Formentin, "Domenico Pizzimenti Vibonense: maestro, interprete, copista del sec. XVI", in Maurizio Baldin, Marialuisa Cecere, and Daria Crismani (eds), *Testi medici latini antichi: le parole della medicina, lessico e storia* (Edizioni e saggi universitari di filologia classica 9). Bologna: Pàtron, 2004, pp. 691–701.

Gabaldón 2017: Jesús García Gabaldón, *Juan Andrés (1740–1817). Ensayo de una biografía intelectual* (Verbum Menor). Madrid: Verbum, 2017.

Giménez López 2021: Enrique Giménez López, *Juan Andrés. Un erudito en el exilio de Italia* (Historia). Alicante: Publicacions Universitat d'Alacant, 2021.

Goldschmidt 1932: Günther Goldschmidt, *Manuscrits d'Allemagne, d'Autriche, de Danemark, de Hollande et de Suisse* (Catalogue des manuscrits alchimiques grecs 4). Brussels: Secrétariat administratif de l'U. A. I., 1932.

González Palencia and Mele 1941–1943: Ángel González Palencia and Eugenio Mele, *Vida y obras de Don Diego Hurtado de Mendoza*, 3 vols. Madrid: Instituto de Valencia de Don Juan, 1941–1943.

Graux 1880: Charles Graux, *Essai sur les origines du fonds grec de l'Escurial: épisode de l'histoire de la renaissance des lettres en Espagne* (Bibliothèque de l'École des Hautes Études, Sciences philologiques et historiques 46). Paris: F. Vieweg, 1880.

Halleux 1980: Robert Halleux, "La controverse sur les origines de la chimie, de Paracelse à Borrichius", in Jean-Claude Margolin (ed.), *Acta conventus neo-latini Turonensis: Troisième congrès international d'études néo-latines*, 2 vols (De Pétrarque à Descartes 38). Paris: Librairie Philosophique J. Vrin, vol. 2, pp. 807–819.

Hase 1858: Eduard Friedrich Hase, "Dr. Thomas Reinesius, Stadtphysikus und Bürgermeister zu Altenburg. Ein Lebensbild aus dem 17. Jahrhundert", *Mittheilungen der Geschichts- und Alterthumsforschenden Gesellschaft des Osterlandes* 4 (1858), pp. 309–348.

Hajdú 2003: Kerstin Hajdú, *Catalogus codicum manu scriptorum Bibliothecae Monacensis*, vol. 2.3: *Codices graecos 110–185 continens*. Wiesbaden: Harrassowitz, 2003.

Hartnup 2004: Karen Hartnup, *"On the Beliefs of the Greeks": Leo Allatios and Popular Orthodoxy* (The Medieval Mediterranean 54). Leiden and Boston, MA: Brill, 2004.

Hutter 1982: Irmgard Hutter, *Corpus der byzantinischen Miniaturenhandschriften*, vol. 3.1: *Oxford Bodleian Library III: Textband*. Stuttgart: A. Hiersemann, 1982.

Jacono 1962: Carmela Jacono, *Bibliografia di Leone Allacci (1588–1669)* (Quaderni dell'Istituto di Filologia greca della Universita di Palermo 2). Palermo: Presso l'Accademia, 1962.

Joly 2013: Bernard Joly, *Histoire de l'alchimie*. Paris: Vuibert and Adapt-Snes, 2013.

Josten 1964: Conrad H. Josten, "A Translation of John Dee's 'Monas Hieroglyphica' (Antwerp, 1564), with an Introduction and Annotations", *Ambix* 12.2–3 (1964), pp. 84–221.

Kahn 2007: Didier Kahn, *Alchimie et paracelsisme en France à la fin de la Renaissance (1567–1625)* (Cahiers d'Humanisme et Renaissance 80). Geneva: Droz, 2007.

Kahn 2016: Didier Kahn, *Le fixe et le volatil. Chimie et alchimie, de Paracelse à Lavoisier*. Paris: CNRS Éditions, 2016.

Kahn and Matton 1995: Didier Kahn and Sylvain Matton (eds), *Alchimie: art, histoire et mythes* (Textes et travaux de *Chrysopoeia* 1). Paris: S. É. H. A.; Milan: Archè, 1995.

Kelly and Stolleis 1983: William Ashford Kelly and Michael Stolleis, "Hermann Conring: Gedruckte Werke 1627–1751", in Michael Stolleis (ed.), *Hermann Conring (1606–1681): Beiträge zu Leben und Werk* (Historische Forschungen 23). Berlin: Duncker & Humblot, 1983, pp. 535–572.

Kennedy 2018: Scott Kennedy, "Bessarion's Date of Birth: A New Assessment of the Evidence", *Byzantinische Zeitschrift* 111 (2018), pp. 641–658.

Kern 1928: Marie Kern, *Daniel Georg Morhof: Inaugural-Dissertation zur Erlangung der Doktorwürde der Hohen Philosophischen Fakultät der Albert-Ludwig-Universität zu Freiburg*. Landau-Pfalz: Vorderpfälzische Genossenschafts-Druckerei, 1928.

Kersaint 1966: Georges Kersaint, *Antoine-François de Fourcroy (1755–1809): sa vie et son œuvre*. Paris: Muséum national d'Histoire naturelle, 1966.

Kidwell 2004: Carol Kidwell, *Pietro Bembo: Lover, Linguist, Cardinal*. Montreal: McGill-Queen's University Press, 2004.

Koch 1866: Emil Ferdinand Koch, *Oluf Borck, en literærhistorisk-biografisk Skildring*. Copenhagen: J. Lund, 1866.

König 1979: Gebhard König, "Peter Lambeck (1628–80), Bibliothekar Kaiser Leopolds I", *Mitteilungen des Instituts für Österreichische Geschichtsforschung* 87 (1979), pp. 121–166.

Kopp 1869–1875: Hermann Kopp, *Beiträge zur Geschichte der Chemie*, 3 vols. Braunschweig: Vieweg, 1869–1875.

Koutalis 2022: Vangelis Koutalis, *Η αλχημεία στο έργο των ελληνόφωνων λογίων της Ιταλίας: 15ος–17ος αιώνας (Βησσαρίων, Angelo Forte, Leone Allacci)*. PhD Dissertation. University of Ioannina, 2022.

Koutalis et al. 2018: Vangelis Koutalis, Matteo Martelli, and Gerasimos Merianos, "Graeco-Egyptian, Byzantine and Post-Byzantine Alchemy: Introductory Remarks", in Efthymios Nicolaidis (ed.), *Greek Alchemy from Late Antiquity to Early Modernity* (De Diversis Artibus 104). Turnhout: Brepols, 2018, pp. 11–43.

Labowsky 1966: Lotte Labowsky, "Il Cardinale Bessarione e gli inizi della Biblioteca Marciana", in Agostino Pertusi (ed.), *Venezia e l'Oriente fra tardo Medioevo e Rinascimento* (Civiltà europea e civiltà veneziana, Aspetti e problemi 4). Florence: Sansoni, 1966, pp. 159–182.

Labowsky 1979: Lotte Labowsky, *Bessarion's Library and the Biblioteca Marciana: Six Early Inventories* (Sussidi eruditi 31). Rome: Storia e letteratura, 1979.

Lagercrantz 1927: Otto Lagercrantz, "Ueber das Verhältnis des Codex Parisinus 2327 (= A) zum Codex Marcianus 299 (= M)", in Zuretti et al. 1927: 341–358.

Lagercrantz 1932: Otto Lagercrantz, "Ueber das Verhältnis des Codex Parisinus 2327 (= A) zum Codex Marcianus 299 (= M) (Fortsetzung von Catalogue II 341–358)", in Goldschmidt 1932: 399–432.

Lebègue 1924: Henri Lebègue, *Catalogue des manuscrits alchimiques grecs*, vol. 1: *Les* Parisini. Brussels: M. Lamertin, 1924.

Legnani 2013: Marco Legnani, *Antonio Perrenot de Granvelle: Politica e diplomazia al servizio dell'imperio spagnolo (1517–1586)*. Milan: Unicopli, 2013.

Legrand 1894–1896: Émile Legrand, *Bibliographie hellénique, ou description raisonnée des ouvrages publiés par des Grecs au dix-septième siècle*, 4 vols. Paris: A. Picard et Fils, 1894–1896.

Lennartson 2020: Anders Lennartson, *Carl Wilhelm Scheele and Torbern Bergman: The Science, Lives and Friendship of Two Pioneers in Chemistry* (Perspectives on the History of Chemistry). Cham: Springer, 2020.

Letrouit 1990: Jean Letrouit, "Datation d'Olympiodore l'alchimiste", *Emerita* 58.2 (1990), pp. 289–292.

Letrouit 1995: Jean Letrouit, "Chronologie des alchimistes grecs", in Kahn and Matton 1995, pp. 11–93.

Letrouit 2002: Jean Letrouit, "Hermétisme et alchimie: contribution à l'étude du Marcianus graecus 299 (= M)", in Carlos Gilly and Cis van Heertum (eds), *Magia, alchimia, scienza dal '400 al '700. L'influsso di Ermete Trismegisto*, 2 vols. Florence: Centro Di, 2002, vol. 1, pp. 85–109.

Lewis 2008: Bernard Lewis, "The Arab Destruction of the Library of Alexandria: Anatomy of a Myth", in Mostafa El-Abbadi and Omnia Mounir Fathallah (eds), *What Happened to the Ancient Library of Alexandria?* (Library of the Written Word 3, The Manuscript World 1). Leiden and Boston, MA: Brill, 2008, pp. 213–217.

Livrea 1976: Enrico Livrea, "Chi è l'autore della 'Blemyomachia' (P. Berol. 5003)?", *Prometheus* 2.2 (1976), pp. 97–123.

Lüthy 2000: Christoph Lüthy, "The Fourfold Democritus on the Stage of Early Modern Science", *Isis* 91 (2000), pp. 443–479.

Lüthy and Nicoli 2023: Christoph Lüthy and Elena Nicoli, "Atoms, Corpuscles, and Minima in the Renaissance: An Overview", in Christoph Lüthy and Elena Nicoli (eds), *Atoms, Corpuscles and Minima in the Renaissance* (Medieval and Early Modern Philosophy and Science 36). Leiden and Boston, MA: Brill, 2023, pp. 1–32.

Madan et al. 1895–1953: Falconer Madan, Herbert Henry Edmund Craster, Noël Denholm-Young, and Richard W. Hunt, *A Summary Catalogue of Western Manuscripts in the Bodleian Library at Oxford, Which Have Not Hitherto Been Catalogued in the Quarto Series*, 6 parts in 7 vols. Oxford: Clarendon Press, 1895–1953.

Malone-Lee 2019: Michael Malone-Lee, "Cardinal Bessarion and the Introduction of Plato to the Latin West", in Giancarlo Abbamonte and Stephen Harrison (eds), *Making and Rethinking the Renaissance: Between Greek and Latin in 15th–16th Century Europe* (Trends in Classics, Supplementary Volumes 77). Berlin and Boston, MA: De Gruyter, 2019, pp. 109–123.

Malone-Lee 2024: Michael Malone-Lee, *Cardinal Bessarion (1403–1472): Most Latin of Greeks, Most Greek of Latins* (Studies in Medieval History and Culture). London: Routledge, 2024.

Manfredini 1994: Mario Manfredini, "I manoscritti plutarchei del Bessarione", *Annali della Scuola Normale Superiore di Pisa: Classe di Lettere e Filosofia*, 3rd Series, 24.1 (1994), pp. 31–48.

Manzi 1974: Pietro Manzi, *La tipografia napoletana nel '500. Annali di Giuseppe Cacchi, Giovanni Battista Cappelli e tipografi minori (1566–1600)* (Biblioteca di bibliografia italiana 77). Florence: Leo S. Olschki, 1974.

Mariev 2013: Sergei Mariev, "Der Traktat *De natura et arte* des Kardinal Bessarion", in Claudia Märtl, Christian Kaiser, and Thomas Ricklin (eds), *"Inter graecos latinissimus, inter latinos graecissimus": Bessarion zwischen den Kulturen* (Pluralisierung und Autorität 39). Berlin and Boston, MA: De Gruyter, 2013, pp. 367–389.

Martelli 2011: Matteo Martelli, *Pseudo-Democrito, Scritti alchemici, con il commentario di Sinesio. Edizione critica del testo greco, traduzione e commento* (Textes et travaux de *Chrysopoeia* 12). Paris: S. É. H. A.; Milan: Archè, 2011.

Martelli 2013: Matteo Martelli, *The* Four Books *of Pseudo-Democritus* (Sources of Alchemy and Chemistry 1; Ambix 60, Supplement 1). Leeds: Maney Publishing, 2013.

Martínez Manzano 2018: Teresa Martínez Manzano, "La biblioteca manuscrita griega de Diego Hurtado de Mendoza: problemas y prospectivas", *Segno e testo* 16 (2018), pp. 315–433.

Martínez Manzano 2021: Teresa Martínez Manzano, "Towards the Reconstruction of a Little-Known Renaissance Library: The Greek Incunabula and Printed Editions of Diego Hurtado de Mendoza", in Rosa Maria Piccione (ed.), *Greeks, Books and Libraries in Renaissance Venice* (Transmissions 1). Berlin and Boston, MA: De Gruyter, 2021, pp. 163–176.

Matton 1995: Sylvain Matton, "L'influence de l'humanisme sur la tradition alchimique", *Micrologus* 3 (1995), Special Issue: *Le crisi dell'alchimia / The Crisis of Alchemy*, pp. 279–345.

McEvoy 2010: John G. McEvoy, *The Historiography of the Chemical Revolution: Patterns of Interpretation in the History of Science*. London: Pickering & Chatto, 2010.

Mioni 1981: Elpidio Mioni, *Bibliothecae Divi Marci Venetiarum codices graeci manuscripti*, 5: *Thesaurus antiquus*, 1: *Codices 1–299* (Ministero per i beni culturali e ambientali, Indici e cataloghi, Nuova Serie 6). Rome: Istituto Poligrafico e Zecca dello Stato, Libreria dello Stato, 1981.

Mioni 1991: Elpidio Mioni, "Vita del Cardinale Bessarione", *Miscellanea Marciana* 6 (1991), pp. 13–197.

Monfasani 1986: John Monfasani, "Platina, Capranica, and Perotti: Bessarion's Latin Eulogists and His Date of Birth", in Augusto Campana and Paola Medioli Masotti (eds), *Bartolomeo Sacchi il Platina (Piadena 1421–Roma 1481)* (Medioevo e Umanesimo 62). Padua: Editrice Antenore, 1986, pp. 97–136.

Monfasani 2013: John Monfasani, "The Pre- and Post- History of Cardinal Bessarion's 1469 *In Calumniatorem Platonis*", in Claudia Märtl, Christian Kaiser, and Thomas Ricklin (eds), *"Inter* *graecos latinissimus, inter latinos graecissimus": Bessarion zwischen den Kulturen* (Pluralisierung und Autorität 39). Berlin and Boston, MA: De Gruyter, 2013, pp. 347–366.

Moström 1957: Birgitta Moström, *Torbern Bergman: A Bibliography of His Works* (New Editions of Torbern Bergman-Literature 2). Stockholm: Almqvist & Wiksell, 1957.

Neville 2006: Roy G. Neville, *The Roy G. Neville Historical Chemical Library: An Annotated Catalogue of Printed Books on Alchemy, Chemistry, Chemical Technology, and Related Subjects*, 2 vols. Philadelphia, PA: Chemical Heritage Foundation, 2006.

Newman 2006: William R. Newman, *Atoms and Alchemy: Chymistry & the Experimental Origins of the Scientific Revolution*. Chicago, IL: University of Chicago Press, 2006.

Newman 2009: William R. Newman, "The Significance of 'Chymical Atomism'", *Early Science and Medicine* 14 (2009), Special Issue: *Evidence and Interpretation: Studies on Early Science and Medicine in Honor of John E. Murdoch*, pp. 248–264.

Nielsen 2001: Hans Toftlund Nielsen, "Hvem skrev den første kemihistorie?", *Dansk Kemi* 82.9 (2001), pp. 40–43.

Papadopoulos 1984–1986: Thomas I. Papadopoulos, *Ἑλληνικὴ βιβλιογραφία (1466 ci. – 1800)*, 2 vols (Πραγματεῖαι τῆς Ἀκαδημίας Ἀθηνῶν 48). Athens: Academy of Athens, 1984–1986.

Papadopoulos 1987: Thomas I. Papadopoulos, "Οἰκογενειακὰ Λέοντος Ἀλλατίου", *Chiaka Chronika* 18 (1987), pp. 17–26.

Papadopoulos 1989: Thomas I. Papadopoulos, "Ὁ Λέων Ἀλλάτιος καὶ ἡ Χίος", *Chiaka Chronika* 20 (1989), pp. 3–144.

Parry 2011: Glyn Parry, *The Arch-Conjuror of England: John Dee*. New Haven, CT: Yale University Press, 2011.

Pausillo 2020: Giorgia Pausillo, "Nuove considerazioni sui manoscritti alchemici di Domenico Pizzimenti", *Scripta* 13 (2020), pp. 141–159.

Pausillo 2022: Giorgia Pausillo, *Un nuovo catalogo dei manoscritti alchemici greci delle biblioteche italiane*. PhD Dissertation. Alma Mater Studiorum – Università di Bologna, 2022.

Petersen 1998: Erik Petersen, *Intellectum liberare: Johann Albert Fabricius: En humanist i Europa*, 2 vols (Danish Humanist Texts and Studies 18). Copenhagen: Det Kongelige Bibliotek and Museum Tusculanums Forlag, 1998.

Petzholdt 1866: Julius Petzholdt, *Bibliotheca bibliographica. Kritisches Verzeichniss der das Gesammtgebiet der Bibliographie betreffenden Litteratur des In- und Auslandes in systematischer Ordnung*. Leipzig: W. Engelmann, 1866.

Principe 2000: Lawrence M. Principe, "D. G. Morhof's Analysis and Defence of Transmutational Alchemy", in Waquet 2000, pp. 138–153.

Principe 2011: Lawrence M. Principe, "Alchemy Restored", *Isis* 102.2 (2011), pp. 305–312.

Principe 2013: Lawrence M. Principe, *The Secrets of Alchemy* (Synthesis). Chicago, IL: University of Chicago Press, 2013.

Qassem 2008: Qassem Abdou Qassem, "The Arab Story of the Destruction of the Ancient Library of Alexandria", in Mostafa El-Abbadi and Omnia Mounir Fathallah (eds), *What Happened to the Ancient Library of Alexandria?* (Library of the Written Word 3, The Manuscript World 1). Leiden and Boston, MA: Brill, 2008, pp. 207–211.

Rampling 2012: Jennifer M. Rampling, "John Dee and the Alchemists: Practising and Promoting English Alchemy in the Holy Roman Empire", *Studies in History and Philosophy of Science* 43 (2012), pp. 498–508.

Rehm 1939: Albert Rehm, "Zur Überlieferung der griechischen Alchemisten", *Byzantinische Zeitschrift* 39 (1939), pp. 394–434.

Reitzenstein 1919: Richard Reitzenstein, "Zur Geschichte der Alchemie und des Mystizismus", *Nachrichten von der Königlichen Gesellschaft der Wissenschaften zu Göttingen: Philologisch-Historische Klasse* (1919), pp. 1–37.

Robert-Nicoud 2018: Vincent Robert-Nicoud, *The World Upside Down in 16th-Century French Literature and Visual Culture* (Faux Titre 426). Leiden: Brill Rodopi, 2018.

Roberts 2019: Alexandre M. Roberts, "Framing a Middle Byzantine Alchemical Codex", *Dumbarton Oaks Papers* 73 (2019), pp. 69–102.

Roberts 2022: Alexandre M. Roberts, "Islamicate Alchemy in Greek Letters on the First Page of *Marcianus graecus* 299", *Byzantinische Zeitschrift* 115 (2022), pp. 341–350.

Rohrbacher 2002: David Rohrbacher, *The Historians of Late Antiquity*. London: Routledge, 2002.

Saffrey 1995: Henri Dominique Saffrey, "Historique et description du manuscrit alchimique de Venise *Marcianus Graecus* 299", in Kahn and Matton 1995, pp. 1–10.

Shackelford 2004: Jole Shackelford, *A Philosophical Path for Paracelsian Medicine: The Ideas, Intellectual Context, and Influence of Petrus Severinus (1540/2–1602)*. Copenhagen: Museum Tusculanum Press and University of Copenhagen, 2004.

Schartau 1994: Bjarne Schartau, *Codices graeci haunienses. Ein deskriptiver Katalog des griechischen Handschriftenbestandes der Königlichen Bibliothek Kopenhagen* (Danish Humanist Texts and Studies 9). Copenhagen: Museum Tusculanum Press and The Royal Library, 1994.

Schepelern 1983: Henrik D. Schepelern, "Introduction", in Henrik D. Schepelern (ed.), *Olai Borrichii Itinerarium 1660–1665: The Journal of the Danish Polyhistor Ole Borch*, 4 vols. Copenhagen: The Danish Society for Language and Literature, 1983, vol. 1, pp. VII–XLIII.

Schütt 1996: Hans-Werner Schütt, *Ein verborgener Schatz Venedigs: Der alchemische Kodex 299 aus der Biblioteca Marciana* (Quaderni 50). Venice: Centro Tedesco di Studi Veneziani, 1996.

Singer et al. 1924: Dorothea Waley Singer, Annie Anderson, and William J. Anderson, *Catalogue des manuscrits alchimiques grecs*, vol. 3: *Les manuscrits des îles Britanniques*. Brussels: M. Lamertin, 1924.

Smeaton 1962: William Arthur Smeaton, *Fourcroy: Chemist and Revolutionary, 1755–1809*. Cambridge: W. Heffer & Sons, 1962.

Sommervogel et al. 1890–1932: Carlos Sommervogel, Aloys de Backer, Augustin de Backer, Pierre Bliard, and Auguste Carayon, *Bibliothèque de la Compagne de Jésus. Nouvelle édition*, 11 vols. Brussels: O. Schepens; Paris: A. Picard, 1890–1932.

Spivakovsky 1970: Erika Spivakovsky, *Son of the Alhambra: Don Diego Hurtado de Mendoza, 1504–1575*. Austin, TX: University of Texas Press, 1970.

Sterck 1936: Johannes Franciscus Maria Sterck, "Calcovius-Blaeu-Vondel (Keulsche boeken te Amsterdam gedruckt)", *Vondel-Kroniek* 7 (1936), pp. 66–76.

Stöger 1855: [Johann Nepomuk Stöger], *Scriptores provinciae Austriacae Societatis Jesu. Collectionis scriptorum ejusdem Societatis universae tomus I*. Vienna: Mekhitarist Congregation, 1855.

Stolleis 1983: Michael Stolleis (ed.), *Hermann Conring (1606–1681): Beiträge zu Leben und Werk*. Berlin: Duncker & Humblot, 1983.

Suarez and Woudhuysen 2010: Michael F. Suarez and H. R. Woudhuysen (eds), *The Oxford Companion to the Book*, 2 vols. Oxford: Oxford University Press, 2010.

Surace 2014: Domenico Surace, "Vita e opere di Leone Allacci", in Claudia Montuschi (ed.), *Storia della Biblioteca Apostolica Vaticana*, vol. 3: *La Vaticana nel Seicento (1590–1700): una biblioteca di biblioteche*. Vatican City: Biblioteca Apostolica Vaticana, 2014, pp. 199–204.

Tarrant 1998: Harold Tarrant, "Introduction", in Robin Jackson, Kimon Lycos, and Harold Tarrant, *Olympiodorus: Commentary on Plato's* Gorgias. *Translated with Full Notes* (Philosophia Antiqua 78). Leiden: Brill, 1998, pp. 1–52.

Touwaide 2016: Alain Touwaide, *A Census of Greek Medical Manuscripts: From Byzantium to the Renaissance* (Medicine in the Medieval Mediterranean). London: Routledge, 2016.

VD17 – Das Verzeichnis der im deutschen Sprachraum erschienenen Drucke des 17. Jahrhunderts, http://www.vd17.de.

Viano 1995: Cristina Viano, "Olympiodore l'alchimiste et les présocratiques: une doxographie de l'unité (*De arte sacra*, § 18–27)", in Kahn and Matton 1995, pp. 95–150.

Viano 2006: Cristina Viano, *La matière des choses. Le livre IV des* Météorologiques *d'Aristote, et son interprétation par Olympiodore* (Tradition de la pensée classique). Paris: Librairie Philosophique J. Vrin, 2006.

Viano 2021: Cristina Viano, "Olympiodorus and Greco-Alexandrian Alchemy", in Albert Joose (ed.), *Olympiodorus of Alexandria: Exegete, Teacher, Platonic Philosopher* (Philosophia Antiqua 159). Leiden and Boston, MA: Brill, 2021, pp. 14–30.

Waquet 2000: Françoise Waquet (ed.), *Mapping the World of Learning: The* Polyhistor *of Daniel Georg Morhof* (Wolfenbütteler Forschungen 91). Wiesbaden: Harrassowitz, 2000.

Zorzi 1987: Marino Zorzi, *La Libreria di San Marco: libri, lettori, società nella Venezia dei Dogi*. Milan: A. Mondadori, 1987.

Zuretti 1928: Carlo Oreste Zuretti, *Catalogue des manuscrits alchimiques grecs*, vol. 5.1: *Les manuscrits d'Espagne*. Brussels: M. Lamertin, 1928.

Zuretti et al. 1927: Carlo Oreste Zuretti, Otto Lagercrantz, Johan Ludvig Heiberg, Ingeborg Hammer-Jensen, Domenico Bassi, and Emidio Martini, *Catalogue des manuscrits alchimiques grecs*, vol. 2: *Les manuscrits italiens*. Brussels: M. Lamertin, 1927.

Index

This index includes references from the main text and the footnotes, without distinguishing between them in the page ranges. References to individual footnotes (e.g., "161n82") are given only when the relevant material appears exclusively in the footnotes.

The *Index nominum et operum anonymorum* encompasses historical, mythological, biblical, and literary figures, as well as modern scholars. Page references for personal names are included only where the individuals concerned are discussed, not where they appear solely in bibliographical citations. Only anonymous works, together with those known by their traditional designations (e.g., *Aetius Arabus*), are indexed as separate entries; works by named authors are accessible through the relevant author entries.

The *Indices verborum* also include specialized expressions (e.g., *Sciant artifices*) and other items, such as names (e.g., Χημία). In the *Index verborum graecorum*, in particular, the following conventions are followed in special cases: (1) entries given in both the singular and the plural (e.g., ἀπορία/ἀπορίαι) indicate that the word has a distinct meaning when used in the plural; (2) obscure words not recorded in the major dictionaries (e.g., λυμπικήου) are indexed in the exact form in which they are attested in the main text; (3) when a term can be associated with a known lexical form with reasonable confidence, both are indexed separately, with the known form cross-referencing the form attested in the main text (e.g., διαβάφω; διβάφον). The terms in the *Index verborum arabicorum* are given in the Latin transliteration adopted in the essays.

In the *Index rerum*, frequent terms such as "Byzantium" and "science" have been omitted as separate entries.

https://doi.org/10.1515/9783111589619-013

Index papyrorum

Index codicum manuscriptorum

Index nominum et operum anonymorum

Index geographicus

Index verborum

Index verborum graecorum

Index verborum arabicorum (in Latin transliteration)

Index verborum latinorum

Index rerum